T0094018

Physics of Relativistic Objects in Compact Binaries: From Birth to Coalescence

Astrophysics and Space Science Library

For other titles published in this series, go to
www.springer.com/series/5664

Physics of Relativistic Objects in Compact Binaries: From Birth to Coalescence

Edited by

Monica Colpi
Piergiorgio Casella
Vittorio Gorini
Ugo Moschella
Andrea Possenti

 Springer

Editors

Monica Colpi
Department of Physics G. Occhialini
University of Milano Bicocca
Piazza della Scienza 3
20126 Milano
Italy

Piergiorgio Casella
Astronomical Institute "Anton Pannekoek"
University of Amsterdam
Kruislaan 403
1098 SJ Amsterdam
The Netherlands

Vittorio Gorini
Department of Physics and Mathematics
University of Insubria
Via Valleggio 11
22100 Como
Italy

Ugo Moschella
Department of Physics and Mathematics
University of Insubria
Via Valleggio 11
22100 Como
Italy

Andrea Possenti
INAF- Osservatorio Astronomico di
Cagliari
Loc. Poggio dei Pini, Strada 54,
09012 Capoterra (CA),
Italy

Cover illustration: Artist's impression (courtesy of Ceravolo Graphic Studio) of a double pulsar binary emitting both radio and gravitational waves during the spiral-in phase. The system is depicted against a background of the population of coalescing double neutron star binaries in the Galaxy.

ISSN 1864-5879
ISBN 978-1-4020-9263-3 (HB)   ISBN 978-1-4020-9264-0 (e-book)

Library of Congress Control Number: 2008936684

Published by Springer Science+Business Media B.V.
P.O. Box 17, 3300 AA Dordrecht, The Netherlands
In association with
Canopus Publishing Limited,
27 Queen Square, Bristol BS1 4ND, UK

www.springer.com and www.canopusbooks.com

Preface

A very attractive feature of the theory of general relativity is that it is a perfect example of a "falsifiable" theory: no tunable parameter is present in the theory and therefore even a single experiment incompatible with a prediction of the theory would immediately lead to its inevitable rejection, at least in the physical regime of application of the aforementioned experiment. This fact provides additional scientific value to one of the boldest and most fascinating achievements of the human intellect ever, and motivates a wealth of efforts in designing and implementing tests aimed at the falsification of the theory.

The first historical test on the theory has been the deflection of light grazing the solar surface (Eddington 1919): the compatibility of the theory with this first experiment together with its ability to explain the magnitude of the perihelion advance of Mercury contributed strongly to boost acceptance and worldwide knowledge. However, technological limitations prevented physicists from setting up more constraining tests for several decades after the formulation of the theory. In fact, a relevant problem with experimental general relativity is that the predicted deviations from the Newtonian theory of gravity are very small when the experiments are carried out in terrestrial laboratories. A rough estimate of the expected magnitude of general relativity corrections can be drawn from a comparison between the classical gravitational potential energy $E_{grav} \sim -GM^2/R$ of a body of mass M and radius R and the rest mass energy of the same body, $E_{rest} = Mc^2$ (G being the gravitational constant and c the speed of light in vacuum). On Earth, the dimensionless ratio $\epsilon = |E_{grav}/E_{rest}|$ is of only $\sim 10^{-10}$ and, as a consequence, very sensitive experimental apparatuses are necessary for detecting the tiny discrepancies resulting from the application of Einstein's theory with respect to the predictions of Newtonian gravity.

After the pioneering experimental era, general relativity tests have reflourished only in the last fifty years, when new generation devices (such as radar and lasers) and the birth of astronautics allowed us to design and carry out much more accurate experiments in space, such as the use of radar technology (Shapiro 1964) for transmitting pulses towards either Venus or Mercury, and for measuring the time delay of the echoes. According to general relativity these delays are enhanced when the pulses pass close to the Sun. Laser ranging to the Moon (Nordtvedt 1968) was used to set upper limits to the amplitude of the lunar orbit oscillations that are expected to occur if the ratio of the gravitational to inertial mass of the Earth differs from the value (equal to unity) predicted by general relativity. More recently, the frequency shift of radio photons to and from the Cassini spacecraft as they pass near the Sun has been measured (Bertotti 2003) probing the degree to which photons are deflected and delayed by the curvature of space-time produced by the Sun. The last big step forward in the field has been the launch of the Relativity Gyroscope Experiment (Gravity Probe B) which collected data orbiting the Earth between 2004 an 2005 (Everitt 2007). The aim was to investigate two well known predictions of general relativity : (i) the occurrence of the geodetic precession effect (resulting from the warp of the local space-time due to the mass of the Earth) and, even more ambitiously, the frame dragging effect (also known as gravitomagnetism), generated by the rotation of Earth, which drags and twists the space-time surrounding it. Final data release of this experiment is expected during 2008, although some preliminary announcements about the measurement of the geodetic effect have already been reported.

All the general relativity tests listed above involve celestial bodies of the solar system, among which the highest value of ϵ is that of the Sun, having $\epsilon \sim 10^{-6}$. Therefore, those tests belong to the class of experiments that investigate the so-called *weak-field limit* of gravitational physics, i.e. a regime where the velocities of the bodies imposed by the gravitational field are small compared to c and the energy associated with the gravitational field of the bodies is a tiny fraction of their rest mass energy. The amazing outcome of a century of experimental physics is that general relativity has been passing all the aforementioned tests – as well as all the others of the same class – with full marks and *cum laude*.

Thus, why bother with further investigations? On the epistemological side, no physical theory can ever be proved, but only falsified; hence, the request for more and more refined experiments can never find an end. However, in the particular case of general relativity, there are still fundamental and unanswered problems urging an extension of the range of tests. At present, the main goal of theoretical physics is to describe all the phenomena in Nature in terms of a unified theory, which should encompass gravity, nuclear and electroweak interactions in a single framework. A knotty step toward this unified model is to combine the classical and deterministic (although often very difficult to calculate) predictions of general relativity with the probabilistic approach of quantum mechanics. Many decades of unsatisfactory conclusions in the field of quantum gravity may be an indication that some of the ingredients in current attempts are not suitable for constructing the desired Grand Unification Theory. In particular, one may wonder whether general relativity – very successful in the condition of *weak-field* gravity – remains the best available theory for describing Nature under extreme physical conditions, such as those holding in the regime where a unified model applies.

In the presence of gravitational effects alone, the notion of extreme physical conditions can be reworded stating that the dimensionless parameter ϵ tends to unity. This is the so-called *strong-field* realm of gravity theories (e.g. Will 2001): it applies for instance to the intense gravitational fields associated with the Planck-scale, at which a quantum formulation of general relativity is unavoidable, or to the nearness of a collapsed object, like the event horizon of a static black hole ($\epsilon \sim 0.5$) or the surface of a neutron star ($\epsilon \sim 0.2$). Under these conditions, general relativity could finally show its limits of applicability, being supplanted by alternative theories of gravity, either already proposed or still to be elaborated. In this perspective, considering the ambitious aims of the physics of the XXI century, it is mandatory to establish the most stringent limits on the falsifiability of the Einsteinian theory in the *strong-field* regime.

How to perform that task? It is really hard to conceive the possibility of testing the *strong-field* regime in a terrestrial laboratory or in the solar system. As a consequence, we have to search for natural laboratories in the Cosmos: in this context, *relativistic objects* hosted in *compact binaries* offer magnificent opportunities.

The term *relativistic objects* refers mostly to black holes and neutron stars, while *compact binaries* to binaries where gravitational wave inspiral influences binary evolution within a Hubble time. When binaries hosting two collapsed stars are in their latest stage of coalescence – driven by gravitational radiation damping – the typical orbital velocity approaches c and the orbital separation becomes of the order of their size, implying that the physical processes occur in the typical *strong-field* regime. Various types of spiralling compact binaries are expected to result from stellar binary evolution: double neutron stars, neutron star plus black hole systems and double black hole systems. These binaries are the most promising targets for the current generation of ground-based interferometers for the detection of gravitational waves, like LIGO and VIRGO. Much more massive binary black hole systems should also develop from cosmological mergers of galaxies and their inspiral is the main objective of LISA, the future space-based gravitational wave interferometer. Direct detection of the gravitational waves released during these inspirals would be an enormous step forward for experimental gravity and a wonderful support to the 100 year old predictions of general relativity. At the same time, these investigations have the potentiality of discovering the existence of limits to the applicability of Einstein's theory, e.g. if a dipole gravitational radiation contribution – null in general relativity, but predicted in the class of the so called scalar-tensor theories – should enter the formulae describing the effects of radiation back-reaction on orbit phasing. Ruinous consequences for general relativity would also derive for example from the detection of more than two transverse modes of propagation of the waves.

Taken at face value, the earlier stages of evolution of close binaries – well before coalescence – may not satisfy the conditions for a *strong-field* regime. In fact, when the orbital separation is large, any binary component moves in the weak gravitational field of the companion and thus $\epsilon \ll 1$. However, in most alternative theories of gravity – e.g. for example the scalar-tensor theories mentioned above – the motion and the gravitational radiation damping depend on the gravitational binding energy of the involved bodies. Since in compact objects, like neutron stars, the gravitational binding energy is very large (i.e. ϵ is close to unity), significant deviations from the orbital motion are expected to be detectable even in the weak field limit. In this framework, a particularly interesting case of study is that of *binary pulsars*, that is binary systems comprising a neutron star which emits collimated beams of radio waves, observed as pulses, once per rotation of the neutron star. Due to evolutionary reasons, the pulsars included in binary systems with other compact objects very often behave as *highly stable clocks* . The measurement of the times of arrival of their pulses allows for an accurate determination of their orbital motion. This provides the observational basis for the search for deviations from the general relativity predictions and, in summary, for using the binary pulsars as a magnificent laboratory for the *strong-field* tests of Einstein's theory.

Distant cosmic laboratories compared to terrestrial or space-based laboratories require a deep understanding of the experimental procedures adopted in astrophysics, for which we are only *passive* observers. In these laboratories we have no opportunity to alter the conditions of the apparatus, nor to selectively modify the experimental parameters in order to highlight and better study the various physical effects contributing to the process under analysis. In the case of astrophysics, all the information must be extracted only from the flow of particles (in the form of photons, gravitons, etc.) supplied, at a non-tunable level, by the cosmic laboratory. In light of that, a first relevant issue is to maximize the number and the usefulness of the collected particles, i.e. to make the detectors and the data analysis as efficient as possible. Selection of the most suitable natural laboratories is another key point, as well as the understanding of how the history and the properties of their constituents can affect the interpretation of the data. In this respect, it is very important to know the origin and the evolutionary path that led to the formation of any natural laboratory, its basic physical features and its cosmic environment.

This book, entitled *Physics of relativistic objects in compact binaries: from birth to coalescence*, has been conceived and assembled with the aim of providing a comprehensive and updated report on the astrophysical approach to the investigation of gravity theories, with particular attention to *strong-field* tests of general relativity and alternate theories, performed using collapsed objects in relativistic binaries as laboratories. The text collects reviews written by scientists at the front-line of this investigation and known worldwide for their major and often still unsurpassed contributions in this field. The book covers various topics, which are reviewed both on the observational and the theoretical side: (i) from binaries as test-beds of gravity theories to binary pulsars as cosmic laboratories, (ii) from binary stars evolution to the formation of relativistic binaries, (iii) from short gamma-ray bursts to low mass X-ray binaries, (iv) from stellar-mass black hole binaries to coalescing super-massive black holes in galaxy mergers.

In particular, the first chapter reviews gravity theories describing the motion of point masses in a binary. The physics presented and the mathematical tools introduced set the basis for a correct and unambiguous interpretation of the large data set provided by the observation of binary pulsars. The procedure of timing this class of pulsars, as well as some of the most remarkable applications of that, are described in detail in Chapter 2, whereas Chapter 3 deals with the concepts driving the experiments which search for additional relativistic binary pulsars.

The following five chapters introduce the reader to the realm of binaries hosting one or two relativistic objects, as observed and interpreted in an astrophysical context. Chapter 4, reviews the critical channels of formation of compact binaries in the galactic field, highlighting the chief steps that signal their formation and evolution. Theory and observations are inter-winded in this chapter, designed to give a rather complete view of the complex phenomenology of neutron stars and black holes in binaries and of the physical tools used for the interpretation and description of these sources. Chapter 5 continues in this study, exploring the formation of binary pulsars, and more generally of binaries with two compact stars – either a neutron star or a black hole, also of intermediate mass – in the dense collisional environment of a globular cluster, where they form preferentially via close gravitational encounters. Short gamma-ray burst sources are the subject of Chapter 6: in the context of this book, they can be viewed as the counterparts of the inspiral and merger of neutron star-black hole or double neutron star binaries. The chapter reviews the impact that short gamma-ray burst sources have on our knowledge of the most energetic events in the universe. Chapter 7 continues by surveying a number of powerful diagnostic tools used to explore and map space-time in the vicinity of accreting neutron stars or black holes in low mass X-ray binary systems. Chapter 8 introduces the reader to a class of systems, the white dwarfs in ultra-compact binaries, that are currently receiving attention, being important sources of low frequency gravitational waves, and so potential systems for testing general relativity.

The book ends with a chapter dedicated to gravitational waves from coalescing black holes, from the perspective that their detection will open a new window into our universe. The recent remarkable advances in numerical relativity in describing the coalescence of black holes have motivated the introduction of this chapter. In particular, Chapter 9 describes again the two body problem and how it is solved in numerical relativity. Focused mainly on massive binary black holes, this chapter reports on the latest achievements in the study of the dynamics of the merger in full general relativity and their impact on physics, astronomy, and cosmology.

This volume is based on the lecture notes of a doctoral school: *A Century from Einstein Relativity: Probing Gravity Theories with Binary Systems*, promoted by SIGRAV, the Italian Society for Relativity and Gravitation, and supported by the University of Milano Bicorra, the University of Insubria, and the National Institute of Nuclear Physics (INFN).

Milano,   *Monica Colpi*
February 2008   *Andrea Possenti*

Contents

List of Contributors

Marta Burgay
INAF- Osservatorio Astronomico di
Cagliari
Loc. Poggio dei Pini, Strada 54,
09012 Capoterra (CA), Italy
burgay@ca.astro.it

Simone Dall'Osso
INAF - Osservatorio Astronomico di
Roma
Via Frascati 33, 00040 Monteporzio
Catone (Roma), Italy
dallosso@mporzio.astro.it

Thibault Damour
Institut des Hautes Etudes Scientifiques
35 route de Chartres, F-91440
Bures-sur-Yvette, France
damour@ihes.fr

Gianluca Israel
INAF - Osservatorio Astronomico di
Roma
Via Frascati 33, 00040 Monteporzio
Catone (Roma), Italy
gianluca@mporzio.astro.it*

Davide Lazzati
JILA, University of Colorado
440 UCB, Boulder, CO 80309-0440,
USA
lazzati@colorado.edu

Monica Colpi
Department of Physics G. Occhialini,
University of Milano Bicocca
Piazza della Scienza 3, 20126 Milano,
Italy
colpi@mib.infn.it

Nichi D'Amico
INAF- Osservatorio Astronomico di
Cagliari
Loc. Poggio dei Pini, Strada 54,
09012 Capoterra (CA), Italy
damico@ca.astro.it

Bernadetta Devecchi
Department of Physics G. Occhialini,
University of Milano Bicocca
Piazza della Scienza 3, 20126 Milano,
Italy
bernadetta.devecchi@mib.infn.it

Michael Kramer
The University of Manchester,
Jodrell Bank Observatory
Jodrell Bank, Cheshire, UK
Michael.Kramer@manchester.ac.uk

Rosalba Perna
JILA, University of Colorado
440 UCB, Boulder, CO 80309-0440,
USA
rosalba@colorado.edu

Andrea Possenti
INAF- Osservatorio Astronomico di
Cagliari
Loc. Poggio dei Pini, Strada 54,
09012 Capoterra (CA), Italy
possenti@ca.astro.it

Luigi Stella
INAF - Osservatorio Astronomico di
Roma

Via Frascati 33, 00040 Monteporzio
Catone (Roma), Italy
stella@mporzio.astro.it

Frans Pretorius
Department of Physics, Princeton
University
Princeton, NJ 08544, USA
fpretori@princeton.edu

Ed van den Heuvel
Astronomical Institute "Anton Pannekoek", University of Amsterdam
Kruislaan 403, 1098SJ Amsterdam,The Netherlands
edvdh@science.uva.nl

Binary Systems as Test-Beds of Gravity Theories

Thibault Damour

Institut des Hautes Etudes Scientifiques, 35 route de Chartres,
F-91440 Bures-sur-Yvette, France
damour@ihes.fr

1 Introduction

The discovery of binary pulsars in 1974 [1] opened up a new testing ground for relativistic gravity. Before this discovery, the only available testing ground for relativistic gravity was the solar system. As Einstein's theory of General Relativity (GR) is one of the basic pillars of modern science, it deserves to be tested, with the highest possible accuracy, in all its aspects. In the solar system, the gravitational field is slowly varying and represents only a very small deformation of a flat spacetime. As a consequence, solar system tests can only probe the quasi-stationary (non-radiative) *weak-field* limit of relativistic gravity. By contrast binary systems containing compact objects (neutron stars or black holes) involve spacetime domains (inside and near the compact objects) where *the gravitational field is strong*. Indeed, the surface relativistic gravitational field $h_{00} \simeq 2\,GM/c^2R$ of a neutron star is of order 0.4, which is close to the one of a black hole $(2\,GM/c^2R = 1)$ and much larger than the surface gravitational fields of solar system bodies: $(2\,GM/c^2R)_{\mathrm{Sun}} \sim 10^{-6}$, $(2\,GM/c^2R)_{\mathrm{Earth}} \sim 10^{-9}$. In addition, the high stability of "pulsar clocks" has made it possible to monitor the dynamics of its orbital motion down to a precision allowing one to measure the small $(\sim (v/c)^5)$ orbital effects linked to the propagation of the gravitational field at the velocity of light between the pulsar and its companion.

The recent discovery of the remarkable *double* binary pulsar PSR J0737−3039 [2, 3] (see also the contributions by M. Kramer and by N. D'Amico and M. Burgay to this volume) has renewed the interest in the use of binary pulsars as test-beds of gravity theories. The aim of this chapter is to provide an introduction to the theoretical frameworks needed for interpreting binary pulsar data as tests of GR and alternative gravity theories.

2 Motion of Binary Pulsars in General Relativity

The traditional (text book) approach to the problem of motion of N separate bodies in GR consists of solving, by successive approximations, Einstein's field equations (we use the signature $-+++$)

$$R_{\mu\nu} - \frac{1}{2} R \, g_{\mu\nu} = \frac{8\pi \, G}{c^4} \, T_{\mu\nu} \,, \tag{1}$$

together with their consequence

$$\nabla_\nu \, T^{\mu\nu} = 0 \,. \tag{2}$$

To do so, one assumes some specific matter model, say a perfect fluid,

$$T^{\mu\nu} = (\varepsilon + p) \, u^\mu \, u^\nu + p \, g^{\mu\nu} \,. \tag{3}$$

One expands (say in powers of Newton's constant)

$$g_{\mu\nu}(x^\lambda) = \eta_{\mu\nu} + h^{(1)}_{\mu\nu} + h^{(2)}_{\mu\nu} + \cdots \,, \tag{4}$$

together with the use of the simplifications brought by the 'Post-Newtonian' approximation ($\partial_0 \, h_{\mu\nu} = c^{-1} \, \partial_t \, h_{\mu\nu} \ll \partial_i \, h_{\mu\nu}$; $v/c \ll 1$, $p \ll \varepsilon$). Then one integrates the local material equation of motion (2) over the volume of each separate body, labelled say by $a = 1, 2, \ldots, N$. In so doing, one must define some 'center of mass' z_a^i of body a, as well as some (approximately conserved) 'mass' m_a of body a, together with some corresponding 'spin vector' S_a^i and, possibly, higher multipole moments.

An important feature of this traditional method is to use a *unique coordinate chart* x^μ to describe the full N-body system. For instance, the center of mass, shape and spin of each body a are all described within this common coordinate system x^μ. This use of a single chart has several inconvenient aspects, even in the case of weakly self-gravitating bodies (as in the solar system case). Indeed, it means for instance that a body which is, say, spherically symmetric in its own 'rest frame' X^α will appear as deformed into some kind of ellipsoid in the common coordinate chart x^μ. Moreover, it is not clear how to construct 'good definitions' of the center of mass, spin vector, and higher multipole moments of body a, when described in the common coordinate chart x^μ. In addition, as we are interested in the motion of strongly self-gravitating bodies, it is not a priori justified to use a simple expansion of the type (4) because $h^{(1)}_{\mu\nu} \sim \sum_a Gm_a/(c^2 \, |\boldsymbol{x} - \boldsymbol{z}_a|)$ will not be uniformly small in the common coordinate system x^μ. It will be small if one stays far away from each object a, but, as recalled above, it will become of order unity on the surface of a compact body.

These two shortcomings of the traditional 'one-chart' approach to the relativistic problem of motion can be cured by using a *'multi-chart' approach*. The multi-chart approach describes the motion of N (possibly, but not necessarily, compact) bodies by using $N + 1$ separate coordinate systems: (i) one *global* coordinate chart x^μ ($\mu = 0, 1, 2, 3$) used to describe the spacetime outside N 'tubes', each containing one body, and (ii) N *local* coordinate charts X_a^α ($\alpha = 0, 1, 2, 3; a = 1, 2, \ldots, N$) used to describe the spacetime in and around each body a. The multi-chart approach was first used to discuss the motion of black holes and other compact objects [4, 5, 6, 7, 8, 9, 10, 11]. Then it was also found to be very convenient for describing, with the high-accuracy required for dealing with modern technologies such as VLBI, systems of N weakly self-gravitating bodies, such as the solar system [12, 13].

The essential idea of the multi-chart approach is to combine the information contained in *several expansions*. One uses both a global expansion of the type (4) and several local expansions of the type

$$G_{\alpha\beta}(X_a^\gamma) = G_{\alpha\beta}^{(0)}(X_a^\gamma; m_a) + H_{\alpha\beta}^{(1)}(X_a^\gamma; m_a, m_b) + \cdots , \tag{5}$$

where $G_{\alpha\beta}^{(0)}(X; m_a)$ denotes the (possibly strong-field) metric generated by an isolated body of mass m_a (possibly with the additional effect of spin).

The separate expansions (4) and (5) are then 'matched' in some overlapping domain of common validity of the type $Gm_a/c^2 \lesssim R_a \ll |\boldsymbol{x} - \boldsymbol{z}_a| \ll d \sim |\boldsymbol{x}_a - \boldsymbol{x}_b|$ (with $b \neq a$), where one can relate the different coordinate systems by expansions of the form

$$x^\mu = z_a^\mu(T_a) + e_i^\mu(T_a)\, X_a^i + \frac{1}{2}\, f_{ij}^\mu(T_a)\, X_a^i\, X_a^j + \cdots \tag{6}$$

The multi-chart approach becomes simplified if one considers *compact* bodies (of radius R_a comparable to $2\, Gm_a/c^2$). In this case, it was shown [9], by considering how the 'internal expansion' (5) propagates into the 'external' one (4) via the matching (6), that, *in General Relativity*, the internal structure of each compact body was *effaced* to a very high degree, when seen in the external expansion (4). For instance, for non-spinning bodies, the internal structure of each body (notably the way it responds to an external tidal excitation) shows up in the external problem of motion only at the *fifth post-Newtonian* (5PN) approximation, i.e. in terms of order $(v/c)^{10}$ in the equations of motion.

This *'effacement of internal structure'* indicates that it should be possible to simplify the rigorous multi-chart approach by skeletonizing each compact body by means of some delta-function source. Mathematically, the use of distributional sources is delicate in a non-linear theory such as GR. However, it was found that one can reproduce the results of the more rigorous matched-multi-chart approach by treating the divergent integrals generated by the use of delta-function sources by means of (complex) analytic continuation [9]. The most efficient method (especially to high PN orders) has been found to use analytic continuation in the dimension of space d [14].

Finally, the most efficient way to derive the general relativistic equations of motion of N compact bodies consists of solving the equations derived from the action (where $g \equiv -\det(g_{\mu\nu})$)

$$S = \int \frac{d^{d+1} x}{c} \sqrt{g} \, \frac{c^4}{16\pi G} \, R(g) - \sum_a m_a c \int \sqrt{-g_{\mu\nu}(z_a^\lambda) \, dz_a^\mu \, dz_a^\nu}, \quad (7)$$

formally using the standard weak-field expansion (4), but considering the space dimension d as an arbitrary complex number which is sent to its physical value $d = 3$ only at the end of the calculation.

Using this method[1] one has derived the equations of motion of *two* compact bodies at the 2.5PN (v^5/c^5) approximation level needed for describing binary pulsars [15, 16, 9]:

$$\begin{aligned}
\frac{d^2 z_a^i}{dt^2} = {} & A_{a0}^i(z_a - z_b) + c^{-2} A_{a2}^i(z_a - z_b, v_a, v_b) \\
& + c^{-4} A_{a4}^i(z_a - z_b, v_a, v_b, S_a, S_b) \\
& + c^{-5} A_{a5}^i(z_a - z_b, v_a - v_b) + \mathcal{O}(c^{-6}).
\end{aligned} \quad (8)$$

Here $A_{a0}^i = -G m_b(z_a^i - z_b^i)/|z_a - z_b|^3$ denotes the Newtonian acceleration, A_{a2}^i its 1PN modification, A_{a4}^i its 2PN modification (together with the spin–orbit effects), and A_{a5}^i the 2.5PN contribution of order v^5/c^5. [See the references above; or the review [17], for more references and the explicit expressions of A_2, A_4 and A_5.] It was verified that the term A_{a5}^i has the effect of decreasing the mechanical energy of the system by an amount equal (on average) to the energy lost in the form of gravitational wave flux at infinity. Note, however, that here A_{a5}^i was derived, in the near zone of the system, as a *direct* consequence of the general relativistic propagation of gravity, at the velocity c, between the two bodies. This highlights the fact that binary pulsar tests of the existence of A_{a5}^i are *direct tests of the reality of gravitational radiation*.

Recently, the equations of motion (8) have been computed to even higher accuracy: 3PN $\sim v^6/c^6$ [18, 19, 20, 21, 22] and 3.5PN $\sim v^7/c^7$ [23, 24, 25] (see also the review [26]). These refinements are, however, not (yet) needed for interpreting binary pulsar data.

3 Timing of Binary Pulsars in General Relativity

In order to extract observational effects from the equations of motion (8) one needs to go through two steps: (i) to solve the equations of motion (8) so as to get the coordinate positions z_1 and z_2 as explicit functions of the coordinate

---

[1] Or, more precisely, an essentially equivalent analytic continuation using the so-called 'Riesz kernels'.

time t, and (ii) to relate the coordinate motion $z_a(t)$ to the pulsar observables, i.e. mainly to the times of arrival of electromagnetic pulses on Earth.

The first step has been accomplished, in a form particularly useful for discussing pulsar timing, in [27]. There (see also [28]) it was shown that, when considering the full (periodic and secular) effects of the $A_2 \sim v^2/c^2$ terms in Eq. (8), together with the secular effects of the $A_4 \sim v^4/c^4$ and $A_5 \sim v^5/c^5$ terms, the relativistic two-body motion could be written in a very simple 'quasi-Keplerian' form (in polar coordinates), namely:

$$\int n \, dt + \sigma = u - e_t \sin u \,, \tag{9}$$

$$\theta - \theta_0 = (1 + k) \, 2 \arctan \left[\left(\frac{1 + e_\theta}{1 - e_\theta} \right)^{\frac{1}{2}} \tan \frac{u}{2} \right] \,, \tag{10}$$

$$R \equiv r_{ab} = a_R (1 - e_R \cos u) \,, \tag{11}$$

$$r_a \equiv |z_a - z_{CM}| = a_r (1 - e_r \cos u) \,, \tag{12}$$

$$r_b \equiv |z_b - z_{CM}| = a_{r'} (1 - e_{r'} \cos u) \,. \tag{13}$$

Here $n \equiv 2\pi/P_b$ denotes the orbital frequency, $k = \Delta\theta/2\pi = \langle \dot{\omega} \rangle / n = \langle \dot{\omega} \rangle P_b/2\pi$ the fractional periastron advance per orbit, u an auxiliary angle ('relativistic eccentric anomaly'), e_t, e_θ, e_R, e_r and $e_{r'}$ various 'relativistic eccentricities' and a_R, a_r and $a_{r'}$ some 'relativistic semi-major axes'. See [27] for the relations between these quantities, as well as their link to the relativistic energy and angular momentum E, J. A direct study [28] of the dynamical effect of the contribution $A_5 \sim v^5/c^5$ in the equations of motion (8) has checked that it led to a secular increase of the orbital frequency $n(t) \simeq n(0) + \dot{n}(t - t_0)$, and thereby to a quadratic term in the 'relativistic mean anomaly' $\ell = \int n \, dt + \sigma$ appearing on the left-hand side (L.H.S.) of Eq. (9):

$$\ell \simeq \sigma_0 + n_0(t - t_0) + \frac{1}{2} \dot{n}(t - t_0)^2 \,. \tag{14}$$

As for the contribution $A_4 \sim v^4/c^4$ it induces several secular effects in the orbital motion: various 2PN contributions to the dimensionless periastron parameter k ($\delta_4 k \sim v^4/c^4+$ spin–orbit effects), and secular variations in the inclination of the orbital plane (due to spin–orbit effects).

The second step in relating (8) to pulsar observations has been accomplished through the derivation of a 'relativistic timing formula' [29, 30]. The 'timing formula' of a binary pulsar is a multi-parameter mathematical function relating the observed time of arrival (at the radio-telescope) of the center of the Nth pulse to the integer N. It involves many different physical effects: (i) dispersion effects, (ii) travel time across the solar system, (iii) gravitational delay due to the Sun and the planets, (iv) time dilation effects between the time measured on the Earth and the solar-system-barycenter time, (v) variations in the travel time between the binary pulsar and the solar-system barycenter (due to relative accelerations, parallax and proper motion), (vi) time delays happening within the binary system. We shall focus here on the time delays which take place within the binary system (see the chapter by M. Kramer for a discussion of the other effects).

For a proper derivation of the time delays occurring within the binary system we need to use the multi-chart approach mentioned above. In the 'rest frame' $(X_a^0 = c\,T_a, X_a^i)$ attached to the pulsar a, the pulsar phenomenon can be modelled by the secularly changing rotation of a beam of radio waves:

$$\Phi_a = \int \Omega_a(T_a)\,dT_a \simeq \Omega_a\,T_a + \frac{1}{2}\,\dot{\Omega}_a\,T_a^2 + \frac{1}{6}\,\ddot{\Omega}_a\,T_a^3 + \cdots, \qquad (15)$$

where Φ_a is the longitude around the spin axis. [Depending on the precise definition of the rest-frame attached to the pulsar, the spin axis can either be fixed, or be slowly evolving, see e.g. [13].] One must then relate the initial direction (Θ_a, Φ_a), and proper time T_a, of emission of the pulsar beam to the coordinate direction and coordinate time of the null geodesic representing the electromagnetic beam in the 'global' coordinates x^μ used to describe the dynamics of the binary system [NB: the explicit orbital motion (9)–(13) refers to such global coordinates $x^0 = ct$, x^i.] This is done by using the link (6) in which z_a^i denotes the global coordinates of the 'center of mass' of the pulsar, T_a the local (proper) time of the pulsar frame, and where, for instance

$$e_i^0 = \frac{v_i}{c}\left(1 + \frac{1}{2}\frac{v^2}{c^2} + 3\frac{Gm_b}{c^2\,r_{ab}} + \cdots\right) + \cdots \qquad (16)$$

Using the link (6) (with expressions such as (16) for the coefficients $e_i^\mu, \dots$) one finds, among other results, that a radio beam emitted in the proper direction N^i in the local frame appears to propagate, in the global frame, in the coordinate direction n^i where

$$n^i = N^i + \frac{v^i}{c} - N^i\frac{N^j\,v^j}{c} + \mathcal{O}\left(\frac{v^2}{c^2}\right). \qquad (17)$$

This is the well known 'aberration effect', which will then contribute to the timing formula.

One must also write the link between the pulsar 'proper time' T_a and the coordinate time $t = x^0/c = z_a^0/c$ used in the orbital motion (9)–(13). This reads

$$-c^2\,dT_a^2 = \tilde{g}_{\mu\nu}(a_a^\lambda)\,dz_a^\mu\,dz_a^\nu \qquad (18)$$

where the 'tilde' denotes the operation consisting (in the matching approach) in discarding in $g_{\mu\nu}$ the 'self contributions' $\sim (Gm_a/R_a)^n$, while keeping the effect of the companion ($\sim Gm_b/r_{ab}$, etc...). One checks that this is equivalent (in the dimensional-continuation approach) in taking $x^\mu = z_a^\mu$ for sufficiently *small* values of the real part of the dimension d. To lowest order this yields the link

$$T_a \simeq \int dt \left(1 - \frac{2\,Gm_b}{c^2\,r_{ab}} - \frac{v_a^2}{c^2}\right)^{\frac{1}{2}} \simeq \int dt \left(1 - \frac{Gm_b}{c^2\,r_{ab}} - \frac{1}{2}\frac{v_a^2}{c^2}\right) \quad (19)$$

which combines the special relativistic and general relativistic time dilation effects. Hence, following [30] we can refer to them as the 'Einstein time delay'.

Then, one must compute the (global) time taken by a light beam emitted by the pulsar, at the proper time T_a (linked to $t_{emission}$ by (19)), in the initial global direction n^i (see (17)), to reach the barycenter of the solar system. This is done by writing that this light beam follows a null geodesic: in particular

$$0 = ds^2 = g_{\mu\nu}(x^\lambda)\,dx^\mu\,dx^\nu \simeq -\left(1 - \frac{2U}{c^2}\right)c^2\,dt^2 + \left(1 + \frac{2U}{c^2}\right)dx^2 \quad (20)$$

where $U = Gm_a/|\boldsymbol{x} - \boldsymbol{z}_a| + Gm_b/|\boldsymbol{x} - \boldsymbol{z}_b|$ is the Newtonian potential within the binary system. This yields (with $t_e \equiv t_{emission}$, $t_a \equiv t_{arrival}$)

$$t_a - t_e = \int_{t_e}^{t_a} dt \simeq \frac{1}{c}\int_{t_e}^{t_a} |d\boldsymbol{x}| + \frac{2}{c^3}\int_{t_e}^{t_a}\left(\frac{Gm_a}{|\boldsymbol{x} - \boldsymbol{z}_a|} + \frac{Gm_b}{|\boldsymbol{x} - \boldsymbol{z}_b|}\right)|d\boldsymbol{x}|. \quad (21)$$

The first term on the last RHS of (21) is the usual 'light crossing time' $\frac{1}{c}|\boldsymbol{z}_{\text{barycenter}}(t_a) - \boldsymbol{z}_a(t_e)|$ between the pulsar and the solar barycenter. It contains the 'Roemer time delay' due to the fact that $\boldsymbol{z}_a(t_e)$ moves on an orbit. The second term on the last RHS of (21) is the 'Shapiro time delay' due to the propagation of the beam in a curved spacetime (only the Gm_b piece linked to the companion is variable).

When inserting the 'quasi-Keplerian' form (9)–(13) of the relativistic motion in the 'Roemer' term in (21), together with all other relativistic effects, one finds that the final expression for the relativistic timing formula can be significantly simplified by doing two mathematical transformations. One can redefine the 'time eccentricity' e_t appearing in the 'Kepler equation' (9), and one can define a new 'eccentric anomaly' angle: $u \to u^{\text{new}}$ [we henceforth drop the superscript 'new' on u]. After these changes, the binary-system part of the general relativistic timing formula [30] takes the form (we suppress the index a on the pulsar proper time T_a)

$$t_{\text{barycenter}} - t_0 = D^{-1}[T + \Delta_R(T) + \Delta_E(T) + \Delta_S(T) + \Delta_A(T)] \quad (22)$$

with

$$\Delta_R = x\sin\omega[\cos u - e(1 + \delta_r)] + x[1 - e^2(1 + \delta_\theta)^2]^{1/2}\cos\omega\sin u, \quad (23)$$
$$\Delta_E = \gamma\sin u, \quad (24)$$
$$\Delta_S = -2r\ln\{1 - e\cos u - s[\sin\omega(\cos u - e) + (1 - e^2)^{1/2}\cos\omega\sin u]\}, \quad (25)$$
$$\Delta_A = A\{\sin[\omega + A_e(u)] + e\sin\omega\} + B\{\cos[\omega + A_e(u)] + e\cos\omega\}, \quad (26)$$

where $x = x_0 + \dot{x}(T - T_0)$ represents the projected light-crossing time ($x = a_{\text{pulsar}} \sin i/c$), $e = e_0 + \dot{e}(T - T_0)$ a certain (relativistically-defined) 'timing eccentricity', $A_e(u)$ the function

$$A_e(u) \equiv 2 \arctan \left[\left(\frac{1+e}{1-e} \right)^{1/2} \tan \frac{u}{2} \right], \tag{27}$$

$\omega = \omega_0 + k\, A_e(u)$ the 'argument of the periastron', and where the (relativistically-defined) 'eccentric anomaly' u is the function of the 'pulsar proper time' T obtained by solving the Kepler equation

$$u - e \sin u = 2\pi \left[\frac{T - T_0}{P_b} - \frac{1}{2} \dot{P}_b \left(\frac{T - T_0}{P_b} \right)^2 \right]. \tag{28}$$

It is understood here that the pulsar proper time T corresponding to the Nth pulse is related to the integer N by an equation of the form

$$N = c_0 + \nu_p\, T + \frac{1}{2} \dot{\nu}_p\, T^2 + \frac{1}{6} \ddot{\nu}_p\, T^3. \tag{29}$$

From these formulas, one sees that δ_θ (and δ_r) measure some relativistic distortion of the pulsar orbit, γ the amplitude of the 'Einstein time delay'[2] Δ_E, and r and s the *range* and *shape* of the 'Shapiro time delay'[3] Δ_S. Note also that the dimensionless PPK parameter k measures the *non-uniform* advance of the periastron. It is related to the often quoted *secular* rate of periastron advance $\dot{\omega} \equiv \langle d\omega/dt \rangle$ by the relation $k = \dot{\omega} P_b/2\pi$. It has been explicitly checked that binary-pulsar observational data do indeed require to model the relativistic periastron advance by means of the non-uniform (and non-trivial) function of u multiplying k on the R.H.S. of (27) [31][4]. Finally, we see from (28) that P_b represents the (periastron to periastron) orbital period at the fiducial epoch T_0, while the dimensionless parameter $\dot{P}_b$ represents the time derivative of P_b (at T_0).

Schematically, the structure of the DD timing formula (22) is

$$t_{\text{barycenter}} - t_0 = F\left[T_N; \{p^K\}; \{p^{PK}\}; \{q^{PK}\} \right], \tag{30}$$

---

[2] The post-Keplerian timing parameter γ, first introduced in [29], has the dimension of time, and should not be confused with the dimensionless post-Newtonian Eddington parameter γ^{PPN} probed by solar-system experiments (see below).

[3] The dimensionless parameter s is numerically equal to the sine of the inclination angle i of the orbital plane, but its real definition within the PPK formalism is the timing parameter which determines the 'shape' of the logarithmic time delay $\Delta_S(T)$.

[4] Alas this function is theory-independent, so that the non-uniform aspect of the periastron advance cannot be used to yield discriminating tests of relativistic gravity theories.

where $t_{\text{barycenter}}$ denotes the solar-system barycentric (infinite frequency) arrival time of a pulse, T the pulsar emission proper time (corrected for aberration), $\{p^K\} = \{P_b, T_0, e_0, \omega_0, x_0\}$ is the set of *Keplerian* parameters, $\{p^{PK} = k, \gamma, \dot{P}_b, r, s, \delta_\theta, \dot{e}, \dot{x}\}$ the set of *separately measurable post-Keplerian* parameters, and $\{q^{PK}\} = \{\delta_r, A, B, D\}$ the set of *not separately measurable post-Keplerian* parameters [31]. [The parameter D is a 'Doppler factor' which enters as an overall multiplicative factor D^{-1} on the right-hand side of Eq. (22).]

A further simplification of the DD timing formula was found possible. Indeed, the fact that the parameters $\{q^{PK}\} = \{\delta_r, A, B, D\}$ are not separately measurable means that they can be absorbed in changes of the other parameters. The explicit formulas for doing that were given in [30] and [31]: they consist in redefining e, x, P_b, δ_θ and δ_r. At the end of the day, it suffices to consider a simplified timing formula where $\{\delta_r, A, B, D\}$ have been set to some given fiducial values, e.g. $\{0, 0, 0, 1\}$, and where one only fits for the remaining parameters $\{p^K\}$ and $\{p^{PK}\}$.

Finally, let us mention that it is possible to extend the general parametrized timing formula (30) by writing a similar parametrized formula describing the effect of the pulsar orbital motion on the directional spectral luminosity $[d(\text{energy})/d(\text{time})\, d(\text{frequency})\, d(\text{solid angle})]$ received by an observer. As discussed in detail in [31] this introduces a new set of 'pulse-structure post-Keplerian parameters'.

4 Phenomenological Approach to Testing Relativistic Gravity with Binary Pulsar Data

As stated in the Introduction, binary pulsars contain strong gravity domains and should therefore allow one to test the strong-field aspects of relativistic gravity. The question we face is then the following: How can one use binary pulsar data to test strong-field (and radiative) gravity?

Two different types of answers can be given to this question: a *phenomenological* (or *theory-independent*) one, or various types of *theory-dependent* approaches. In this section we shall consider the phenomenological approach.

The phenomenological approach to binary-pulsar tests of relativistic gravity is called the *parametrized post-Keplerian formalism* [32, 31]. This approach is based on the fact that the mathematical form of the multi-parameter DD timing formula (30) was found to be applicable not only in General Relativity, but also in a wide class of alternative theories of gravity. Indeed, any theory in which gravity is mediated not only by a metric field $g_{\mu\nu}$ but by a general combination of a metric field and of one or several scalar fields $\varphi^{(a)}$ will induce relativistic timing effects in binary pulsars which can still be parametrized by the formulas (22)–(29). Such general 'tensor-multi-scalar' theories of gravity contain arbitrary functions of the scalar fields. They have been studied in full generality in [33]. It was shown that, under certain conditions, such tensor–scalar gravity theories could lead, because of strong-field effects, to very different predictions from those of General Relativity in binary pulsar timing observations [34, 35, 36]. However, the point which is important for this section, is that even when such strong-field effects develop one can still use the universal DD timing formula (30) to fit the observed pulsar times of arrival.

The basic idea of the phenomenological, parametrized post-Keplerian (PPK) approach is then the following: By least-square fitting the observed sequence of pulsar arrival times t_N to the parametrized formula (30) (in which T_N is defined by (29) which introduces the further parameters $\nu_p, \dot{\nu}_p, \ddot{\nu}_p$) one can phenomenologically extract from raw observational data the (best fit) values of all the parameters entering (29) and (30). In particular, one so determines both the set of Keplerian parameters $\{p^K\} = \{P_b, T_0, e_0, \omega_0, x_0\}$, and the set of post-Keplerian (PK) parameters $\{p^{PK}\} = \{k, \gamma, \dot{P}_b, r, s, \delta_\theta, \dot{e}, \dot{x}\}$. In extracting these values, we did not have to assume any theory of gravity. However, each specific theory of gravity will make specific predictions relating the PK parameters to the Keplerian ones, and to the two (a priori unknown) masses m_a and m_b of the pulsar and its companion. [For certain PK parameters one must also consider other variables related to the spin vectors of a and b.] In other words, the measurement (in addition of the Keplerian parameters) of each PK parameter defines, for each given theory, a *curve in the* (m_a, m_b) *mass plane*. For any given theory, the measurement of two PK parameters determines two curves and thereby generically determines the values of the two masses m_a and m_b (as the point of intersection of these two curves). Therefore, as soon as one measures *three* PK parameters one obtains a *test* of the considered gravity theory. The test is passed only if the three curves meet at one point. More generally, the measurement of n PK timing parameters yields $n - 2$ independent tests of relativistic gravity. Any one of these tests, i.e. any simultaneous measurement of three PK parameters can either confirm or put in doubt any given theory of gravity.

As General Relativity is our current most successful theory of gravity, it is clearly the prime target for these tests. We have seen above that the timing data of each binary pulsar provides a maximum of 8 PK parameters: $k, \gamma, \dot{P}_b, r, s, \delta_\theta, \dot{e}$ and $\dot{x}$. Here, we were talking about a normal 'single line' binary pulsar where, among the two compact objects a and b only one of the two, say a is observed as a pulsar. In this case, one binary system can provide up to $8 - 2 = 6$ tests of GR. In practice, however, it has not yet been possible to measure the parameter δ_θ (which measures a small relativistic deformation of the elliptical orbit), nor the secular parameters $\dot{e}$ and $\dot{x}$. The original Hulse-Taylor system PSR 1913+16 has allowed one to measure 3 PK parameters: $k \equiv \langle \dot{\omega} \rangle P_b/2\pi$, γ and $\dot{P}_b$. The two parameters k and γ involve (non-radiative) strong-field effects, while, as explained above, the orbital period derivative $\dot{P}_b$ is a direct consequence of the term $A_5 \sim v^5/c^5$ in the binary-system equations of motion (5). The term A_5 is itself directly linked to the retarded propagation, at the velocity of light, of the gravitational interaction between the two strongly self-gravitating bodies a and b. Therefore, any test involving $\dot{P}_b$ will be a *mixed radiative strong-field* test.

Let us explain in this example what information one needs to implement a phenomenological test such as the $(k - \gamma - \dot{P}_b)_{1913+16}$ one. First, we need to know the predictions made by the considered target theory for the PK parameters k, γ and $\dot{P}_b$ as functions of the two masses m_a and m_b. These predictions have been worked out, for General Relativity, in [29, 28, 30]. Introducing the notation (where $n \equiv 2\pi/P_b$)

$$M \equiv m_a + m_b \tag{31}$$

$$X_a \equiv m_a/M ; \quad X_b \equiv m_b/M ; \quad X_a + X_b \equiv 1 \tag{32}$$

$$\beta_O(M) \equiv \left(\frac{GMn}{c^3}\right)^{1/3} , \tag{33}$$

they read

$$k^{\mathrm{GR}}(m_a, m_b) = \frac{3}{1-e^2}\, \beta_O^2 , \tag{34}$$

$$\gamma^{\mathrm{GR}}(m_a, m_b) = \frac{e}{n}\, X_b(1 + X_b)\, \beta_O^2 , \tag{35}$$

$$\dot{P}_b^{\mathrm{GR}}(m_a, m_b) = -\frac{192\pi}{5} \frac{1 + \frac{73}{24} e^2 + \frac{37}{96} e^4}{(1 - e^2)^{7/2}}\, X_a X_b \beta_O^5 . \tag{36}$$

However, if we use the three predictions (34)–(36), together with the best current observed values of the PK parameters $k^{\mathrm{obs}}, \gamma^{\mathrm{obs}}, \dot{P}_b^{\mathrm{obd}}$ [37] we shall find that the three curves $k^{\mathrm{GR}}(m_a, m_b) = k^{\mathrm{obs}}$, $\gamma^{\mathrm{GR}}(m_a, m_b) = \gamma^{\mathrm{obs}}$, $\dot{P}_b^{\mathrm{GR}}(m_a, m_b) = \dot{P}_b^{\mathrm{obs}}$ in the (m_a, m_b) mass plane *fail to meet* at about the $13\,\sigma$ level! Should this put in doubt General Relativity? No, because [38] has shown that the time variation (notably due to galactic acceleration effects) of the Doppler factor D entering (22) entailed an extra contribution to the 'observed' period derivative $\dot{P}_b^{\mathrm{obs}}$. We need to subtract this non-GR contribution before drawing the corresponding curve: $\dot{P}_b^{\mathrm{GR}}(m_a, m_b) = \dot{P}_b^{\mathrm{obs}} - \dot{P}_b^{\mathrm{galactic}}$. Then one finds that the three curves *do meet* within one σ. This yields a deep confirmation of General Relativity, and a direct observational proof of the reality of gravitational radiation.

We said several times that this test is also a probe of the strong-field aspects of GR. How can one see this? A look at the GR predictions (34)–(36) does not exhibit explicit strong-field effects. Indeed, the derivation of (34)–(36) used in a crucial way the 'effacement of internal structure' that occurs in the general relativistic dynamics of compact objects. This non-trivial property is rather specific of GR and means that, in this theory, all the strong-field effects can be *absorbed* in the definition of the masses m_a and m_b. One can, however, verify that strong-field effects do enter the observable PK parameters

$k, \gamma, \dot{P}_b$ etc... by considering how the theoretical predictions (34)–(36) get modified in alternative theories of gravity. The presence of such strong-field effects in PK parameters was first pointed out in [7] (see also [39]) for the Jordan–Fierz–Brans–Dicke theory of gravity, and in [8] for Rosen's bi-metric theory of gravity. A detailed study of such strong-field deviations was then performed in [33, 34, 35] for general tensor-(multi-)scalar theories of gravity. In the following section we shall exhibit how such strong-field effects enter the various post-Keplerian parameters.

Continuing our historical review of phenomenological pulsar tests, let us come to the binary system which was the first one to provide several 'pure strong-field tests' of relativistic gravity, without mixing of radiative effects: PSR 1534+12. In this system, it was possible to measure the four (non radiative) PK parameters k, γ, r and s. [We see from (25) that r and s measure, respectively, the *range* and the *shape* of the 'Shapiro time delay' Δ_S.] The measurement of the 4 PK parameters k, γ, r, s define 4 curves in the (m_a, m_b) mass plane, and thereby yield 2 strong-field tests of GR. It was found in [40] that GR passes these two tests. For instance, the ratio between the measured value s^{obs} of the phenomenological parameter[5] s and the value $s^{\mathrm{GR}}[k^{\mathrm{obs}}, \gamma^{\mathrm{obs}}]$ predicted by GR on the basis of the measurements of the two PK parameters k and γ (which determine, via (34), (35), the GR-predicted value of m_a and m_b) was found to be $s^{\mathrm{obs}}/s^{\mathrm{GR}}[k^{\mathrm{obs}}, \gamma^{\mathrm{obs}}] = 1.004 \pm 0.007$ [40]. The most recent data [41] yield $s^{\mathrm{obs}}/s^{\mathrm{GR}}[k^{\mathrm{obs}}, \gamma^{\mathrm{obs}}] = 1.000 \pm 0.007$. We see that we have here a confirmation of the strong-field regime of GR at the 1% level.

Another way to get phenomenological tests of the strong field aspects of gravity concerns the possibility of a violation of the strong equivalence principle. This is parametrized by phenomenologically assuming that the ratio between the gravitational and the inertial mass of the pulsar differs from unity (which is its value in GR): $(m_{\mathrm{grav}}/m_{\mathrm{inert}})_a = 1 + \Delta_a$. Similar to what happens in the Earth-Moon-Sun system [42], the three-body system made of a binary pulsar and of the Galaxy exhibits a 'polarization' of the orbit which is proportional to $\Delta \equiv \Delta_a - \Delta_b$, and which can be constrained by considering certain quasi-circular neutron-star-white-dwarf binary systems [43]. See [44] for recently published improved limits[6] on the phenomenological equivalence-principle violation parameter Δ.

The Parkes multibeam survey has recently discovered several new interesting 'relativistic' binary pulsars, thereby giving a huge increase in the number of phenomenological tests of relativistic gravity. Among those new binary pul-

---

[5] As already mentioned the dimensionless parameter s is numerically equal (in all theories) to the sine of the inclination angle i of the orbital plane, but it is better thought, in the PPK formalism, as a phenomenological timing parameter determining the 'shape' of the logarithmic time delay $\Delta_S(T)$.

[6] Note, however, that these limits, as well as those previously obtained in [45], assume that the (a priori pulsar-mass dependent) parameter $\Delta \simeq \Delta_a$ is the same for all the analyzed pulsars.

sar systems, two stand out as superb testing grounds for relativistic gravity: (i) PSR J1141−6545 [46, 47], and (ii) the remarkable double binary pulsar PSR J0737−3039A and B [2, 3, 48, 49] (see also the following chapter by M. Kramer).

The PSR J1141−6545 timing data have led to the measurement of 3 PK parameters: k, γ, and $\dot{P}_b$ [47]. As in PSR 1913+16 this yields one mixed radiative-strong-field test[7].

The timing data of the millisecond binary pulsar PSR J0737−3039A have led to the direct measurement of 5 PK parameters: k, γ, r, s and $\dot{P}_b$ [3, 48, 49]. In addition, the 'double line' nature of this binary system (i.e. the fact that one observes both components, A and B, as radio pulsars) allows one to perform new phenomenological tests by using *Keplerian* parameters. Indeed, the simultaneous measurement of the Keplerian parameters x_a and x_b representing the projected light crossing times of both pulsars (A and B) gives access to the combined Keplerian parameter

$$R^{\mathrm{obs}} \equiv \frac{x_b^{\mathrm{obs}}}{x_a^{\mathrm{obs}}} . \tag{37}$$

On the other hand, the general derivation of [30] (applicable to any Lorentz-invariant theory of gravity, and notably to any tensor–scalar theory) shows that the theoretical prediction for the the ratio R, considered as a function of the masses m_a and m_b, is

$$R^{\mathrm{theory}} = \frac{m_a}{m_b} + \mathcal{O}\left(\frac{v^4}{c^4}\right) . \tag{38}$$

The absence of any *explicit* strong-field-gravity effects in the theoretical prediction (38) (to be contrasted, for instance, with the predictions for PK parameters in tensor–scalar gravity discussed in the next section) is mainly due to the convention used in [30] and [31] for *defining* the masses m_a and m_b. These are always defined so that the Lagrangian for two non interacting compact

---

[7] In addition, scintillation data have led to an estimate of the sine of the orbital inclination, $\sin i$ [50]. As said above, $\sin i$ numerically coincides with the PK parameter s measuring the 'shape' of the Shapiro time delay. Therefore, one could use the scintillation measurements as an indirect determination of s, thereby obtaining two independent tests from PSR J1141−6545 data. A caveat, however, is that the extraction of $\sin i$ from scintillation measurements rests on several simplifying assumptions whose validity is unclear. In fact, in the case of PSR J0737−3039 the direct timing measurement of s disagrees with its estimate via scintillation data [49]. It is therefore safer not to use scintillation estimates of $\sin i$ on the same footing as direct timing measurements of the PK parameter s. On the other hand, a safe way of obtaining an s-related gravity test consists in using the necessary mathematical fact that $s = \sin i \leq 1$. In GR the definition $x_a = a_a \sin i / c$ leads to $\sin i = n\,x_a / (\beta_0\,X_b)$. Therefore we can write the inequality $n\,x_a / (\beta_0(M)\,X_b) \leq 1$ as a phenomenological test of GR.

objects reads $L_0 = \sum_a - m_a c^2 (1 - \boldsymbol{v}_a^2/c^2)^{1/2}$. In other words, $m_a c^2$ represents the total energy of body a. This means that one has *implicitly* lumped in the definition of m_a many strong-self-gravity effects. [For instance, in tensor–scalar gravity m_a includes not only the usual Einsteinian gravitational binding energy due to the self-gravitational field $g_{\mu\nu}(x)$, but also the extra binding energy linked to the scalar field $\varphi(x)$.] Anyway, what is important is that, when performing a phenomenological test from the measurement of a triplet of parameters, e.g. $\{k, \gamma, R\}$, *at least one* parameter among them be a priori sensitive to strong-field effects. This is enough for guaranteeing that the crossing of the three curves $k^{\mathrm{theory}}(m_a, m_b) = k^{\mathrm{obs}}$, $\gamma^{\mathrm{theory}}(m_a, m_b) = \gamma^{\mathrm{obs}}$, $R^{\mathrm{theory}}(m_a, m_b) = R^{\mathrm{obs}}$ is really a probe of strong-field gravity.

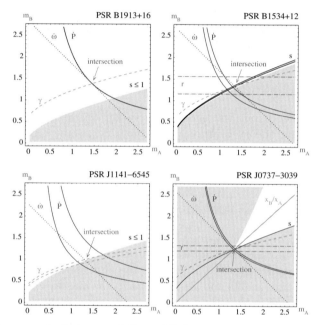

Fig. 1. Phenomenological tests of General Relativity obtained from Keplerian and post-Keplerian timing parameters of four relativistic pulsars. Figure taken from [51].

In conclusion, the two recently discovered binary pulsars PSR J1141−6545 and PSR J0737−3039 have more than *doubled* the number of phenomenological tests of (radiative and) strong-field gravity. Before their discovery, the 'canonical' relativistic binary pulsars PSR 1913+16 and PSR 1534+12 had given us *four* such tests: one $(k - \gamma - \dot{P}_b)$ test from PSR 1913+16 and three $(k - \gamma - r - s - \dot{P}_b[8])$ tests from PSR 1534+12. The two new binary systems

---

[8] The timing measurement of $\dot{P}_b^{\mathrm{obs}}$ in PSR 1534+12 is even more strongly affected by kinematic corrections ($\dot{D}$ terms) than in the PSR 1913+16 case. In absence of a

have given us *five*[9] more phenomenological tests: one $(k - \gamma - \dot{P}_b)$ (or two, $k - \gamma - \dot{P}_b - s$) tests from PSR J1141−6545 and four $(k - \gamma - r - s - \dot{P}_b - R)$ tests from PSR J0737−3039[10]. As illustrated in Fig. 1, these *nine* phenomenological tests of strong-field (and radiative) gravity are all in beautiful agreement with General Relativity.

In addition, let us recall that several quasi-circular wide binaries, made of a neutron star and a white dwarf, have led to high-precision phenomenological confirmations [44] (in strong-field conditions) of one of the deep predictions of General Relativity: the 'strong' equivalence principle, i.e. the fact that various bodies fall with the same acceleration in an external gravitational field, independently of the strength of their self-gravity.

Finally, let us mention that Ref. [31] has extended the philosophy of the phenomenological (parametrized post-Keplerian) analysis of timing data, to a similar phenomenological analysis of *pulse-structure data*. Ref. [31] showed that, in principle, one could extract up to 11 'post-Keplerian pulse-structure parameters'. Together with the 8 post-Keplerian timing parameters of a (single-line) binary pulsar, this makes a total of 19 phenomenological PK parameters. As these parameters depend not only on the two masses m_a, m_b but also on the two angles λ, η determining the direction of the spin axis of the pulsar, the maximum number of tests one might hope to extract from one (single-line) binary pulsar is $19 - 4 = 15$. However, the present accuracy with which one can model and measure the pulse structure of the known pulsars has not yet allowed one to measure any of these new pulse-structure parameters in a theory-independent and model-independent way.

Nonetheless, it has been possible to confirm the reality (and order of magnitude) of the spin–orbit coupling in GR which was pointed out [52, 53] to be observable via a secular change of the intensity profile of a pulsar signal. Confirmations of general relativistic spin–orbit effects in the evolution of pulsar profiles were obained in several pulsars: PSR 1913+16 [54, 55], PSR B1534+12 [56] and PSR J1141−6545 [57]. In this respect, let us mention that the spin–orbit interaction affects also several PK parameters, either by inducing a secular evolution in some of them (see [31]) or by contributing to their value. For instance, the spin–orbit interaction contributes to the observed value of the periastron advance parameter k an amount which is significant for the pulsars (such as 1913+16 and 0737−3039) where k is measured with high-accuracy. It was then pointed out [58] that this gives, in principle, and indirect way of measuring the moment of inertia of neutron stars (a useful quantity for probing the equation of state of nuclear matter [59, 60]). However, this can

---

precise, independent measurement of the distance to PSR 1534+12, the $k - \gamma - \dot{P}_b$ test yields, at best, a $\sim$ 15% test of GR.

[9] Or even six, if we use the scintillation determination of s in PSR J1141−6545.

[10] The companion pulsar 0737−3039B being non recycled, and being visible only during a small part of its orbit, cannot be timed with sufficient accuracy to allow one to measure any of its post-Keplerian parameters.

be done only if one measures, besides k, *two other* PK parameters with 10^{-5} accuracy. A rather tall order which will be a challenge to meet.

The phenomenological approach to pulsar tests has the advantage that it can confirm or invalidate a specific theory of gravity without making assumptions about other theories. Moreover, as General Relativity has no free parameters, any test of its predictions is a potentially lethal test. From this point of view, it is remarkable that GR has passed with flying colours all the pulsar tests if has been submitted to. [See, notably, Fig. 1.] As argued above, these tests have probed strong-field aspects of gravity which had not been probed by solar-system (or cosmological) tests. On the other hand, a disadvantage of the phenomenological tests is that they do not tell us in any precise way which strong-field structures have been actually tested. For instance, let us imagine that one day one specific PPK test fails to be satisfied by GR, while the others are OK. This leaves us in a quandary: If we trust the problematic test, we must conclude that GR is wrong. However, the other tests say that GR is OK. This example shows that we would like to have some idea of what physical effects, linked to strong-field gravity, enter in each test, or even better in each PK parameter. The 'effacement of internal structure' which takes place in GR does not allow one to discuss this issue. This gives us a motivation for going beyond the phenomenological PPK approach by considering *theory-dependent* formalisms in which one embeds GR within a *space of alternative gravity theories.*

5 Theory-Space Approach to Testing Relativistic Gravity with Binary Pulsar Data

A complementary approach to testing gravity with binary pulsar data consists in embedding General Relativity within a *multi-parameter space of alternative theories of gravity.* In other words, we want to *contrast* the predictions of GR with the predictions of continuous families of alternative theories. In so doing we hope to learn more about which structures of GR are actually being probed in binary pulsar tests. This is a bit similar to the well-known psycho-physiological fact that the best way to appreciate a nuance of colour is to surround a given patch of colour by other patches with slightly different colours. This makes it much easier to detect subtle differences in colour. In the same way, we hope to learn about the probing power of pulsar tests by seeing how the phenomenological tests summarized in Fig. 1 fail (or continue) to be satisfied when one continuously deform, away from GR, the gravity theory which is being tested.

Let us first recall the various ways in which this *theory-space approach* has been used in the context of the solar-system tests of relativistic gravity.

5.1 Theory-Space Approaches to Solar-System Tests of Relativistic Gravity

In the quasi-stationary weak-field context of the solar-system, this theory-space approach has been implemented in two different ways. First, the parametrized post-Newtonian (PPN) formalism [61, 62, 63, 42, 64, 65, 11, 66] describes many 'directions' in which generic alternative theories of gravity might differ in their *weak-field* predictions from GR. In its most general versions the PPN formalism contains 10 'post-Einstein' PPN parameters, $\bar{\gamma} \equiv \gamma^{\mathrm{PPN}} - 1$[11], $\bar{\beta} \equiv \beta^{\mathrm{PPN}} - 1, \xi, \alpha_1, \alpha_2, \alpha_3, \zeta_1, \zeta_2, \zeta_3, \zeta_4$. Each one of these dimensionless quantities parametrizes a certain class of slow-motion, weak-field gravitational effects which deviate from corresponding GR predictions. For instance, $\bar{\gamma}$ parametrizes modifications both of the effect of a massive body (say, the Sun) on the light passing near it, and of the terms in the two-body gravitational Lagrangian which are proportional to $(G m_a m_b / r_{ab}) \cdot (\boldsymbol{v}_a - \boldsymbol{v}_b)^2 / c^2$.

A second way of implementing the theory-space philosophy consists in considering some explicit, parameter-dependent family of alternative relativistic theories of gravity. For instance, the simplest tensor–scalar theory of gravity put forward by Jordan [67], Fierz [68] and Brans and Dicke [69] has a unique free parameter, say $\alpha_0^2 = (2\omega_{\mathrm{BD}} + 3)^{-1}$. When $\alpha_0^2 \to 0$, this theory reduces to GR, so that α_0^2 (or $1/\omega_{\mathrm{BD}}$) measures all the deviations from GR. When considering the weak-field limit of the Jordan–Fierz–Brans–Dicke (JFBD) theory, one finds that it can be described within the PPN formalism by choosing $\bar{\gamma} = -2\,\alpha_0^2(1 + \alpha_0^2)^{-1}, \bar{\beta} = 0$ and $\xi = \alpha_i = \zeta_j = 0$.

Having briefly recalled the two types of theory-space approaches used to discuss solar-system tests, let us now consider the case of binary-pulsar tests.

5.2 Theory-Space Approaches to Binary-Pulsar Tests of Relativistic Gravity

There exist generalizations of these two different theory-space approaches to the context of strong-field gravity and binary pulsar tests. First, the PPN formalism has been (partially) extended beyond the 'first post-Newtonian' (1PN) order deviations from GR ($\sim v^2/c^2 + Gm/c^2\,r$) to describe 2PN order deviations from GR $\left(\sim \left(\frac{v^2}{c^2} + \frac{Gm}{c^2 r} \right)^2 \right)$ [70]. Remarkably, there appear only

---

[11] The PPN parameter γ^{PPN} is usually denoted simply as γ. To distinguish it from the Einstein-time-delay PPK timing parameter γ used above we add the superscript PPN. In addition, as the value of γ^{PPN} in GR is 1, we prefer to work with the parameter $\bar{\gamma} \equiv \gamma^{\mathrm{PPN}} - 1$ which vanishes in GR, and therefore measures a 'deviation' from GR in a certain 'direction' in theory-space. Similarly with $\bar{\beta} \equiv \beta^{\mathrm{PPN}} - 1$.

two new parameters at the 2PN level[12]: ϵ and ζ. Also, by expanding in powers of the self-gravity parameters of body a and b the predictions for the PPK timing parameters in generic tensor-multi-scalar theories, one has shown that these predictions depended on several 'layers' of new dimensionless parameters [33]. Early among these parameters one finds, the 1PN parameters $\bar{\beta}, \bar{\gamma}$ and then the basic 2PN parameters ϵ and ζ, but one also finds further parameters $\beta_3, (\beta\beta'), \beta'', \ldots$ which would not enter usual 2PN effects. The two approaches that we have just mentioned can be viewed as generalizations of the PPN formalism.

There exist also useful generalizations to the strong-field context of the idea of considering some explicit parameter-dependent family of alternative theories of relativistic gravity. Early studies [7, 8, 39] focussed either on the one-parameter JFBD tensor–scalar theory, or on some theories which are not continuously connected to GR, such as Rosen's bimetric theory of gravity. Though the JFBD theory exhibits a marked difference from GR in that it predicts the existence of dipole radiation, it has the disadvantage that the weak field, solar-system constraints on its unique parameter α_0^2 are so strong that they drastically constrain (and essentially forbid) the presence of any non-radiative, strong-field deviations from GR. In view of this, it is useful to consider other 'mini-spaces' of alternative theories.

A two-parameter mini-space of theories, that we shall denote[13] here as $T_2(\beta', \beta'')$, was introduced in [33]. This two-parameter family of tensor-bi-scalar theories was constructed so as to have exactly the same first post-Newtonian limit as GR (i.e. $\bar{\gamma} = \bar{\beta} = \cdots = 0$), but to differ from GR in its predictions for the various observables that can be extracted from binary pulsar data. Let us give one example of this behaviour of the $T_2(\beta', \beta'')$ class of theories. For a general theory of gravity we expect to have violations of the strong equivalence principle in the sense that the ratio between the gravitational mass of a self-gravitating body to its inertial mass will admit an expansion of the type

$$\frac{m_a^{\text{grav}}}{m_a^{\text{inert}}} \equiv 1 + \Delta_a = 1 - \frac{1}{2}\eta_1\, c_a + \eta_2\, c_a^2 + \ldots \qquad (39)$$

where $c_a \equiv -2\frac{\partial \ln m_a}{\partial \ln G}$ measures the 'gravitational compactness' (or fractional gravitational binding energy, $c_a \simeq -2\, E_a^{\text{grav}}/m_a c^2$) of body a. The numerical coefficient η_1 of the contribution linear in c_a is a combination of the first post-Newtonian order PPN parameters, namely $\eta_1 = 4\bar{\beta} - \bar{\gamma}$ [42]. The numerical coefficient η_2 of the term quadratic in c_a is a combination of the 1PN *and*

---

[12] When restricting oneself to the general class of tensor-multi-scalar theories. At the 1PN level, this restriction would imply that only the 'directions' $\bar{\gamma}$ and $\bar{\beta}$ are allowed.

[13] We add here an index 2 to T as a reminder that this is a class of tensor-*bi*-scalar theories, i.e. that they contain *two* independent scalar fields φ_1, φ_2 besides a dynamical metric $g_{\mu\nu}$.

2PN parameters. When working in the context of the $T_2(\beta', \beta'')$ theories, the 1PN parameters vanish exactly ($\bar{\beta} = 0 = \bar{\gamma}$) and the coefficient of the quadratic term becomes simply proportional to the theory parameter $\beta' : \eta_2 = \frac{1}{2}B\beta'$, where $B \approx 1.026$. This example shows explicitly how binary pulsar data (here the data constraining the equivalence principle violation parameter $\Delta = \Delta_a - \Delta_b$, see above) can go beyond solar-system experiments in probing certain strong-self-gravity effects. Indeed, solar-system experiments are totally insensitive to 2PN parameters because of the smallness of $c_a \sim Gm_a/c^2 R_a$ and of the structure of 2PN effects [70]. In contrast, the 'compactness' of neutron stars is of order $c_a \sim 0.21\, m_a/M_\odot \sim 0.3$ [33] so that the pulsar limit $|\Delta| < 5.5 \times 10^{-3}$ [44] yields, within the $T_2(\beta', \beta'')$ framework, a significant limit on the dimensionless (2PN order) parameter $\beta' : |\beta'| < 0.12$.

Ref. [35] introduced a new two-parameter mini-space of gravity theories, denoted here as $T_1(\alpha_0, \beta_0)$, which, from the point of view of theoretical physics, has several advantages over the $T_2(\beta', \beta'')$ mini-space mentioned above. First, it is technically simpler in that it contains only *one* scalar field φ besides the metric $g_{\mu\nu}$ (hence the index 1 on $T_1(\alpha_0, \beta_0)$). Second, it contains only positive-energy excitations (while one combination of the two scalar fields of $T_2(\beta', \beta'')$ carried negative-energy waves). Third, it is the minimal way to parametrize the huge class of tensor-mono-scalar theories with a 'coupling function' $a(\varphi)$ satisfying some very general requirements (see below).

Let us now motivate the use of tensor–scalar theories of gravity as alternatives to general relativity.

5.3 Tensor–Scalar Theories of Gravity

Let us start by recalling (essentially from [35]) why tensor-(mono)-scalar theories define a natural class of alternatives to GR. First, and foremost, the existence of scalar partners to the graviton is a simple theoretical possibility which has surfaced many times in the development of unified theories, from Kaluza-Klein to superstring theory. Second, they are general enough to describe many interesting deviations from GR (both in weak-field and in strong field conditions), but simple enough to allow one to work out their predictions in full detail.

Let us therefore consider a general tensor–scalar action involving a metric $\tilde{g}_{\mu\nu}$ (with signature 'mostly plus'), a scalar field Φ, and some matter variables ψ_m (including gauge bosons):

$$S = \frac{c^4}{16\pi G_*} \int \frac{d^4x}{c}\, \tilde{g}^{1/2} \left[F(\Phi)\tilde{R} - Z(\Phi)\tilde{g}^{\mu\nu}\partial_\mu\Phi\,\partial_\nu\Phi - U(\Phi) \right] + S_m[\psi_m; \tilde{g}_{\mu\nu}].$$
(40)

For simplicity, we assume here that the weak equivalence principle is satisfied, i.e., that the matter variables ψ_m are all coupled to the same 'physical metric'[14] $\tilde{g}_{\mu\nu}$. The general model (40) involves three arbitrary functions: a function $F(\Phi)$ coupling the scalar Φ to the Ricci scalar of $\tilde{g}_{\mu\nu}$, $\tilde{R} \equiv R(\tilde{g}_{\mu\nu})$, a function $Z(\Phi)$ renormalizing the kinetic term of Φ, and a potential function $U(\Phi)$. As we have the freedom of arbitrary redefinitions of the scalar field, $\Phi \to \Phi' = f(\Phi)$, only two functions among F, Z and U are independent. It is often convenient to rewrite (40) in a *canonical* form, obtained by redefining both Φ and $\tilde{g}_{\mu\nu}$ according to

$$g^*_{\mu\nu} = F(\Phi)\,\tilde{g}_{\mu\nu}\,, \tag{41}$$

$$\varphi = \pm \int d\Phi \left[\frac{3}{4} \frac{F'^2(\Phi)}{F^2(\Phi)} + \frac{1}{2} \frac{Z(\Phi)}{F(\Phi)} \right]^{1/2} . \tag{42}$$

This yields

$$S = \frac{c^4}{16\pi\, G_*} \int \frac{d^4x}{c}\, g_*^{1/2} \left[R_* - 2g_*^{\mu\nu}\partial_\mu\varphi\,\partial_\nu\varphi - V(\varphi) \right] + S_m \left[\psi_m; A^2(\varphi)\, g^*_{\mu\nu} \right], \tag{43}$$

where $R_* \equiv R(g^*_{\mu\nu})$, where the potential

$$V(\varphi) = F^{-2}(\Phi)\, U(\Phi)\,, \tag{44}$$

and where the *conformal coupling function* $A(\varphi)$ is given by

$$A(\varphi) = F^{-1/2}(\Phi)\,, \tag{45}$$

with $\Phi(\varphi)$ obtained by inverting the integral (42).

The two arbitrary functions entering the canonical form (43) are: (i) the conformal coupling function $A(\varphi)$, and (ii) the potential function $V(\varphi)$. Note that the 'physical metric' $\tilde{g}_{\mu\nu}$ (the one measured by laboratory clocks and rods) is conformally related to the 'Einstein metric' $g^*_{\mu\nu}$, being given by $\tilde{g}_{\mu\nu} = A^2(\varphi)\, g^*_{\mu\nu}$. The canonical representation is technically useful because it decouples the two irreducible propagating excitations: the spin-0 excitations are described by φ, while the pure spin-2 excitations are described by the *Einstein metric* $g^*_{\mu\nu}$ (with kinetic term the usual Einstein–Hilbert action $\propto R(g^*_{\mu\nu})$).

---

[14] Actually, most unified models suggest that there are violations of the weak equivalence principle. However, the study of general string-inspired tensor–scalar models [71] has found that the composition-dependent effects would be negligible in the gravitational physics of neutron stars that we consider here. The experimental limits on tests of the equivalence principle would, however, bring a strong additional constraint of order $10^{-5}\,\alpha_0^2 \sim \Delta a/a \lesssim 10^{-12}$. As this constraint is strongly model-dependent, we will not use it in our exclusion plots below. One should, however, keep in mind that a limit on the scalar coupling strength α_0^2 of order $\alpha_0^2 \lesssim 10^{-7}$ [71, 72] is likely to exist in many, physically-motivated, tensor–scalar models.

In many technical developments it is useful to work with the *logarithmic coupling function* $a(\varphi)$ such that:

$$a(\varphi) \equiv \ln A(\varphi)\,; \;\; A(\varphi) \equiv e^{a(\varphi)}\,. \tag{46}$$

In the case of the general model (40) this logarithmic[15] coupling function is given by

$$a(\varphi) = -\frac{1}{2} \ln F(\Phi)\,,$$

where $\Phi(\varphi)$ must be obtained from (42).

In the following, we shall assume that the potential $V(\varphi)$ is a slowly varying function of φ which, in the domain of variation we shall explore, is roughly equivalent to a very small mass term $V(\varphi) \sim 2\, m_\varphi^2 (\varphi - \varphi_0)^2$ with m_φ^2 of cosmological order of magnitude $m_\varphi^2 = \mathcal{O}(H_0^2)$, or, at least, with a range $\lambda_\varphi = m_\varphi^{-1}$ much larger than the typical length scales that we shall consider (such as the size of the binary orbit, or the size of the Galaxy when considering violations of the strong equivalence principle). Under this assumption[16] the potential function $V(\varphi)$ will only serve the role of fixing the value of φ far from the system (to $\varphi(r = \infty) = \varphi_0$), and its effect on the propagation of φ within the system will be negligible. In the end, the tensor–scalar phenomenology that we shall explore only depends on *one function*: the coupling function $a(\varphi)$.

Let us consider some examples to see what kind of coupling functions might naturally arise. First, the simplest case is the Jordan–Fierz–Brans–Dicke action, which is of the general type (40) with

$$F(\Phi) = \Phi \tag{47}$$
$$Z(\Phi) = \omega_{\mathrm{BD}}\, \Phi^{-1}\,, \tag{48}$$

where ω_{BD} is an arbitrary constant. Using Eqs. (42), (45) above, one finds that $-2\,\alpha_0\,\varphi = \ln \Phi$ and that the (logarithmic) coupling function is simply

$$a(\varphi) = \alpha_0\,\varphi + \mathrm{const.}\,, \tag{49}$$

where $\alpha_0 = \mp(2\omega_{\mathrm{BD}} + 3)^{-1/2}$, depending on the sign chosen in Eq. (42). Independently of this sign, one has the link

$$\alpha_0^2 = \frac{1}{2\,\omega_{\mathrm{BD}} + 3}\,. \tag{50}$$

---

[15] As we shall mostly work with $a(\varphi)$ below, we shall henceforth drop the adjective 'logarithmic'.

[16] Note, however, that, as was recently explored in [73, 74, 75], a sufficiently fast varying potential $V(\varphi)$ can change the tensor–scalar phenomenology by endowing φ with a mass term $m_\varphi^2 = \frac{1}{4}\partial^2 V/\partial\varphi^2$ which strongly depends on the local value of φ and, thereby can get large in sufficiently dense environments.

Note that $2\,\omega_{\rm BD} + 3$ must be positive for the spin-0 excitations to have the correct (non ghost) sign.

Let us now discuss the often considered case of a massive scalar field having a non-minimal coupling to curvature

$$S = \frac{c^4}{16\pi\, G_*} \int \frac{d^4x}{c}\, \tilde{g}^{1/2} \left(\tilde{R} - \tilde{g}^{\mu\nu} \partial_\mu \Phi\, \partial_\nu \Phi - m_\Phi^2 \Phi^2 + \xi \tilde{R}\, \Phi^2 \right) + S_m[\psi_m; \tilde{g}_{\mu\nu}] .$$
(51)

This is of the form (40) with

$$F(\Phi) = 1 + \xi \Phi^2 , \quad Z(\Phi) = 1 , \quad U(\Phi) = m_\Phi^2 \Phi^2 .$$
(52)

The case $\xi = -\frac{1}{6}$ is usually referred to as that of 'conformal coupling'. With the variables (51) the theory is ghost-free only if $2\,(1 + \xi \Phi^2)^2\,(d\varphi/d\Phi)^2 = 1 + \xi(1 + 6\,\xi)\,\Phi^2$ is everywhere positive. If we do not wish to restrict the initial values of Φ, we must have $\xi(1 + 6\,\xi) > 0$. Introducing then the notation $\chi \equiv \sqrt{\xi(1 + 6\,\xi)}$, we get the following link between Φ and φ:

$$2\sqrt{2}\,\varphi = \frac{\chi}{\xi}\, \ln\left[1 + 2\,\chi\,\Phi\left(\sqrt{1 + \chi^2\,\Phi^2} + \chi\,\Phi\right) \right]$$

$$+ \sqrt{6}\, \ln\left[1 - 2\sqrt{6}\,\xi\,\Phi\, \frac{\sqrt{1 + \chi^2\,\Phi^2} - \sqrt{6}\,\xi\,\Phi}{1 + \xi\,\Phi^2} \right] .$$
(53)

For small values of Φ, this yields $\varphi = \Phi/\sqrt{2} + \mathcal{O}(\Phi^3)$. The potential and the coupling functions are given by

$$V(\varphi) = \frac{m_\Phi^2\,\Phi^2}{1 + \xi \Phi^2} ,$$
(54)

$$a(\varphi) = -\frac{1}{2}\, \ln(1 + \xi \Phi^2) .$$
(55)

These functions have singularities when $1 + \xi \Phi^2$ vanishes. If we do not wish to restrict the initial value of Φ we must assume $\xi > 0$ (which then implies our previous assumption $\xi(1 + 6\,\xi) > 0$). Then there is a one-to-one relation between Φ and φ over the entire real line. Small values of Φ correspond to small values of φ and to a coupling function

$$a(\varphi) = -\,\xi\,\varphi^2 + \mathcal{O}(\varphi^4) .$$
(56)

On the other hand, large values of $|\Phi|$ correspond to large values of $|\varphi|$, and to a coupling function of the asymptotic form

$$a(\varphi) \simeq -\sqrt{2}\,\frac{\xi}{\chi}\, |\varphi| + \text{const.}$$
(57)

The potential $V(\varphi)$ has a minimum at $\varphi = 0$, as well as other minima at $\varphi \to \pm\infty$. If we assume, for instance, that m_Φ^2 and the cosmological dynamics are such that the cosmological value of φ is currently attracted towards zero, the value of φ at large distances from the local gravitating systems we shall consider will be $\varphi_0 \ll 1$.

As a final example of a possible tensor–scalar gravity theory, let us discuss the string-motivated dilaton-runaway scenario considered in [76]. The starting action (a functional of $\bar{g}_{\mu\nu}$ and Φ) was taken of the general form

$$S = \int d^4x \sqrt{\bar{g}} \left(\frac{B_g(\Phi)}{\alpha'} \bar{R} + \frac{B_\Phi(\Phi)}{\alpha'} [2 \bar{\Box} \Phi \right.$$

$$\left. - (\bar{\nabla}\Phi)^2] - \frac{1}{4} B_F(\Phi)\bar{F}^2 - V(\Phi) + \cdots \right),$$

and it was assumed that all the functions $B_i(\Phi)$ have a regular asymptotic behaviour when $\Phi \to +\infty$ of the form $B_i(\Phi) = C_i + \mathcal{O}(e^{-\Phi})$. Under this assumption the early cosmological evolution can push Φ towards $+\infty$ (hence the name 'runaway dilaton'). In the canonical, 'Einstein frame' representation (43), one has, for large values of Φ, $\Phi \simeq c\varphi$, where c is a numerical constant, and the coupling function to hadronic matter is given by

$$e^{a(\varphi)} \propto \Lambda_{\text{QCD}}(\varphi) \propto B_g^{-1/2}(\varphi) \exp[-8\pi^2 b_3^{-1} B_F(\varphi)]$$

where b_3 is the one-loop rational coefficient entering the renormalization-group running of the gauge field coupling g_F^2. This finally yields a coupling function of the approximate form (for large values of φ):

$$a(\varphi) \simeq k\, e^{-c\varphi} + \text{const.},$$

where the dimensionless constants k and c are both expected to be of order unity. (The constant c must be positive, but the sign of k is not *a priori* restricted.)

Summarizing: the JFBD model yields a coupling function which is a linear function of φ, Eq. (49), a non-minimally coupled scalar yields a coupling function which interpolates between a quadratic function of φ, Eq. (56), and a linear one, Eq. (57), and the dilaton-runaway scenario of Ref. [76] yields a coupling function of a decaying exponential type.

5.4 The Role of the Coupling Function $a(\varphi)$; Definition of the Two-Dimensional Space of Tensor–Scalar Gravity Theories $T_1(\alpha_0, \beta_0)$

Let us now discuss how the coupling function $a(\varphi)$ enters the observable predictions of tensor–scalar gravity at the first post-Newtonian (1PN) level, i.e., in the weak-field conditions appropriate to solar-system tests. It was shown in previous work that, if one uses appropriate units in the asymptotic region far from the system, namely units such that the asymptotic value $a(\varphi_0)$ of

$a(\varphi)$ vanishes[17], all observable quantities at the 1PN level depend only on the values of the first two derivatives of the $a(\varphi)$ at $\varphi = \varphi_0$. More precisely, if one defines

$$\alpha(\varphi) \equiv \frac{\partial\, a(\varphi)}{\partial\, \varphi}\,; \ \beta(\varphi) \equiv \frac{\partial\, \alpha(\varphi)}{\partial\, \varphi} = \frac{\partial^2\, a(\varphi)}{\partial\, \varphi^2}\,, \tag{58}$$

and denotes by $\alpha_0 \equiv \alpha(\varphi_0)$, $\beta_0 \equiv \beta(\varphi_0)$ their asymptotic values, one finds (see, e.g., [33]) that the effective gravitational constant between two bodies (as measured by a Cavendish experiment) is given by

$$G = G_*(1 + \alpha_0^2)\,, \tag{59}$$

while, among the PPN parameters, only the two basic Eddington ones, $\bar{\gamma} \equiv \gamma^{\mathrm{PPN}} - 1$, and $\bar{\beta} \equiv \beta^{\mathrm{PPN}} - 1$, do not vanish, and are given by

$$\bar{\gamma} \equiv \gamma^{\mathrm{PPN}} - 1 = -2\frac{\alpha_0^2}{1 + \alpha_0^2}\,, \tag{60}$$

$$\bar{\beta} \equiv \beta^{\mathrm{PPN}} - 1 = \frac{1}{2}\frac{\alpha_0\,\beta_0\,\alpha_0}{(1 + \alpha_0^2)^2}\,. \tag{61}$$

The structure of the results (60) and (61) can be transparently expressed by means of simple (Feynman-like) diagrams (see, e.g., [77]). Eqs. (59) and (60) correspond to diagrams where the interaction between two worldlines (representing two massive bodies) is mediated by the sum of the exchange of one graviton and one scalar particle. The scalar couples to matter with strength $\sim \alpha_0\,\sqrt{G_*}$. The exchange of a scalar excitation then leads to a term $\propto \alpha_0^2$. On the other hand, Eq. (61) corresponds to a non-linear interaction between three worldlines involving: (i) the 'generation' of a scalar excitation on a first worldline (factor α_0), (ii) a non-linear vertex on a second worldline associated to the quadratic piece of $a(\varphi)$ ($a_{\mathrm{quad}}(\varphi) = \frac{1}{2}\beta_0(\varphi - \varphi_0)^2$; so that one gets a factor β_0), and (iii) the final 'absorption' of a scalar excitation on a third worldline (second factor α_0).

Eqs. (60) and (61) can be summarized by saying that the first two coefficients in the Taylor expansion of the coupling function $a(\varphi)$ around $\varphi = \varphi_0$ (after setting $a(\varphi_0) = 0$)

$$a(\varphi) = \alpha_0(\varphi - \varphi_0) + \frac{1}{2}\beta_0(\varphi - \varphi_0)^2 + \cdots \tag{62}$$

suffice to determine the quasi-stationary, weak-field (1PN) predictions of any tensor–scalar theory. In other words, the solar-system tests only explore the 'osculating approximation' (62) (slope and local curvature) to the function $a(\varphi)$. Note that GR corresponds to a vanishing coupling function $a(\varphi) = 0$ (so that $\alpha_0 = \beta_0 = \cdots = 0$), the JFBD model corresponds to keeping only

---

[17] In these units the Einstein metric $g^*_{\mu\nu}$ and the physical metric $\tilde{g}_{\mu\nu}$ asymptotically coincide.

the first term on the R.H.S. of (62), while, for instance, the non-minimally coupled scalar field (with asymptotic value $\varphi_0 \ll 1$) does indeed lead to non-zero values for both α_0 and β_0, namely

$$\alpha_0 \simeq -2\,\xi\,\varphi_0 \,;\ \beta_0 \simeq -2\,\xi\,. \tag{63}$$

Finally the dilaton-runaway scenario considered above leads also to non-zero values for both α_0 and β_0, namely

$$\alpha_0 \simeq -k\,c\,e^{-c\varphi_0} \,;\ \beta_0 \simeq +k\,c^2\,e^{-c\varphi_0}\,, \tag{64}$$

for a largish value of φ_0. Note that the dilaton-runaway model naturally predicts that $\alpha_0 \ll 1$, and that β_0 is of the same order of magnitude as $\alpha_0 : \beta_0 \simeq -c\,\alpha_0$ with c being (positive and) of order unity. The interesting outcome is that such a model is well approximated by the usual JFBD model (with $\beta_0 = 0$). This shows that a JFBD-like theory could come out from a model which is initially quite different from the usual exact JFBD theory.

As we shall discuss in detail below, solar-system tests constrain α_0^2 and $\alpha_0^2 |\beta_0|$ to be both small. This immediately implies that $|\alpha_0|$ must be small, i.e., that the scalar field is linearly weakly coupled to matter. On the other hand, the quadratic coupling parameter β_0 is not directly constrained. Both its magnitude and its sign can be more or less arbitrary. Note that there are no *a priori* sign restrictions on β_0. The conformal factor $A^2(\varphi) = \exp(2\,a(\varphi))$ entering Eq. (43) had to be positive, but this leads to no restrictions on the sign of $a(\varphi)$ and of its various derivatives[18]. For instance, in the non-minimally coupled scalar field case, it seemed more natural to require $\xi > 0$, which leads to a negative β_0 in view of Eq. (63).

Let us summarize the results above: (i) the most general tensor–scalar theory[19] is described by one arbitrary function $a(\varphi)$; and (ii) weak-field tests depend only on the first two terms, parametrized by α_0 and β_0, in the Taylor expansion (62) of $a(\varphi)$ around its asymptotic value φ_0.

From this follows a rather natural way to define a simple *mini space of tensor–scalar theories*. It suffices to consider the two-dimensional space of theories, say $T_1(\alpha_0, \beta_0)$, defined by the coupling function which is a quadratic polynomial in φ [34, 35], say

$$a_{\alpha_0,\beta_0}(\varphi) = \alpha_0(\varphi - \varphi_0) + \frac{1}{2}\,\beta_0(\varphi - \varphi_0)^2\,. \tag{65}$$

As indicated, this class of theories depends only on two parameters: α_0 and β_0. The asymptotic value φ_0 of φ does not count as a third parameter (when

---

[18] As explained above, we assume here the presence of a potential term $V(\varphi)$ to fix the asymptotic value φ_0 of φ. If the potential $V(\varphi)$ is absent (or negligible), the 'attractor mechanism' of Refs. [78, 71] would attract φ to a *minimum* of the coupling function $a(\varphi)$, thereby favoring a *positive* value of β_0.

[19] Under the assumption that the potential $V(\varphi)$ is a slowly-varying function of φ, which modifies the propagation of φ only on very large scales.

using the form (65)) because one can always work with the shifted field $\bar{\varphi} \equiv \varphi - \varphi_0$, with asymptotic value $\bar{\varphi}_0 = 0$ and coupling function $a_{\alpha_0,\beta_0}(\bar{\varphi}) = \alpha_0 \bar{\varphi} + \frac{1}{2} \beta_0 \bar{\varphi}^2$. Moreover, as already said, the asymptotic value $a(\varphi_0)$ of $a(\varphi)$ has also no physical meaning, because one can always use units such that it vanishes (as done in (65)).

Note also that an alternative way to represent the same class of theories is to use a coupling function of the very simple form

$$a_\beta(\varphi) = \frac{1}{2} \beta \varphi^2 , \tag{66}$$

but to keep the asymptotic value φ_0 as an independent parameter. This class of theories is clearly equivalent to $T_1(\alpha_0, \beta_0)$, Eq. (65), with the dictionary: $\alpha_0 = \beta \varphi_0$, $\beta_0 = \beta$.

5.5 Tensor–Scalar Gravity, Strong-Field Effects, and Binary-Pulsar Observables

Having chosen some mini-space of gravity theories, we now wish to derive what predictions these theories make for the timing observables of binary pulsars. To do this we need to generalize the general relativistic treatment of the motion and timing of binary systems comprising strongly self-gravitating bodies summarized above. Let us recall that this treatment was based on a multi-chart method, using a *matching* between two separate problems: (i) the 'internal problem' considers each strongly self-gravitating body in a suitable approximately freely falling frame where the influence of its companion is small, and (ii) the 'external problem' where the two bodies are described as effective point masses which interact via the various fields they are coupled to. Let us first consider the *internal problem*, i.e., the description of a neutron star in an approximately freely falling frame where the influence of the companion is reduced to imposing some boundary conditions on the tensor and scalar fields with which it interacts [7, 8, 33, 34, 35]. The field equations of a general tensor–scalar theory, as derived from the canonical action (43) (neglecting the effect of $V(\varphi)$) read

$$R^*_{\mu\nu} = 2\, \partial_\mu \varphi\, \partial_\nu \varphi + 8\pi\, G_* \left(T^*_{\mu\nu} - \frac{1}{2} T^* g^*_{\mu\nu} \right) , \tag{67}$$

$$\Box_{g_*} \varphi = -4\pi\, G_*\, \alpha(\varphi)\, T_* , \tag{68}$$

where $T^{\mu\nu}_* \equiv 2\, c\, (g_*)^{-1/2}\, \delta S_m / \delta g^*_{\mu\nu}$ denotes the material stress-energy tensor in 'Einstein units', and $\alpha(\varphi)$ the φ-derivative of the coupling function, see Eq. (58). All tensorial operations in Eqs. (67) and (68) are performed by using the Einstein metric $g^*_{\mu\nu}$.

Explicitly writing the field equations (67) and (68) for a slowly rotating (stationary, axisymmetric) neutron star, labelled[20] A, leads to a coupled set

---

[20] We henceforth use the labels A and B for the (recycled) pulsar and its companion, instead of the labels a and b used above. We henceforth use the label a to denote

of ordinary differential equations constraining the radial dependence of $g^*_{\mu\nu}$ and φ [35, 79]. Imposing the boundary conditions $g^*_{\mu\nu} \to \eta_{\mu\nu}$, $\varphi \to \varphi_a$ at large radial distances, finally determines the crucial 'form factors' (in Einstein units) describing the effective coupling between the neutron star A and the fields to which it is sensitive: total mass $m_A(\varphi_a)$, total scalar charge $\omega_A(\varphi_a)$, and inertia moment $I_A(\varphi_a)$. As indicated, these quantities are functions of the *asymptotic* value φ_a of φ felt by the considered neutron star[21]. They satisfy the relation $\omega_A(\varphi_a) = -\partial\, m_A(\varphi_a)/\partial\,\varphi_a$. From them, one defines other quantities that play an important role in binary pulsar physics, notably

$$\alpha_A(\varphi_a) \equiv -\frac{\omega_A}{m_A} \equiv \frac{\partial \ln m_A}{\partial \varphi_a} \, , \tag{69}$$

$$\beta_A(\varphi_a) \equiv \frac{\partial \alpha_A}{\partial \varphi_a} \, , \tag{70}$$

as well as

$$k_A(\varphi_a) \equiv -\frac{\partial \ln I_A}{\partial \varphi_a} \, . \tag{71}$$

The quantity α_A, Eq. (69), plays a crucial role. It measures the *effective coupling strength* between the neutron star and the ambient scalar field. If we formally let the self-gravity of the neutron A tend toward zero (i.e., if we consider a weakly self-gravitating object), the function $\alpha_A(\varphi_a)$ becomes replaced by $\alpha(\varphi_a)$ where $\alpha(\varphi) \equiv \partial\, a(\varphi)/\partial\,\varphi$ is the coupling strength appearing in the R.H.S. of Eq. (68). Roughly speaking, we can think of $\alpha_A(\varphi_a)$ as a (suitable defined) average value of the local coupling strength $\alpha(\varphi(r))$ over the radial profile of the neutron star A.

It was pointed out in Refs. [34, 35] that the strong self-gravity of a neutron star can cause the effective coupling strength $\alpha_A(\varphi_a)$ to become of order unity, even when its weak-field counterpart $\alpha_0 = \alpha(\varphi_a)$ is extremely small (as is implied by solar-system tests that put strong constraints on the PPN combination $\bar{\gamma} = -2\,\alpha_0^2/(1 + \alpha_0^2)$). This is illustrated, in the minimal context of the $T_1(\alpha_0, \beta_0)$ class of theories, in Fig. 2.

Note that when the baryonic mass $\bar{m}_A$ of the neutron star is smaller than the critical mass $\bar{m}_{cr} \simeq 1.24\, M_\odot$ the effective scalar coupling strength α_A of the star is quite small (because it is proportional to its weak-field limit $\alpha_0 = \alpha(\varphi_a)$). In contrast, when $\bar{m}_A > \bar{m}_{cr}$, $|\alpha_A|$ becomes of order unity,

---

the *asymptotic* value of some quantity (at large radial distances within the local frame, X^i_A or X^i_B, of the considered neutron star A or B).

[21] This φ_a is a combination of the cosmological background value φ_0 and of the scalar influence of the companion of the considered neutron star. It varies with the orbital period and is determined as part of the 'external problem' discussed below. Note that, strictly speaking, the label a (for *asymptotic*) should be indexed by the label of the considered neutron star: i.e. one should use a label a_A (and a *locally asymptotic* value φ_{a_A}) when considering the neutron star A, and a label a_B (with a corresponding φ_{a_B}) when considering the neutron star B.

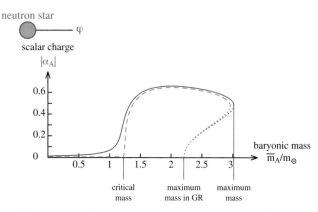

Fig. 2. Dependence upon the baryonic mass $\bar{m}_A$ of the coupling parameter α_A in the theory $T_1(\alpha_0, \beta_0)$ with $\alpha_0 = -0.014$, $\beta_0 = -6$. Figure taken from [80].

nearly independently of the externally imposed $\alpha_0 = \alpha_a = \alpha(\varphi_a)$. This interesting *non-perturbative* behaviour was related in [34, 35] to a mechanism of *spontaneous scalarization*, akin to the well-known mechanism of spontaneous magnetization of ferromagnets. See also [51] for a simple analytical description of the behaviour of α_A.

Let us also mention in passing that, in the case where A is a black hole, the effective coupling strength α_A actually *vanishes* [33]. This result is related to the impossibility of having (regular) 'scalar hair' on a black hole.

We have sketched above the first part of the matching approach to the motion and timing of strongly self-gravitating bodies: the 'internal problem'. It remains to describe the remaining 'external problem'. As already mentioned (and emphasized, in the present context, by Eardley [7, 11]), the most efficient way to describe the external problem is, instead of matching in detail the external fields $(g^*_{\mu\nu}, \varphi)$ to the fields generated by each body in its co-moving frame, to 'skeletonize' the bodies by point masses. Technically this means working with the action

$$S = \frac{c^4}{16\pi\, G_*} \int \frac{d^D x}{c}\, g_*^{1/2}[R_* - 2\, g_*^{\mu\nu}\, \partial_\mu \varphi\, \partial_\nu \varphi]$$
$$- \sum_A c \int m_A(\varphi(z_A))(-g^*_{\mu\nu}(z_A)\, dz_A^\mu\, dz_A^\nu)^{1/2}\,, \tag{72}$$

where the function $m_A(\varphi)$ in the last term on the R.H.S. is the function $m_A(\varphi_a)$ obtained above by solving the internal problem. Eq. (72) indicates that the argument of this function is taken to be $\varphi_a = \varphi(z_A)$, i.e., the value that the scalar field (as viewed in the external problem) takes at the location z_A^μ of the center of mass of body A. However, as body A is described, in the external problem, as a point mass this causes a technical difficulty: the externally determined field $\varphi(x)$ becomes formally singular at the location of the point sources, so that $\varphi(z_A)$ is *a priori* undefined. One can either deal with this problem by coming back to the physically well-defined matching approach (which shows that $\varphi(z_A)$ should be replaced by φ_a, the value of φ in an intermediate domain $R_A \ll r \ll |z_A - z_B|$), or use the efficient technique of dimensional regularization. This means that the spacetime dimension D in Eq. (72) is first taken to have a complex value such that $\varphi(z_A)$ is finite, before being analytically continued to its physical value $D = 4$.

One then derives from the action (72) two important consequences for the motion and timing of binary pulsars. First, one derives the Lagrangian describing the relativistic interaction between N strongly self-gravitating bodies (including orbital $\sim (v/c)^2$ effects, and neglecting $\mathcal{O}(v^4/c^4)$ ones) [11, 7, 33, 39]. It is the sum of one-body, two-body and three-body terms.

The one-body action has the usual form of the sum (over the label A) of the kinetic term of each point mass:

$$L_A^{\text{one-body}} = -m_A c^2 \sqrt{1 - v_A^2/c^2}$$
$$= -m_A c^2 + \frac{1}{2} m_A v_A^2 + \frac{1}{8} m_A \frac{(v_A^2)^2}{c^2} + \mathcal{O}\left(\frac{1}{c^4}\right). \tag{73}$$

Here, we use Einstein units, and the inertial mass m_A entering Eq. (73) is $m_A \equiv m_A(\varphi_0)$, where φ_0 is the asymptotic value of φ far away from the considered N-body system.

The two-body action is a sum over the pairs A, B of a term $L_{AB}^{\text{2-body}}$ which differs from the GR-predicted 2-body Lagrangian in two ways: (i) the usual gravitational constant G appearing as an overall factor in $L_{AB}^{\text{2-body}}$ must be replaced by an effective (body-dependent) gravitational constant (in the appropriate units mentioned above) given by

$$G_{AB} = G_*(1 + \alpha_A \alpha_B), \tag{74}$$

and (ii) the relativistic $(\mathcal{O}(v^2/c^2))$ terms in $L_{AB}^{\text{2-body}}$ contain, in addition to those predicted by GR, new velocity-dependent terms of the form

$$\delta^\gamma L_{AB}^{\text{2-body}} = (\bar{\gamma}_{AB}) \frac{G_{AB} m_A m_B}{r_{AB}} \frac{(v_A - v_B)^2}{c^2}, \tag{75}$$

with

$$\bar{\gamma}_{AB} \equiv \gamma_{AB} - 1 = -2 \frac{\alpha_A \alpha_B}{1 + \alpha_A \alpha_B}. \tag{76}$$

In these expressions $\alpha_A \equiv \alpha_A(\varphi_0) \equiv \partial \ln m_A(\varphi_0)/\partial \varphi_0$ (see Eq. (69) with $\varphi_a \to \varphi_0$).

Finally, the 3-body action is a sum over the pairs B, C and over A (with $A \neq B$, $A \neq C$, but the possibility of having $B = C$) of

$$L_{ABC}^{\text{3-body}} = -(1 + 2\bar{\beta}_{BC}^A) \frac{G_{AB} G_{AC} m_A m_B m_C}{c^2 r_{AB} r_{AC}} \tag{77}$$

where

$$\bar{\beta}^A_{BC} \equiv \beta^A_{BC} - 1 = \frac{1}{2} \frac{\alpha_B \, \beta_A \, \alpha_C}{(1 + \alpha_A \, \alpha_B)(1 + \alpha_A \, \alpha_C)} , \tag{78}$$

with $\beta_A = \partial \alpha_A(\varphi_0)/\partial \varphi_0$ (see Eq. (70) with $\varphi_a \to \varphi_0$).

When comparing the strong-field results (74), (76), (78) to their weak-field counterparts (59), (60), (61) one sees that the body-dependent quantity α_A replaces the weak-field coupling strength α_0 in all quantities which are linked to a scalar effect generated by body A. Note also that, in keeping with the '3-body' nature of Eq. (77), the quantity $\beta^A_{BC} - 1$ is linked to scalar interactions which are generated in bodies B and C and which non-linearly interact on body A. The notation used above has been chosen to emphasize that γ_{AB} and β^A_{BC} are strong-field analogs of the usual Eddington parameters γ^{PPN}, β^{PPN}, so that $\bar{\gamma}_{AB}$ and $\bar{\beta}^A_{BC}$ are strong-field analogs of the 'post-Einstein' 1PN parameters $\bar{\gamma}$ and $\bar{\beta}$ (which vanish in GR). Indeed the usual PPN results for the post-Einstein terms in the $\mathcal{O}(1/c^2)$ 2-body and 3-body Lagrangians are obtained by replacing in Eqs. (75) and (77) $\bar{\gamma}_{AB} \to \bar{\gamma}$, $\bar{\beta}^A_{BC} \to \bar{\beta}$ and $G_{AB} \to G$.

The non-perturbative strong-field effects discussed above show that the strong self-gravity of neutron stars can cause γ_{AB} and β^A_{BC} to be significantly different from their GR values $\gamma^{GR} = 1$, $\beta^{GR} = 1$, in some scalar-tensor theories having a small value of the basic coupling parameter α_0 (so that $\gamma^{PPN} - 1 \propto \alpha_0^2$ and $\beta^{PPN} - 1 \propto \beta_0 \, \alpha_0^2$ are both small). For instance, Fig. 2 shows that it is possible to have $\alpha_A \sim \alpha_B \sim \pm 0.6$ which implies $\gamma_{AB} - 1 \sim -0.53$, i.e., a 50% deviation from GR! Even larger effects can arise in $\beta^A_{BC} - 1$ because of the large values that $\beta_A = \partial \alpha_A/\partial \varphi_0$ can reach near the spontaneous scalarization transition [35].

Those possible strong-field modifications of the effective Eddington parameters γ_{AB}, β^A_{BC}, which parametrize the 'first post-Keplerian' (1PK) effects (i.e., the orbital effects $\sim v^2/c^2$ smaller than those entailed by the Lagrangian $\sum_A \frac{1}{2} m_A \, v_A^2 + \frac{1}{2} \sum_{A \neq B} G_{AB} \, m_A \, m_B/r_{AB}$), can then significantly modify the usual GR predictions relating the directly observable parametrized post-Keplerian (PPK) parameters to the values of the masses of the pulsar and its companion. As worked out in Refs. [11, 31, 33, 35] one finds the following modified predictions for the PPK parameters $k \equiv \langle \dot{\omega} \rangle/n$, r and s:

$$k^{th}(m_A, m_B) = \frac{3}{1-e^2} \left(\frac{G_{AB}(m_A + m_B) \, n}{c^3} \right)^{2/3}$$

$$\left[\frac{1 - \frac{1}{3} \alpha_A \, \alpha_B}{1 + \alpha_A \, \alpha_B} - \frac{X_A \, \beta_B \, \alpha_A^2 + X_B \, \beta_A \, \alpha_B^2}{6 \, (1 + \alpha_A \, \alpha_B)^2} \right] , \tag{79}$$

$$r^{th}(m_A, m_B) = G_{0B} \, m_B , \tag{80}$$

$$s^{th}(m_A, m_B) = \frac{n \, x_A}{X_B} \left[\frac{G_{AB}(m_A + m_B) \, n}{c^3} \right]^{-1/3} . \tag{81}$$

Here, the label A refers to the object which is timed ('the pulsar'[22]), the label B refers to its companion, $x_A = a_A \sin i/c$ denotes the projected semi-major axis of the orbit of A (in light seconds), $X_A \equiv m_A/(m_A + m_B)$ and $X_B \equiv m_B/(m_A + m_B) = 1 - X_A$ the mass ratios, $n \equiv 2\pi/P_b$ the orbital frequency and $G_{0B} = G_*(1 + \alpha_0 \alpha_B)$ the effective gravitational constant measuring the interaction between B and a test object (namely electromagnetic waves on their way from the pulsar toward the Earth). In addition one must replace the unknown bare Newtonian G_* by its expression in terms of the one measured in Cavendish experiments, i.e., $G_* = G/(1 + \alpha_0^2)$ as deduced from Eq. (59).

The modified theoretical prediction for the PPK parameter γ entering the 'Einstein time delay' Δ_E, Eq. (24), is more complicated to derive because one must take into account the modulation of the proper spin period of the pulsar caused by the variation of its moment of inertia I_A under the (scalar) influence of its companion [11, 7, 35]. This leads to

$$\gamma^{\text{th}}(m_A, m_B) = \frac{e}{n}\frac{X_B}{1 + \alpha_A \alpha_B}\left(\frac{G_{AB}(m_A + m_B)n}{c^3}\right)^{2/3}$$
$$[X_B(1 + \alpha_A \alpha_B) + 1 + k_A \alpha_B], \tag{82}$$

where $k_A(\varphi_0) = -\partial \ln I_A(\varphi_0)/\partial \varphi_0$ (see Eq. (71) with $\varphi_a \to \varphi_0$). Numerical studies [35] show that k_A can take quite large values. Actually, the quantity $k_A \alpha_B$ entering (82) blows up near the scalarization transition when $\alpha_0 \to 0$ (keeping $\beta_0 < 0$ fixed). In other words a theory which is closer to GR in weak-field conditions predicts larger deviations in the strong-field regime.

The structure dependence of the effective gravitational constant G_{AB}, Eq. (74), also has the consequence that the object A does not fall in the same way as B in the gravitational field of the Galaxy. As most of the mass of the Galaxy is made of non strongly-self-gravitating bodies, A will fall toward the Galaxy with an acceleration $\propto G_{A0}$, while B will fall with an acceleration $\propto G_{B0}$. Here, as above, $G_{A0} = G_{0A} = G_*(1 + \alpha_0 \alpha_A)$ is the effective gravitational constant between A and any weakly self-gravitating body. As pointed out in Ref. [43] this possible violation of the universality of free fall of self-gravitating bodies can be constrained by using observational data on the class of small-eccentricity long-orbital-period binary pulsars. More precisely, the quantity which can be observationally constrained is not exactly the violation $\Delta_{AB} = (G_{0A} - G_{0B})/G$ of the strong equivalence principle [which simplifies to $\Delta_{A0} = (G_{0A} - G)/G = (1 + \alpha_0^2)^{-1}(\alpha_0 \alpha_A - \alpha_0^2)$ in the case of observational relevance where one neglects the self-gravity of the white-dwarf companion] but rather[23] [33]

---

[22] In the double binary pulsar, both the first discovered pulsar and its companion are pulsars. However, the companion B is a non-recycled, slow pulsar whose motion is well described by Keplerian parameters only.

[23] This refinement is given here for pedagogical completeness. However, in practice, the lowest-order result $\Delta \simeq (1 + \alpha_0^2)^{-1}(\alpha_0 \alpha_A - \alpha_0^2) \simeq \alpha_0 \alpha_A - \alpha_0^2$ is accurate enough.

$$\Delta^{\text{effective}} \equiv \left(\frac{2\,\gamma_{AB} - (X_A\,\beta^B_{AA} + X_B\,\beta^A_{BB}) + 2}{3} \right)^{-1}$$

$$(1 + \alpha_A\,\alpha_B)^{-3/2}(\alpha_0\,\alpha_A - \alpha_0^2). \tag{83}$$

Here, the index B (= white-dwarf companion) can be replaced by 0 (weakly self-gravitating body) so that, for instance, $\gamma_{AB} = \gamma_{A0} = 1 - 2\,\alpha_A\,\alpha_0/(1 + \alpha_A\,\alpha_0) = (1 - \alpha_A\,\alpha_0)/(1 + \alpha_A\,\alpha_0)$, as deduced from Eq. (76).

It remains to discuss the possible strong-field modifications of the theoretical prediction for the orbital period derivative $\dot{P}_b = \dot{P}_b^{\text{th}}(m_A, m_B)$. This is obtained by deriving from the effective action (72) the energy lost by the binary system in the form of fluxes of spin-2 and spin-0 waves at infinity. The needed results in a generic tensor–scalar theory were derived in Refs. [33, 39] (in addition one must take into account the tensor–scalar modification of the additional 'varying-Doppler' contribution to the observed $\dot{P}_b$ due to the galactic acceleration [38]). The final result for $\dot{P}_b$ is of the form

$$\dot{P}_b^{\text{th}}(m_A, m_B) = \dot{P}_{b\varphi}^{\text{monopole}} + \dot{P}_{b\varphi}^{\text{dipole}} + \dot{P}_{b\varphi}^{\text{quadrupole}} + \dot{P}_{bg*}^{\text{quadrupole}}$$

$$+ \dot{P}_{b\,\text{GR}}^{\text{galactic}} + \delta^{\text{th}}\,\dot{P}_b^{\text{galactic}}, \tag{84}$$

where, for instance, $\dot{P}_{b\varphi}^{\text{monopole}}$ is (heuristically[24]) related to the monopolar flux of spin-0 waves at infinity. The term $\dot{P}_{bg*}^{\text{quadrupole}}$ corresponds to the usual quadrupolar flux of spin-2 waves at infinity. It reads:

$$\dot{P}_{bg*}^{\text{quadrupole}}(m_A, m_B) = -\frac{192\pi}{5(1 + \alpha_A\,\alpha_B)}\,\frac{m_A\,m_B}{(m_A + m_B)^2} \tag{85}$$

$$\left(\frac{G_{AB}(m_A + m_B)\,n}{c^3} \right)^{5/3}\,\frac{1 + 73\,e^2/24 + 37\,e^4/96}{(1 - e^2)^{7/2}},$$

with $G_{AB} = G_*(1 + \alpha_A\,\alpha_B) = G(1 + \alpha_A\,\alpha_B)/(1 + \alpha_0^2)$, where G_* is the 'bare' gravitational constant appearing in the action, while G is the gravitational constant measured in Cavendish experiments. The flux (85) is the only one which survives in GR (although without any α_A-related modifications). Among the several other contributions which arise in tensor–scalar theories, let us only write down the explicit expression of the contribution to (84) coming from the *dipolar flux of scalar waves*. Indeed, this contribution is, in most cases, the dominant one [7] because it scales as $(v/c)^3$, while the monopolar and quadrupolar contributions scale as $(v/c)^5$. It reads

$$\dot{P}_{b\varphi}^{\text{dipole}}(m_A, m_B) = -2\pi\,\frac{G_*\,m_A\,m_B\,n}{c^3(m_A + m_B)}\,\frac{1 + e^2/2}{(1 - e^2)^{5/2}}\,(\alpha_A - \alpha_B)^2. \tag{86}$$

---

[24] Contrary to the GR case where a lot of effort was spent to show how the observed $\dot{P}_b$ was directly related to the GR predictions for the $(v/c)^5$-accurate orbital equations of motion of a binary system [9], we use here the indirect and less rigorous argument that the energy flux at infinity should be balanced by a corresponding decrease of the mechanical energy of the binary system.

Note that the dipolar effect (86) vanishes when $\alpha_A = \alpha_B$. Indeed, a binary system made of two identical objects ($A = B$) cannot select a preferred direction for a dipole vector, and cannot therefore emit any dipolar radiation. This also implies that double neutron star systems (which tend to have $m_A \approx m_B \sim 1.35\,M_\odot$) will be rather poor emitters of dipolar radiation (though (86) still tends to dominate over the other terms in (84), because of the remaining difference $(m_A - m_B)/(m_A + m_B) \neq 0$). In contrast, very dissymmetric systems such as a neutron-star and a white-dwarf (or a neutron-star and a black hole) will be very efficient emitters of dipolar radiation, and will potentially lead to very strong constraints on tensor–scalar theories. See below.

5.6 Theory-Space Analyses of Binary Pulsar Data

Having reviewed the theoretical results needed to discuss the predictions of alternative gravity theories, let us end by summarizing the results of various theory-space analyses of binary pulsar data.

Let us first recall what are the best, current solar-system limits on the two 1PN 'post-Einstein' parameters $\bar{\gamma} \equiv \gamma^{\mathrm{PPN}} - 1$ and $\bar{\beta} \equiv \beta^{\mathrm{PPN}} - 1$. They are:

$$\bar{\gamma} = (2.1 \pm 2.3) \times 10^{-5}, \tag{87}$$

from frequency shift measurements made with the Cassini spacecraft [81], which supersedes the constraint

$$\bar{\gamma} = (-1.7 \pm 4.5) \times 10^{-4} \tag{88}$$

from VLBI measurements [82],

$$|2\,\bar{\gamma} - \bar{\beta}| < 3 \times 10^{-3}, \tag{89}$$

from Mercury's perihelion shift [66, 83], and

$$4\,\bar{\beta} - \bar{\gamma} = (4.4 \pm 4.5) \times 10^{-4}, \tag{90}$$

from Lunar laser ranging measurements [84].

Concerning binary pulsar data, we can make use of the published measurements of various Keplerian and post-Keplerian timing parameters in the binary pulsars: PSR 1913+16 [37], PSR B1534+12 [41], PSR J1141−6545 [47] and PSR J0737−3039A+B [3, 48, 49]. In addition, we can use[25] the recently

---

[25] There is, however, a caveat in the theoretical use one can make of the phenomenological limits on Δ. Indeed, in the small-eccentricity long-orbital-period binary pulsar systems used to constrain Δ one does not have access to enough PK parameters to measure the pulsar mass m_A directly. As the theoretical expression of $\Delta \simeq \alpha_0 \alpha_A - \alpha_0^2$ depends on m_A (through α_A), one needs to assume some fiducial value of m_A (say $m_A \simeq 1.35\,M_\odot$).

updated limit on the parameter Δ measuring a possible violation of the strong equivalence principle (SEP), namely $|\Delta| < 5.5 \times 10^{-3}$ at the 95% confidence level [44].

This ensemble of solar-system and binary-pulsar data can then be analyzed within any given parametrized theoretical framework. For instance, one might work within

(i) the 4-parameter framework $T_0(\bar{\gamma}, \bar{\beta}; \epsilon, \zeta)$ [70] which defines the 2PN extension of the original (Eddington) PPN framework $T_0(\bar{\gamma}, \bar{\beta})$; or
(ii) the 2-parameter class of tensor–mono-scalar theories $T_1(\alpha_0, \beta_0)$ [34]; or
(iii) the 2-parameter class of tensor–bi-scalar theories $T_2(\beta', \beta'')$ [33].

Here, the index 0 on $T_0(\bar{\gamma}, \bar{\beta}; \epsilon, \zeta)$ is a reminder of the fact that this framework is not a family of specific theories (it contains *zero* explicit dynamical fields), but is a parametrization of 2PN deviations from GR. As a consequence, its use for analyzing binary pulsar data is somewhat ill-defined because one needs to truncate the various timing observables (which are functions of the compactness of the two bodies A and B, say $P^{\text{PK}} = f(c_A, c_B)$) at the 2PN order (i.e. essentially at the quadratic order in c_A and/or c_B). For some observables (or for product of observables) there might be several ways of defining this truncation. In spite of this slight inconvenience, the use of the $T_0(\bar{\gamma}, \bar{\beta}; \epsilon, \zeta)$ framework is conceptually useful because it shows very clearly why and how binary-pulsar data can probe the behaviour of gravitational theories beyond the usual 1PN regime probed by solar-system tests.

For instance, the parameter $\Delta_A \equiv m_A^{\text{grav}}/m_A^{\text{inert}} - 1$ measuring the strong equivalence principle (SEP) violation in a neutron star has, within the $T_0(\bar{\gamma}, \bar{\beta}; \epsilon, \zeta)$ framework, a 2PN-order expansion of the form [33, 70]

$$\Delta_A = -\frac{1}{2}\left(4\bar{\beta} - \bar{\gamma}\right)c_A + \left(\frac{\epsilon}{2} + \zeta + \mathcal{O}(\bar{\beta})\right)b_A\,, \qquad (91)$$

where $c_A = -2\frac{\partial \ln m_A}{\partial \ln G} \simeq \frac{1}{c^2}\langle U\rangle_A$, $b_A = \frac{1}{c^4}\langle U^2\rangle_A \simeq B\,c_A^2$, with $B \simeq 1.026$ and $c_A \simeq k\,m_A/M_\odot$ with $k \sim 0.21$. The general result (91) is compatible with the result quoted in Sect. 5.2 within the context of the theory $T_2(\beta', \beta'')$ when taking into account the fact that, within $T_2(\beta', \beta'')$, one has $\bar{\beta} = \bar{\gamma} = 0$, $\epsilon = \beta'$ and $\zeta = 0$ [and that β'' parametrizes some effects beyond the 2PN level].

In the example of Eq. (91) one sees that, after having used solar-system tests to constrain the first contribution on the RHS to a very small value, one can use binary-pulsar tests of the SEP to set a significant limit on the combination $\frac{1}{2}\epsilon + \zeta$ of 2PN parameters. Other pulsar data then yield significant limits on other combinations of the two 2PN parameters ϵ and ζ. The final conclusion is that binary-pulsar data allow one to set significant limits (around or better than the 1% level) on the possible 2PN deviations from GR (in contrast to solar-system tests which are unable to yield any limit on ϵ and ζ) [70]. For a recent update of the limits on ϵ and ζ, which makes use of recent pulsar data see [51].

Let us now briefly discuss the use of mini-space of theories, such as $T_1(\alpha_0, \beta_0)$ or $T_2(\beta', \beta'')$, for analyzing solar-system and binary-pulsar data. The basic methodology is to compute, for each given theory (e.g. for each given values of α_0 and β_0 if one chooses to work in the $T_1(\alpha_0, \beta_0)$ theory space) a goodness-of-fit statistics $\chi^2(\alpha_0, \beta_0)$ measuring the quality of the agreement between the experimental data and the considered theory. For instance, when considering the timing data of a particular pulsar, for which one has measured several PK parameters p_i ($i = 1, \ldots, n$) with some standard deviations $\sigma_{p_i}^{\text{obs}}$, one defines, for this pulsar

$$\chi^2(\alpha_0, \beta_0) = \min_{m_A, m_B} \sum_{i=1}^{n} (\sigma_{p_i}^{\text{obs}})^{-2} (p_i^{\text{theory}}(\alpha_0, \beta_0; m_A, m_B) - p_i^{\text{obs}})^2, \qquad (92)$$

where 'min' denotes the result of minimizing over the unknown masses m_A, m_B and where $p_i^{\text{theory}}(\alpha_0, \beta_0; m_A, m_B)$ denotes the theoretical prediction (within $T_1(\alpha_0, \beta_0)$) for the PK observable p_i (given also the observed values of the Keplerian parameters).

The goodness-of-fit quantity $\chi^2(\alpha_0, \beta_0)$ will reach its minimum $\chi^2_{\min}$ for some values, say $\alpha_0^{\min}, \beta_0^{\min}$, of α_0 and β_0. Then, one focusses, for each pulsar, on the level contours of the function

$$\Delta \chi^2(\alpha_0, \beta_0) \equiv \chi^2(\alpha_0, \beta_0) - \chi^2_{\min}. \qquad (93)$$

Each choice of level contour (e.g. $\Delta \chi^2 = 1$ or $\Delta \chi^2 = 2.3$) defines a certain region in theory space, which contains, with a certain corresponding 'confidence level', the 'correct' theory of gravity (if it belongs to the considered mini-space of theories). When combining together several independent data sets (e.g. solar-system data, and different pulsar data) we can define a total goodness-of-fit statistics $\chi^2_{\text{tot}}(\alpha_0, \beta_0)$, by adding together the various individual $\chi^2(\alpha_0, \beta_0)$. This leads to a corresponding combined contour $\Delta \chi^2_{\text{tot}}(\alpha_0, \beta_0)$.

Let us end by briefly summarizing the results of the theory-space approach to relativistic gravity tests. For detailed discussions the reader should consult Refs. [33, 40, 35, 36, 80], and especially the recent update [51] which uses the latest binary-pulsar data.

Regarding the two-parameter class of tensor–bi-scalar theories $T_2(\beta', \beta'')$ the recent analysis [51] has shown that the $\Delta \chi^2(\beta', \beta'')$ corresponding to the double binary pulsar PSR J0737$-$3039 was defining quite a small elliptical allowed region in the (β', β'') plane. By contrast the other pulsar data define much wider allowed regions, while the strong equivalence principle tests define (in view of the theoretical result $\Delta \simeq 1 + \frac{1}{2} B\beta'(c_A^2 - c_B^2)$) a thin, but infinitely long, strip $|\beta'| <$ cst. in the (β', β'') plane. This highlights the power of the double binary pulsar in probing certain specific strong-field deviations from GR.

Contrary to the $T_2(\beta', \beta'')$ tensor–bi-scalar theories, which were constructed to have exactly the same first post-Newtonian limit as GR[26] (so that solar-system tests put no constraints on β' and β''), the class of tensor–mono-scalar theories $T_1(\alpha_0, \beta_0)$ is such that its parameters α_0 and β_0 parametrize *both* the weak-field 1PN regime (see Eqs. (60) and (61) above) and the strong-field regime (which plays an important role in compact binaries). This means that each class of solar-system data (see Eqs. (87)–(90) above) will define, via a corresponding goodness-of-fit statistics of the type, say

$$\chi^2_{\text{Cassini}}(\alpha_0, \beta_0) = (\sigma_\gamma^{\text{Cassini}})^{-2} \left(\bar{\gamma}^{\text{theory}}(\alpha_0, \beta_0) - \bar{\gamma}^{\text{Cassini}} \right)^2$$

a certain allowed region[27] in the (α_0, β_0) plane. As a consequence, the analysis in the framework of the $T_1(\alpha_0, \beta_0)$ space of theories allows one to compare and contrast the probing powers of solar-system tests versus binary-pulsar tests (while also comparing solar-system tests among themselves and binary-pulsar ones among themselves). The result of the recent analysis [51] is shown in Fig. 3.

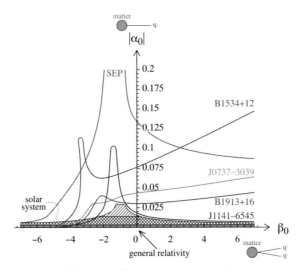

Fig. 3. Solar-system and binary-pulsar constraints on the two-parameter family of tensor–mono-scalar theories $T_1(\alpha_0, \beta_0)$. Figure taken from [51].

---

[26] However, this could be achieved only at the cost of allowing some combination of the two scalar fields to carry a negative energy flux.

[27] Actually, in the case of the Cassini data, as it is quite plausible that the *positive* value of the published central value $\bar{\gamma}^{\text{Cassini}} = +2.1 \times 10^{-5}$ is due to unsubtracted systematic effects, we use $\sigma_\gamma^{\text{Cassini}} = 2.3 \times 10^{-5}$ but $\bar{\gamma}^{\text{Cassini}} = 0$. Otherwise, we would get unreasonably strong 1σ limits on α_0^2 because tensor–scalar theories predict that $\bar{\gamma}$ must be *negative*, see Eqs. (60) and (61).

In Fig. 3, the various solar-system constraints (87)–(90) are concentrated around the horizontal β_0 axis. In particular, the high-precision Cassini constraint is the lower small grey strip. The various pulsar constraints are labelled by the name of the pulsar, except for the strong equivalence principle constraint which is labelled SEP. Note that General Relativity corresponds to the origin of the (α_0, β_0) plane, and is compatible with all existing tests.

The global constraint obtained by combining all the pulsar tests would, to a good accuracy, be obtained by intersecting the various pulsar-allowed regions. One can then see in Fig. 3 that it would be comparable to the pre-Cassini solar-system constraints and that its boundaries would be defined successively (starting from the left) by 1913+16, 1141−6545, 0737−3039, 1913+16 again and 1141−6545 again.

A first conclusion is therefore that, at the *quantitative level*, binary-pulsar tests constrain tensor–scalar gravity theories as strongly as most solar-system tests (excluding the exceptionally accurate Cassini result which constrains α_0^2 to be smaller than 1.15×10^{-5}, i.e. $|\alpha_0| < 3.4 \times 10^{-3}$). A second conclusion is obtained by comparing the behaviour of the solar-system exclusion plots and of the binary-pulsar ones around the negative β_0 axis. One sees that binary-pulsar tests exclude a whole domain of the theory space (located on the left of $\beta_0 < -4$) which is compatible with all solar-system experiments (even when including the very tight Cassini constraint). This remarkable *qualitative* feature of pulsar tests is a direct consequence of the existence of (non-perturbative) *strong-field effects* which start developing when the product $-\beta_0\, c_A$ (with c_A denoting, as above, the compactness of the pulsar) becomes of order unity.

6 Conclusion

In conclusion, we hope to have convinced the reader of the superb opportunities that binary pulsar data offer for testing gravity theories. In particular, they have been able to go qualitatively beyond solar-system experiments in probing two physically important regimes of relativistic gravity: the radiative regime and the strong-field one. Up to now, General Relativity has passed with flying colours all the radiative and strong-field tests provided by pulsar data. However, it is important to continue testing General Relativity in all its aspects (weak-field, radiative and strong-field). Indeed, history has taught us that physical theories have a limited range of validity, and that it is quite difficult to predict in which regime a theory will cease to be an accurate description of nature. Let us look forward to new results, and possibly interesting surprises, from binary pulsar data.

Acknowledgments
It is a pleasure to thank my long-term collaborator Gilles Esposito-Farèse for his useful remarks on the text, and for providing the figures. I wish also

to thank the organizers of the 2005 Sigrav School, and notably Monica Colpi and Ugo Moschella, for organizing a warm and intellectually stimulating meeting. This work was partly supported by the European Research and Training Network "Forces Universe" (contract number MRTN-CT-2004-005104)

References

1. R. A. Hulse and J. H. Taylor: Discovery of a pulsar in a binary system, Astrophys. J. **195**, L51 (1975).
2. M. Burgay et al.: An increased estimate of the merger rate of double neutron stars from observations of a highly relativistic system, Nature **426**, 531 (2003).
3. A. G. Lyne et al.: A double-pulsar system: A rare laboratory for relativistic gravity and plasma physics, Science **303**, 1153 (2004).
4. F. K. Manasse: J. Math. Phys. **4**, 746 (1963).
5. P. D. D'Eath: Phys. Rev. D **11**, 1387 (1975).
6. R. E. Kates: Phys. Rev. D **22**, 1853 (1980).
7. D. M. Eardley: Astrophys. J. **196**, L59 (1975).
8. C. M. Will, D. M. Eardley: Astrophys. J. **212**, L91 (1977).
9. T. Damour: Gravitational radiation and the motion of compact bodies, in *Gravitational Radiation*, edited by N. Deruelle and T. Piran, North-Holland, Amsterdam, pp. 59-144 (1983).
10. K. S. Thorne and J. B. Hartle: Laws of motion and precession for black holes and other bodies, Phys. Rev. D **31**, 1815 (1984).
11. C. M. Will: *Theory and Experiment in Gravitational Physics*, Cambridge University Press (1993) 380 p.
12. V. A. Brumberg and S. M. Kopejkin: Nuovo Cimento B **103**, 63 (1988)
13. T. Damour, M. Soffel and C. M. Xu: General relativistic celestial mechanics. 1. Method and definition of reference system, Phys. Rev. D **43**, 3273 (1991); General relativistic celestial mechanics. 2. Translational equations of motion, Phys. Rev. D **45**, 1017 (1992); General relativistic celestial mechanics. 3. Rotational equations of motion, Phys. Rev. D **47**, 3124 (1993); General relativistic celestial mechanics. 4. Theory of satellite motion, Phys. Rev. D **49**, 618 (1994).
14. G. 't Hooft and M. J. G. Veltman: Regularization and renormalization of gauge fields, Nucl. Phys. B **44**, 189 (1972).
15. T. Damour and N. Deruelle: Radiation reaction and angular momentum loss in small angle gravitational scattering, Phys. Lett. A **87**, 81 (1981).
16. T. Damour: Problème des deux corps et freinage de rayonnement en relativité générale, C.R. Acad. Sci. Paris, Série II, **294**, 1355 (1982).
17. T. Damour: The problem of motion in Newtonian and Einsteinian gravity, in *Three Hundred Years of Gravitation*, edited by S.W. Hawking and W. Israel, Cambridge University Press, Cambridge, pp. 128-198 (1987).
18. P. Jaranowski, G. Schäfer: Third post-Newtonian higher order ADM Hamilton dynamics for two-body point-mass systems, Phys. Rev. D **57**, 7274 (1998).
19. L. Blanchet, G. Faye: General relativistic dynamics of compact binaries at the third post-Newtonian order, Phys. Rev. D **63**, 062005-1-43 (2001).
20. T. Damour, P. Jaranowski, G. Schäfer: Dimensional regularization of the gravitational interaction of point masses, Phys. Lett. B **513**, 147 (2001).

21. Y. Itoh, T. Futamase: New derivation of a third post-Newtonian equation of motion for relativistic compact binaries without ambiguity, Phys. Rev. D **68**, 121501(R), (2003).

22. L. Blanchet, T. Damour, G. Esposito-Farèse: Dimensional regularization of the third post-Newtonian dynamics of point particles in harmonic coordinates, Phys. Rev. D **69**, 124007 (2004).

23. M. E. Pati, C. M. Will: Post-Newtonian gravitational radiation and equations of motion via direct integration of the relaxed Einstein equations. II. Two-body equations of motion to second post-Newtonian order, and radiation-reaction to 3.5 post-Newtonian order, Phys. Rev. D **65**, 104008-1-21 (2001).

24. C. Königsdörffer, G. Faye, G. Schäfer: Binary black-hole dynamics at the third-and-a-half post-Newtonian order in the ADM formalism, Phys. Rev. D **68**, 044004-1-19 (2003).

25. S. Nissanke, L. Blanchet: Gravitational radiation reaction in the equations of motion of compact binaries to 3.5 post-Newtonian order, Class. Quantum Grav. **22**, 1007 (2005).

26. L. Blanchet: Gravitational radiation from post-Newtonian sources and inspiralling compact binaries, Living Rev. Rel. **5**, 3 (2002); Updated article: http://www.livingreviews.org/lrr-2006-4

27. T. Damour, N. Deruelle: General relativitic celestial mechanics of binary system I. The post-Newtonian motion, Ann. Inst. Henri Poincaré **43**, 107 (1985).

28. T. Damour: Gravitational radiation reaction in the binary pulsar and the quadrupole formula controversy, Phys. Rev. Lett. **51**, 1019 (1983).

29. R. Blandford, S. A. Teukolsky: Astrophys. J. **205**, 580 (1976).

30. T. Damour, N. Deruelle: General relativitic celestial mechanics of binary system II. The post-Newtonian timing formula, Ann. Inst. Henri Poincaré **44**, 263 (1986).

31. T. Damour, J. H. Taylor: Strong field tests of relativistic gravity and binary pulsars, Phys. Rev. D **45**, 1840 (1992).

32. T. Damour: Strong-field tests of general relativity and the binary pulsar, in *Proceedings of the 2nd Canadian Conference on General Relativity and Relativistic Astrophysics*, edited by A. Coley, C. Dyer, T. Tupper, World Scientific, Singapore, pp. 315-334 (1988).

33. T. Damour, G. Esposito-Farèse: Tensor-multi-scalar theories of gravitation, Class. Quant. Grav. **9**, 2093 (1992).

34. T. Damour, G. Esposito-Farèse: Non-perturbative strong-field effects in tensor–scalar theories of gravitation, Phys. Rev. Lett. **70**, 2220 (1993).

35. T. Damour, G. Esposito-Farèse: Tensor–scalar gravity and binary-pulsar experiments, Phys. Rev. D **54**, 1474 (1996), arXiv:gr-qc/9602056.

36. T. Damour, G. Esposito-Farèse: Gravitational-wave versus binary-pulsar tests of strong-field gravity, Phys. Rev. D **58**, 042001 (1998).

37. J. M. Weisberg, J. H. Taylor: Relativistic binary pulsar B1913+16: thirty years of observations and analysis, To appear in the proceedings of Aspen Winter Conference on Astrophysics: Binary Radio Pulsars, Aspen, Colorado, 11-17 Jan 2004., arXiv:astro-ph/0407149.

38. T. Damour, J. H. Taylor: On the orbital period change of the binary pulsar Psr-1913+16, The Astrophysical Journal **366**, 501 (1991).

39. C. M. Will, H. W. Zaglauer: Gravitational radiation, close binary systems, and the Brans-Dicke theory of gravity, Astrophys. J. **346**, 366 (1989).

40. J. H. Taylor, A. Wolszczan, T. Damour, J. M. Weisberg: Experimental constraints on strong field relativistic gravity, Nature **355**, 132 (1992).
41. I. H. Stairs, S. E. Thorsett, J. H. Taylor, A. Wolszczan: Studies of the relativistic binary pulsar PSR B1534+12: I. Timing analysis, Astrophys. J. **581**, 501 (2002).
42. K. Nordtvedt: Equivalence principle for massive bodies. 2. Theory, Phys. Rev. **169**, 1017 (1968).
43. T. Damour and G. Schäfer: New tests of the strong equivalence principle using binary pulsar data, Phys. Rev. Lett. **66**, 2549 (1991).
44. I. H. Stairs *et al.*: Discovery of three wide-orbit binary pulsars: implications for binary evolution and equivalence principles, Astrophys. J. **632**, 1060 (2005).
45. N. Wex: New limits on the violation of the Strong Equivalence Principle in strong field regimes, Astronomy and Astrophysics **317**, 976 (1997), gr-qc/9511017.
46. V. M. Kaspi *et al.*: Discovery of a young radio pulsar in a relativistic binary orbit, Astrophys. J. **543**, 321 (2000).
47. M. Bailes, S. M. Ord, H. S. Knight, A. W. Hotan: Self-consistency of relativistic observables with general relativity in the white dwarf-neutron star binary pulsar PSR J1141-6545, Astrophys. J. **595**, L49 (2003).
48. M. Kramer *et al.*: eConf **C041213**, 0038 (2004), astro-ph/0503386.
49. M. Kramer *et al.*: Tests of general relativity from timing the double pulsar, Science **314**, 97-102 (2006).
50. S. M. Ord, M. Bailes and W. van Straten: The Scintillation Velocity of the Relativistic Binary Pulsar PSR J1141-6545, arXiv:astro-ph/0204421.
51. T. Damour, G. Esposito-Farèse: Binary-pulsar versus solar-system tests of tensor–scalar gravity, 2007, in preparation.
52. T. Damour, R. Ruffini: Sur certaines vérifications nouvelles de la relativité générale rendues possibles par la découverte d'un pulsar membre d'un système binaire, C.R. Acad. Sci. Paris (Série A) **279**, 971 (1974).
53. B. M. Barker, R. F. O'Connell: Gravitational two-body problem with arbitrary masses, spins, and quadrupole moments, Phys. Rev. D **12**, 329 (1975).
54. M. Kramer: Astrophys. J. **509**, 856 (1998).
55. J. M. Weisberg and J. H. Taylor: Astrophys. J. **576**, 942 (2002).
56. I. H. Stairs, S. E. Thorsett, Z. Arzoumanian: Measurement of gravitational spin–orbit coupling in a binary pulsar system, Phys. Rev. Lett. **93**, 141101 (2004).
57. A. W. Hotan, M. Bailes, S. M. Ord: Geodetic Precession in PSR J1141-6545, Astrophys. J. **624**, 906 (2005).
58. T. Damour, G. Schäfer: Higher order relativistic periastron advances and binary pulsars, Nuovo Cim. B **101**, 127 (1988).
59. J.M. Lattimer, B.F. Schutz: Constraining the equation of state with moment of inertia measurements, Astrophys. J. **629**, 979 (2005).
60. I. A. Morrison, T. W. Baumgarte, S. L. Shapiro, V. R. Pandharipande: The moment of inertia of the binary pulsar J0737-3039A: constraining the nuclear equation of state, Astrophys. J. **617**, L135 (2004).
61. A. S. Eddington: *The Mathematical Theory of Relativity*, Cambridge University Press, London (1923).
62. L. I. Schiff: Am. J. Phys. **28**, 340 (1960).
63. R. Baierlein: Phys. Rev. **162**, 1275 (1967).

64. C. M. Will: Astrophys. J. **163**, 611 (1971).
65. C. M. Will, K. Nordtvedt: Astrophys. J. **177**, 757 (1972).
66. C. M. Will: The confrontation between general relativity and experiment, Living Rev. Rel. **4**, 4 (2001) arXiv:gr-qc/0103036; update (2005) in arXiv:gr-qc/0510072.
67. P. Jordan, Nature (London) **164**, 637 (1949); Schwerkraft und Weltall (Vieweg, Braunschweig, 1955); Z. Phys. **157**, 112 (1959).
68. M. Fierz: Helv. Phys. Acta **29**, 128 (1956).
69. C. Brans, R. H. Dicke: Mach's principle and a relativistic theory of gravitation, Phys. Rev. **124**, 925 (1961).
70. T. Damour, G. Esposito-Farèse: Testing gravity to second postNewtonian order: A Field theory approach, Phys. Rev. D **53**, 5541 (1996), arXiv:gr-qc/9506063.
71. T. Damour, A. M. Polyakov: The string dilaton and a least coupling principle, Nucl. Phys. B **423**, 532 (1994) arXiv:hep-th/9401069; String theory and gravity, Gen. Rel. Grav. **26**, 1171 (1994), arXiv:gr-qc/9411069.
72. T. Damour, D. Vokrouhlicky: The equivalence principle and the moon, Phys. Rev. D **53**, 4177 (1996), arXiv:gr-qc/9507016.
73. J. Khoury, A. Weltman: Chameleon fields: Awaiting surprises for tests of gravity in space, Phys. Rev. Lett. **93**, 171104 (2004).
74. J. Khoury, A. Weltman: Chameleon cosmology, Phys. Rev. D **69**, 044026 (2004).
75. P. Brax, C. van de Bruck, A. C. Davis, J. Khoury, A. Weltman: Detecting dark energy in orbit: The cosmological chameleon, Phys. Rev. D **70**, 123518 (2004).
76. T. Damour, F. Piazza, G. Veneziano: Runaway dilaton and equivalence principle violations, Phys. Rev. Lett. **89**, 081601 (2002), arXiv:gr-qc/0204094; Violations of the equivalence principle in a dilaton-runaway scenario, Phys. Rev. D **66**, 046007 (2002).
77. T. Damour, G. Esposito-Farèse: Testing gravity to second postNewtonian order: A Field theory approach, Phys. Rev. D **53**, 5541 (1996).
78. T. Damour, K. Nordtvedt: General relativity as a cosmological attractor of tensor scalar theories, Phys. Rev. Lett. **70**, 2217 (1993); Tensor–scalar cosmological models and their relaxation toward general relativity, Phys. Rev. D **48**, 3436 (1993).
79. J. B. Hartle: Slowly rotating relativistic stars. 1. Equations of structure, Astrophys. J. **150**, 1005 (1967).
80. G. Esposito-Farèse: Binary-pulsar tests of strong-field gravity and gravitational radiation damping, in Proceedings of the tenth Marcel Grossmann Meeting, July 2003, edited by M. Novello et al., World Scientific (2005), p. 647, arXiv:gr-qc/0402007.
81. B. Bertotti, L. Iess, P. Tortora: A test of general relativity using radio links with the Cassini spacecraft, Nature **425**, 374 (2003).
82. S. S. Shapiro *et al*: Phys. Rev. Lett **92**, 121101 (2004).
83. I. I. Shapiro, in *General Relativity and Gravitation 12*, edited by N. Ashby, D. F. Bartlett, and W. Wyss, Cambridge University Press, Cambridge, p. 313 (1990).
84. J. G. Williams, S. G. Turyshev, D. H. Boggs: Progress in lunar laser ranging tests of relativistic gravity, Phys. Rev. Lett. **93**, 261101 (2004), arXiv:gr-qc/0411113.

Exploiting Binary Pulsars as Laboratories of Gravity Theories

Michael Kramer

The University of Manchester, Jodrell Bank Observatory, Jodrell Bank, Cheshire, UK `Michael.Kramer@manchester.ac.uk`

1 Introduction

Four decades have passed since the discovery of pulsars [1]. These unique objects have been proven to be invaluable in the study of a wide variety of physical and astrophysical problems. Most notable are studies of gravitational physics, the interior of neutron stars, the structure of the Milky Way and stellar and binary evolution. A number of these studies utilize the pulsar emission properties and/or the interaction of the radiation with the ambient medium. Most applications, however, are enabled by a technique known as pulsar timing. Here, pulsar astronomers make use of pulsars as accurate cosmic clocks where a number of fast-rotating pulsars, so called millisecond pulsars, show long-term stabilities that rival the best atomic clocks on Earth. Being compact massive objects with the most extreme states of matter in the present-day Universe, a number of pulsars are also moving in the gravitational field of a companion star, hence providing ideal conditions for tests of general relativity and alternative theories of gravity. In this review, I first discuss why a continuing challenge of Einstein's theory of gravitation, the theory of general relativity, with new observational data is still necessary. Then I describe pulsars and their use as clocks, in particular for their use as cosmic gravitational laboratories. Finally, I review some classical tests and report on the recent progress such as the discovery of the first double pulsar and look ahead to the future. This text should be read in close comparison to the contribution of Thibault Damour who provides much of the theoretical framework and motivation for the observations described here.

2 General Relativity – The Last Word in Our Understanding of Gravity?

Almost a hundred years after Einstein formulated his theory of general relativity (GR), efforts in testing GR and its concepts are still being made by many colleagues around the world, using many different approaches. These tests started promptly after GR's formulation.[1] The theory of GR was an immediate success when it was confronted with experimental data. Whilst being a bold move away from Newton's theory of gravity, which had reigned supreme for nearly three centuries, GR could naturally explain the perihelion advance of Mercury and predicted the deflection of light by the Sun and other massive bodies. Its predictive power motivated physicists to come up with new and more precise tests. Eventually, tests could be performed in space, using, for instance, Lunar laser ranging [2] or radar reflection measurements of inner planets [3]. These tests, providing some of the most stringent limits on GR, are joined by recent modern experiments using space probes such as Cassini [4]. So far, GR has passed all these tests with flying colours.

However, in light of recent progress in observational cosmology in particular, the question of whether alternative theories of gravity need to be considered, is as topical as ever. Indeed, Einstein may not have had the final word after all. GR is a classical relativistic theory that describes the gravitational interaction of bodies on large scales. One aim of today's physicists is to formulate a theory of quantum gravity which would fuse this classical world of gravitation with the strange world of quantum mechanics. Quantum mechanics (QM) describes nature on small, atomic scales. The success of both GR and QM suggests that we know physics on both extreme scales rather well, but it seems difficult or even impossible to extrapolate one theory into the other's regime. This problem is particularly demonstrated when trying to reconcile the very successful particle physics framework with the also very successful "Lambda-CDM" model which combines cold dark matter (CDM) with a cosmological constant, Λ, to impressively describe the results of cosmological observations. The problem is that most quantum field theories predict a huge cosmological constant from the energy of the quantum vacuum, up to 120 orders of magnitude too large. So, what is the solution? What are we missing?

Since the aim of quantum gravity is to account for all of the known particles and interactions of the physical world, it is conceivable that the correct theory of quantum gravity will predict some deviations from GR, eventually occurring in some limit. If our efforts can determine to what extend the as yet accurate theory of GR describes the gravitational interaction of the macroscopic world correctly, the current approaches to use GR as the basis of quantum gravity would be justified. On the other hand, a flaw discovered in GR would imply that other alternative lines of investigations may have to be followed.

How can we find a flaw in GR, should it exist? Most experimental tests of GR are made in the solar system, providing exciting limits on the validity of GR and alternative theories of gravity like tensor–scalar theories. However, solar-system experiments are made in the gravitational weak-field

---

[1] Expeditions to observe the next total solar eclipse started immediately after World War I ended.

regime, while deviations from GR may appear only in strong gravitational fields. Therefore, no test can be considered to be complete without probing the *strong-field* realm of gravitational physics. Fortunately, nature provides us with an almost perfect laboratory to test the strong-field regime using binary radio pulsars.

In the *weak-field* limit of gravity we deal with small velocities and weak gravitational potentials. Strictly speaking, currently known binary pulsars are separated well enough from their companion to move in their weak gravitational field. However, their movement still provides us with precision tests of the strong-field regime. This becomes clear when considering strong self-field effects which are predicted by the majority of alternative theories. Such effects would, for instance, clearly affect the pulsars' orbital motion, allowing us to search for these effects and hence providing us with a unique precision strong-field test of gravity.

The strength of self-field effects in compact bodies can be estimated from computing the body's gravitational self-energy, ϵ. For a mass M with radius R, ϵ can expressed in a dimensionless way in units of its rest-mass energy, i.e. $\epsilon = E_{\mathrm{grav}}/Mc^2 \sim -GM/Rc^2$, where G is the gravitational constant and c is the speed of light. In the solar system we find $\epsilon \sim -10^{-6}$ for the Sun, $\epsilon \sim -10^{-10}$ for the Earth and $\epsilon \sim -10^{-11}$ for the Moon, so that we are performing solar system tests clearly in the weak-field limit. The problem is that we cannot rule out that the true theory of gravity "looks and feels" like GR in the weak field, but that it reveals deviations as soon as the strong-field limit is encountered where velocities and self-energies are large.

For instance, for a neutron star we expect $\epsilon \sim -0.2$ whilst for a BH $\epsilon = -0.5$, i.e. five to nine orders of magnitude larger than encountered in the solar system. Indeed, Damour & Esposito-Farèse [5] have shown that in the family of tensor–scalar theories, alternative solutions to GR exist that would pass all solar system tests but that would be violated as soon as the strong-field limit is reached. Even though in all metric theories like GR, matter and non-gravitational fields respond only to a spacetime metric, scalar-tensor theories assume that gravity is mediated by a tensor field and one or more scalar field. It is possible that deviations from GR manifest themselves in a "spontaneous scalarization" [5] if the strong field limit is approached. Therefore, solar system tests will not be able to replace the strong field tests provided by radio pulsars [6].

Radio pulsars therefore probe different regimes of the parameter space, and even allow us to test the radiative aspects of gravitational theories by either experiencing gravitational wave damping of their orbital motion or by acting as gravitational wave detectors. We summarize the probed parameter space in Fig. 1 where we concentrate on tests with binary pulsars. Whilst the recent discovery of the first double pulsar system, PSR J0737−3039 (see Sect. 7) pushes the probed limits into previously unexplored regimes, there is still a huge fraction of parameter space that needs studying. These areas can be explored by finding and studying radio pulsars in even more compact

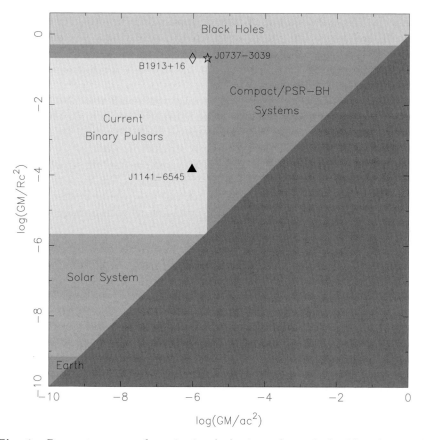

Fig. 1. Parameter space of gravitational physics to be probed with pulsars and black holes. Some theories of gravity predict effects that depend on the compactness of the gravitating body which is shown here (y-axis) as a function of orbital size and probed gravitational potential (x-axis). Note that the lower right half of the diagram is excluded as it implies an orbit smaller than the size of the body.

orbits, in particular in orbits about stellar and massive black holes. We discuss the exciting possibilities arising from such pulsar-black hole systems in some detail in Sect. 8.

3 Pulsars

Pulsars are highly magnetized, rotating neutron stars which emit a narrow radio beam along their magnetic dipole axis (Fig. 2). As the magnetic axis is inclined to the rotation axis, the pulsar acts like a cosmic light-house emitting a radio pulse that can be detected once per rotation period when the beam is directed towards Earth.

3.1 Neutron Star Properties

Pulsars are born as neutron stars in supernova explosions of massive stars, created in the collapse of the progenitor's core. Neutron stars are the most compact objects next to black holes. From timing measurements of binary pulsars, we can determine the masses of pulsars to be within a narrow range around $1.35 M_\odot$ [7], where the smallest value is found for pulsar B in the double pulsar system with $M_B = 1.250 \pm 0.005 M_\odot$ [8]. Modern calculations for different equations of state [9] produce results for the size of a neutron star quite similar to the very first calculations by Oppenheimer & Volkov [10], i.e. about 20 km in diameter. Such sizes are consistent with independent estimates derived from modelling light-curves and luminosities of pulsars observed at X-rays [11]. Underneath a solid crust, the interior is super-conducting and super-liquid at the same time [12].

As rotating magnets, pulsars lose rotational energy due to magnetic dipole radiation and a particle wind. This leads to an increase in rotation period, P, described by $\dot{P} > 0$. Equating the corresponding energy output of the dipole to the loss rate in rotational energy, we obtain an estimate for the magnetic field strength at the pulsar surface,

$$B = 3.2 \times 10^{19} \sqrt{P\dot{P}} \text{ G}, \tag{1}$$

with P measured in s and $\dot{P}$ in s s^{-1}. Typical values are of order 10^{12} G, although field strengths up to 10^{14} G have been observed [13]. Millisecond pulsars have lower field strengths of the order of 10^8 to 10^{10} G which appear to be a result of their evolutionary history (see Sect. 3.3).

3.2 Radio Emission

The radio signal of a pulsar is of coherent nature but usually weak, both because the pulsar is distant and the size of the actual emission region is small. Despite intensive research, the responsible radiation processes are still unidentified. We believe that the neutron star's rotating magnetic field induces an electric quadrupole field which is strong enough to pull out charges from the stellar surface (the electrical force exceeds the gravitational force by a factor of $\sim 10^{12}$!). The magnetic field forces the resulting dense plasma to co-rotate with the pulsar. This *magnetosphere* can only extend up to a distance where the co-rotation velocity reaches the speed of light. This distance defines the so-called *light cylinder* which separates the magnetic field lines into two distinct groups, i.e. *open and closed field lines* (Fig. 2). The plasma on the closed field lines is trapped and co-rotates with the pulsar forever. In contrast, plasma on the open field lines reaches highly relativistic velocities, produces the observed radio beam at some distance to the pulsar surface, and finally leaves the magnetosphere in a particle wind that contributes to the pulsar spin-down [14].

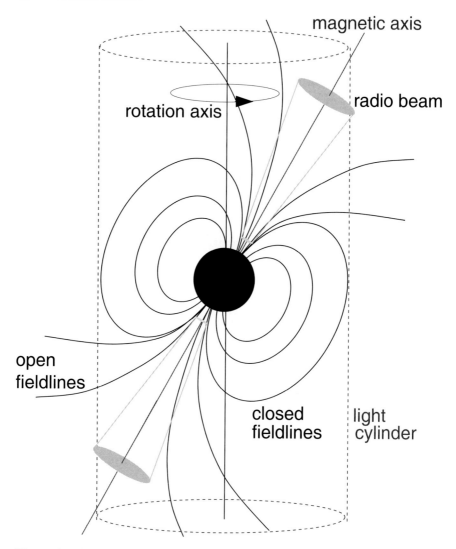

Fig. 2. A pulsar is a rotating, highly magnetised neutron star. A radio beam centred on the magnetic axis is created at some height above the surface. The tilt between the rotation and magnetic axes makes the pulsar in effect a cosmic lighthouse when the beam sweeps around in space.

Individual radio pulses provide a "snapshot" of the plasma processes occurring in the pulsar magnetosphere at an instant, resulting in varying, often seemingly random looking pulses (see Fig. 3). Averaging over all dynamical processes in the pulsar magnetosphere forms an integrated pulse shape, the *pulse profile*. The average pulse profile reflects the global constraints given by geometrical factors and a conal beam structure with, possibly, a beam pattern.

Apart from a frequency dependence, the profile is stable and independent of the particular pulses added (see Fig. 3). It is this profile stability which allows us to time pulsars to high precision.

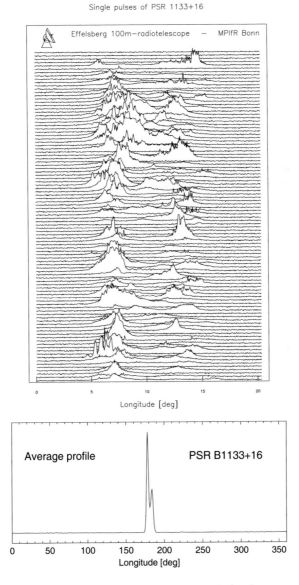

Fig. 3. Individual pulses vary in shapes and strength (top), average profiles are stable (bottom). The typical pulse width is only ∼4% of the period.

3.3 Evolution

The observed rotational slow-down, $\dot{P}$, and the resulting evolution in pulsar period, P, can be used to describe the life of a pulsar. This is usually done in a (logarithmic) P-$\dot{P}$-diagram as shown in Fig. 4 where we can draw lines of constant loss in rotational energy, magnetic field (Eq. (1)), and constant *characteristic age*, estimated from

$$\tau = \frac{P}{2\dot{P}}. \tag{2}$$

This quantity is a valid estimate for the true age under the assumption that the initial spin period is much smaller than the present period and that the spin-down is fully determined by magnetic dipole braking. Pulsars are born in the upper left area of Fig. 4 and move into the central part where they spend most of their lifetime.

Most known pulsars have spin periods between 0.1 and 1.0 s with period derivatives of typically $\dot{P} = 10^{-15}$ s s^{-1}. Selection effects are only partly responsible for the limited number of pulsars known with very long periods, the longest known period being 8.5 s [15]. The dominant effect is due to the "death" of pulsars when their slow-down has reached a critical state. This state seems to depend on a combination of P and $\dot{P}$ known as the *pulsar death-line*. The normal life of radio pulsars is limited to a few tens or hundreds million years or so.

The described evolution does not explain the about 100 pulsars in the lower left of the $P - \dot{P}$-diagram. These pulsars have simultaneously small periods (few milliseconds) and small period derivatives ($\dot{P} \leq 10^{-18}$ s s^{-1}). These *millisecond pulsars* (MSPs) are much older than ordinary pulsars with ages up to $\sim 10^{10}$ yr. It is believed that MSPs are born when mass and thereby angular momentum is transferred from an evolving binary companion while it overflows its Roche lobe [16]. In this model, MSPs are recycled from a dead binary pulsar via an accreting X-ray binary phase. This model implies a number of observational consequences: a) most normal pulsars do not develop into a MSP as they have long lost a possible companion during their violent birth event; b) for surviving binary systems, X-ray binary pulsars represent the progenitor systems for MSPs; c) the final spin period of recycled pulsars depends on the mass of the binary companion. A more massive companion evolves faster, limiting the duration of the accretion process; d) the ram-pressure of the magnetic field limits the accretion flow onto the neutron star surface. This results in the limiting *spin-up line* shown in Fig. 4; e) the majority of MSPs have low-mass white-dwarf companions as the remnant of the binary star. These systems evolve from low-mass X-ray binary systems; f) high-mass X-ray binary systems represent the progenitors for double neutron star systems (DNSs). DNSs are rare since these systems need to survive a second supernova explosion. The resulting MSP is only mildly recycled with a period of tens of milliseconds. The properties of MSPs and X-ray binaries are consistent with the described picture. For instance, it is striking that $\sim 80\%$ of all MSPs are in a binary orbit while this is true for only less than 1% of the non-recycled population. For MSPs with a low-mass white dwarf companion the orbit is nearly circular. In case of DNSs, the orbit is affected by the unpredictable nature of the kick imparted onto the newly born neutron star in the asymmetric supernova explosion of the companion. If the system survives, the result is typically an eccentric orbit with an orbital period of a few hours.

4 Observations of Pulsars

An important complication when finding and observing pulsars is the impact of the interstellar medium (ISM). As the pulsar signal propagates through the ISM, it is affected in a number of ways, such as frequency dispersion, Faraday rotation, scintillation and scattering. Once the amount of dispersion is determined for each pulsar, this effect, described in more detail below, can be corrected for. In contrast, scattering, arising from inhomogeneities in the ISM's electron density distribution, cannot be compensated for by instrumental means. Scintillation is a variation of the pulsar intensity with frequency and time, and its impact can be largely minimized by observing with large bandwidths. Observations with large bandwidths of course also improve the sensitivity, but also make the correction for dispersion, the so-called "de-dispersion", usually more sophisticated.

4.1 Dispersion

The ISM is a cold, ionized plasma. Electromagnetic radiation from pulsars and other sources will experience a frequency-dependent index of refraction as they propagate through the ISM. The refractive index (ignoring magnetic fields) is

$$\mu = \sqrt{1 - \left(\frac{f_p}{f}\right)^2} < 1, \tag{3}$$

so that the group velocity of a propagating electromagnetic wave is less than the speed of light c. Consequently, the propagation of a radio signal along a path of length d from the pulsar to Earth will be delayed in time with respect to a signal of infinite frequency by an amount

$$t = \frac{1}{c}\int_0^d \left[1 + \frac{f_p^2}{2f^2}\right] dl - \frac{d}{c} = \frac{e^2}{2\pi m_e c}\frac{\int_0^d n_e\, dl}{f^2} \equiv \mathcal{D} \times \frac{\mathrm{DM}}{f^2}, \tag{4}$$

where the *dispersion measure*

$$\mathrm{DM} = \int_0^d n_e\, dl \tag{5}$$

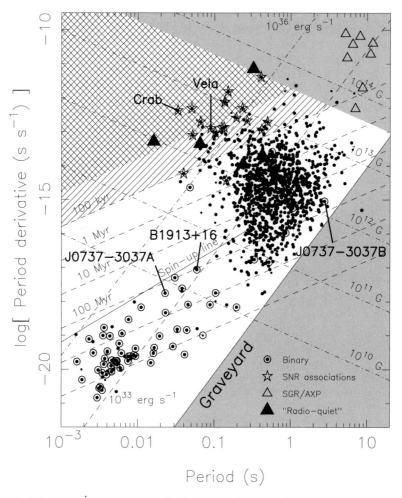

Fig. 4. The $P - \dot{P}$–diagram for the known pulsar population. Lines of constant characteristic age, surface magnetic field and spin-down luminosity are shown. Binary pulsars are marked by a circle. The lower solid line represents the pulsar "death line" enclosing the "pulsar graveyard" where pulsars are expected to switch off radio emission. The gray area in the top right corner indicates the region where the surface magnetic field appears to exceed the quantum critical field of 4.4×10^{13} Gauss. For such values, some theories expect the quenching of radio emission in order to explain the radio-quiet "magnetars" (i.e. Soft-gamma ray repeaters, SGRs, and Anomalous X-ray pulsars, AXPs). The upper solid line is the "spin-up" line which is derived for the recycling process as the period limit for millisecond pulsars.

is usually expressed in cm^{-3} pc and the *dispersion constant* $(\mathcal{D})$

$$\mathcal{D} \equiv \frac{e^2}{2\pi m_e c} = (4.148808 \pm 0.000003) \times 10^3 \ \text{MHz}^2 \, \text{pc}^{-1} \, \text{cm}^3 \, \text{s}. \qquad (6)$$

The delay between two frequencies, f_1 and f_2 both in MHz, is

$$\Delta t \simeq 4.15 \times 10^6 \text{ ms } \times (f_1^{-2} - f_2^{-2}) \times \text{DM}. \qquad (7)$$

From a measurement of the pulse arrival time at two or more different frequencies (see Fig. 5), we can infer the DM along the line of sight to the pulsar. The DM then can be used to estimate the distance to a pulsar by numerically integrating Eq. (5) assuming a model for the galactic electron density distribution, n_e.

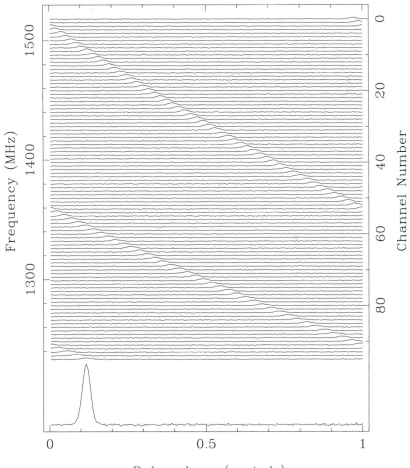

Fig. 5. Pulse dispersion shown in this Parkes observation of the 128 ms pulsar B1356–60. The dispersion measure is 295 cm^{-3} pc. The quadratic frequency dependence of the dispersion delay is clearly visible. Figure provided by Andrew Lyne; for details see [17].

As pulsars are observed with a finite bandwidth, the pulsar signal would be smeared if it is detected without taking into account that the radio pulses first arrive at the upper end of the bandwidth before they finally reach the lower end. In the extreme case, the dispersion delay between the the upper and lower frequencies exceeds the pulse period and the pulses become un-detectable. Two methods exist to correct for dispersion. In the "incoherent de-dispersion" technique, a filterbank is used to divide the bandwidth into a large number of small frequency channels, in which the signal is detected sep-arately. Before adding the output of these filterbank channels, each channel is delayed electronically by an appropriate amount until the signal has arrived in the lowest-frequency channel (see Fig. 6). This method is straightforward and easy to implement, but it is not perfect, as a dispersion smearing in each small filterbank channel remains.

The ideal de-dispersion method is a "coherent" method, as it completely corrects the effect of the ISM on the phases of the propagating electromagnetic radiation[2]. This method requires access to the raw (undetected) voltages of the incident electromagnetic wave and involves a number of transforms of the signal between the time and frequency domain. As a result, this method is technically and computationally more challenging but improvements in com-puting power in the recent years have meant that this method is essentially routine at the major pulsar observatories.

4.2 Finding Pulsars

When searching for pulsars, the dispersion measure of a new pulsar is a priori unknown, so that the DM has to be determined during the search process. This is done by using a filterbank system. The outputs of the individual channels are combined for a large number of trial DM values, and each of the resulting time-series is searched for a periodic signal. Indeed, the pulsars that we are mostly interested in for our gravitational studies, are those which are precise cosmic clocks. The most effective way to find such highly-periodic signals is to employ Fourier transforms. Including a number of specialized tricks (see [17]), this is done very effectively, having led to the discovery of almost 1800 known radio pulsars.

When employing Fourier transforms, however, the assumption is made that the spin frequency is constant. While this is essentially correct for isolated pul-sars[3], this assumption is not fulfilled for pulsars in fast binary orbits. In these cases, the orbital motion can modify the pulsar spin frequency observed on Earth by such an amount that the signal is spread over a number or many

---

[2] A delay of a pulsed signal in the time domain corresponds to a change of the phases of the wave signal in the frequency domain.

[3] The small variation in observed spin-frequency due to a relative Doppler-motion between the telescope on Earth and the pulsar is usually negligible during the observing time.

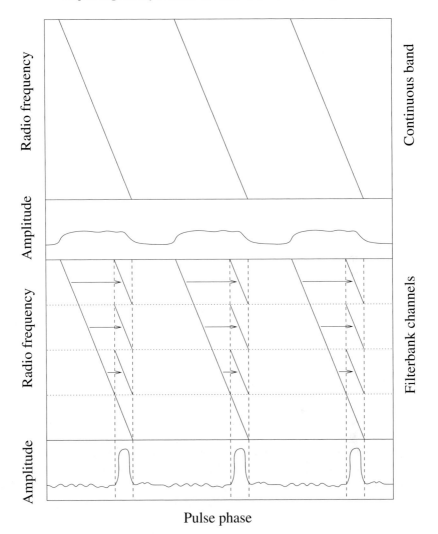

Fig. 6. Pulse dispersion and the process of incoherent de-dispersion. Simply detecting the pulse over a finite bandwidth results in a significantly broadened profile (top panel). Dividing the passband into smaller bandwidth channels and applying the appropriate delay to each channel considerably reduces the broadening and increases the pulse S/N ratio (lower panel). Figure provided by Duncan Lorimer.

adjacent Fourier frequency bins, effectively reducing the overall signal-to-noise ratio until the signal is undetectable. In order to overcome this problem, "acceleration searches" need to be applied. In an ideal world (i.e. a world where infinite computing power is available), we would transform the observed time-series into the rest-frame of the pulsar, assuming all possible binary orbits and orbital phases. Eventually, we would find the right combination of orbital parameters for the correct transformation into the rest frame, so that the Fourier transform would recover the full pulsar signal. However, such an ideal world does not exist, and searching all possible orbital parameters is impossible. Therefore, the searched parameter space is usually restricted, e.g. by assuming that the acceleration is constant during the observing time.

Fig. 7 gives an example of the improvement from the use of an acceleration search in case of the Hulse–Taylor pulsar B1913+16. The left panel shows that the pulsar is strong enough to be detectable without any acceleration search, but folding the data at the nominal period from the standard search results in a heavily smeared profile. During an acceleration search the data are re-sampled assuming a constant acceleration value. With the correct value, the acceleration-corrected time series as shown on the right effectively has removed the deleterious effects of Doppler smearing and the true pulse shape is seen with greater significance A detailed description of this and other varieties of acceleration searches can be found in Lorimer & Kramer (2005) [17].

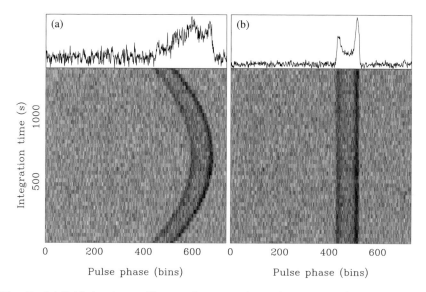

Fig. 7. (a) Folded pulse profiles as a function of time for a 22 min Arecibo observation of PSR B1913+16 showing the effects of a changing apparent pulse period. (b) The same time series now corrected for a constant acceleration during the observing time. Figure provided by Duncan Lorimer.

4.3 Observing Pulsars

We will concentrate here on the observational procedures that are relevant for pulsar timing. These include in particular the detection and de-dispersion of the pulsar signal, the forming of an average pulse profile, and the determination of a pulse time of arrival (TOA). It is the latter which ultimately forms the data set that is studied for tests of theories of gravity.

Fig. 8 summarizes the basic observational setup required for pulsar timing. As pulsars are typically very weak radio sources, and because single pulses differ significant in shape and amplitude while average profiles are typically stable, the observed pulses are de-dispersed and a few hundred or thousand pulses are folded and added to form a mean pulse profile.

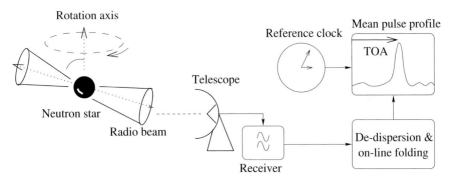

Fig. 8. Diagram showing the basic concept of a pulsar timing observation. Figure provided by Duncan Lorimer.

Usually, the TOA is defined as the arrival time of the nearest pulse to the mid-point of the observation. As the pulses have a certain width, the TOA refers to some *fiducial point* on the profile. Ideally, this point coincides with the plane defined by the rotation and magnetic axes of the pulsar and the line of sight to the observer. The determination of the TOA makes use of the stability of the pulse profile by using a cross-correlation of the observed profile with a high signal-to-noise "template" obtained from the addition of many earlier observations. Assuming that the discretely sampled profile, $\mathcal{P}(t)$, is a scaled and shifted version of the template, $\mathcal{T}(t)$, with added noise, $\mathcal{N}(t)$, we may write

$$\mathcal{P}(t) = a + b\mathcal{T}(t - \tau) + \mathcal{N}(t) \tag{8}$$

where a is an arbitrary offset and b a scaling factor. The time shift between the profile and the template, τ, yields the TOA relative to the fiducial point of the template and the start time of the observation [18]. We expect the uncertainty of a TOA measurement, σ_{TOA}, to scale as

$$\sigma_{\text{TOA}} \simeq \frac{W}{\text{S/N}} \propto \frac{S_{\text{sys}}}{\sqrt{t_{\text{obs}}\Delta f}} \times \frac{P\delta^{3/2}}{S_{\text{mean}}}, \tag{9}$$

where S_{sys} is the system equivalent flux density, Δf is the available observing bandwidth, t_{obs} is the integration time, P is the pulse period, $\delta = W/P$ is the pulse duty cycle and S_{mean} is the mean flux density of the pulsar. Best results obviously are obtained with sensitive wide-band systems (low S_{sys} and large Δf) for bright (high S_{mean}) short-period pulsars with narrow pulses (i.e. small duty cycles). For millisecond pulsars, a few thousand pulses can be added easily in a few minutes of observing time. This usually results in extremely stable profiles. In addition to their higher rotational stability, this represents an important factor in explaining the superior timing stability of millisecond pulsars when compared to normal pulsars.

5 Pulsar Timing

By measuring the arrival time of the pulsar signals clock very precisely, we can study effects that determine the propagation of the pulses in four-dimensional spacetime. Millisecond pulsars are the most useful objects for these investigations: their pulse arrival times can not only be measured much more accurately than for normal but their rotation is also much smoother, making them intrinsically better clocks. For both types we use pulsar timing to account for *every single* rotation of the neutron star between two different observations. Hence, we aim to determine the number of an observed pulse, counting from some reference epoch, t_0. We can write

$$N = N_0 + \nu_0 \times (t - t_0) + \frac{1}{2}\dot{\nu}_0 \times (t - t_0)^2 + ..., \qquad (10)$$

where N_0 is the pulse number and ν_0 the spin frequency at the reference time, respectively. Whilst for most millisecond pulsars a second derivative, $\ddot{\nu}$, is usually too small to be measured, we expect ν and $\dot{\nu}$ to be related via the physics of the braking process. For magnetic dipole braking we find

$$\dot{\nu} = -\mathrm{const.} \times \nu^n, \qquad (11)$$

where the *braking index* takes the value $n = 3$. If ν and its derivatives are accurately known and if t_0 coincides with the arrival of a pulse, all following pulses should appear at integer values of N — when observed in an inertial reference frame. However, our observing frame is not inertial, as we are using telescopes that are located on a rotating Earth orbiting the Sun. Therefore, we need to transfer the pulse times-of-arrival (TOAs) measured with the observatory clock (*topocentric arrival times*) to the centre of mass of the solar system as the best approximation to an inertial frame available. The transformation of a topocentric TOA to such *barycentric arrival times*, t_{SSB}, is given by

$$t_{\mathrm{SSB}} = \quad t_{\mathrm{topo}} - t_0 + t_{\mathrm{corr}} - D/f^2 , \qquad (12)$$
$$+ \Delta_{\mathrm{Roemer},\odot} + \Delta_{\mathrm{Shapiro},\odot} + \Delta_{\mathrm{Einstein},\odot} , \qquad (13)$$
$$+ \Delta_{\mathrm{Roemer,Bin}} + \Delta_{\mathrm{Shapiro,Bin}} + \Delta_{\mathrm{Einstein,Bin}}. \qquad (14)$$

We have split the transformation into three lines. The first two lines apply to every pulsar whilst the third line is only applicable to binary pulsars.

5.1 Clock and Frequency Corrections

The observatory time is typically maintained by local Hydrogen-maser clocks monitored by GPS signals. In a process involving a number of steps, clock corrections, t_{corr}, are retroactively applied to the arrival times in order to transfer them to a uniform atomic time that would be kept by an ideal atomic clock on the geoid. It is published retroactively by the *Bureau International des Poids et Mesures* (BIPM).

As the pulses are delayed due to dispersion in the interstellar medium, the arrival time depends on the observing frequency, f. The TOA is therefore corrected for a pulse arrival at an infinitely high frequency (last term in Eq. (12)). For some pulsars, "interstellar weather" causes small changes in DM, which would cause time-varying drifts in the TOAs. In such cases, the above term is extended to include time-derivatives of DM which can be measured using multi-frequency observations.

5.2 Barycentric Corrections

The *Roemer delay*, $\Delta_{\mathrm{Roemer},\odot}$, is the classical light-travel time between the phase centre of the telescope and the solar system barycentre (SSB). Given a unit vector, $\hat{s}$, pointing from the SSB to the position of the pulsar and the vector connecting the SSB to the observatory, $\boldsymbol{r}$, we find:

$$\Delta_{\mathrm{Roemer},\odot} = -\frac{1}{c}\,\boldsymbol{r}\cdot\hat{s} = -\frac{1}{c}\left(\boldsymbol{r}_{\mathrm{SSB}} + \boldsymbol{r}_{\mathrm{EO}}\right)\cdot\hat{s}. \tag{15}$$

Here c is the speed of light and we have split $\boldsymbol{r}$ into two parts. The vector $\boldsymbol{r}_{\mathrm{SSB}}$ points from the SSB to the centre of the Earth (geocentre). Computation of this vector requires accurate knowledge of the locations of all major bodies in the solar system and uses solar system ephemerides. The second vector $\boldsymbol{r}_{\mathrm{EO}}$, connects the geocentre with the phase centre of the telescope. In order to compute this vector accurately, the non-uniform rotation of the Earth has to be taken into account, so that the correct relative position of the observatory is derived.

The *Shapiro delay*, $\Delta_{\mathrm{Shapiro},\odot}$, is a relativistic correction that corrects for an extra delay due to the curvature of spacetime in the solar system [3]. It is largest for a signal passing the Sun's limb ($\sim 120~\mu$s) while Jupiter can contribute as much as 200 ns. In principle one has to sum over all bodies in the solar system, but in practice only the Sun is usually taken into account.

The last term in Eq. (13), $\Delta_{\mathrm{Einstein},\odot}$, is called *Einstein delay* and it describes the combined effect of gravitational redshift and time dilation due to motion of the Earth and other bodies, taking into account the variation of an atomic clock on Earth in the varying gravitational potential as it follows its elliptical orbit around the Sun [19].

5.3 Relative Motion and Shklovskii Effect

If the pulsar is moving relative to the SSB, the transverse component of the velocity, v_t, can be measured as the vector $\hat{s}$ in Eq. (15) changes with time. Present day timing precision is not sufficient to measure a radial motion although it is theoretically possible. This leaves Doppler corrections to observed periods, masses etc. undetermined. The situation changes if the pulsar has an optically detectable companion such as a white dwarf for which Doppler shifts can be measured from optical spectra.

Another effect arising from a transverse motion is the *Shklovskii effect*, also known in classical astronomy as *secular acceleration*. With the pulsar motion, the projected distance of the pulsar to the SSB is increasing, leading to an increase in any observed change of periodicity, such as pulsar spin-down or orbital decay. The observed pulse period derivative, for instance, is increased over the intrinsic value by

$$\left(\frac{\dot{P}}{P}\right)_{obs} = \left(\frac{\dot{P}}{P}\right)_{int} + \frac{1}{c}\frac{v_t^2}{d}. \tag{16}$$

For millisecond pulsars where $\dot{P}_{int}$ is small, a significant fraction of the observed change in period can be due to the Shklovskii effect.

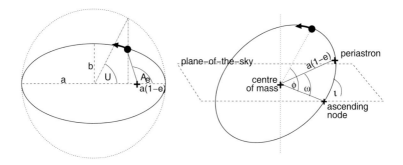

Fig. 9. Definition of the orbital elements in a Keplerian orbit and the angles relating both the orbit and the pulsar to the observer's coordinate system and line-of-sight. (left) The closest approach of the pulsar to the centre-of-mass of the binary system marks periastron, given by the longitude ω and a chosen epoch T_0 of its passage. The distance between centre of mass and periastron is given by $a(1-e)$ where a is the semi-major axis of the orbital ellipse and e its eccentricity. (right) Usually, only the projection on the plane of the sky, $a\sin i$, is measurable, where i is the orbital inclination defined as the angle between the orbital plane and the plane of the sky. The *true anomaly*, A_E, and *eccentric anomaly*, U are related to the *mean anomaly*. The orbital phase of the pulsar Φ is measured relative to the ascending node.

5.4 Binary Pulsars

Equations (12) and (13) are used to transfer the measured TOAs to the SSB. If the pulsar has a binary companion, the light-travel time across the orbit and further relativistic effects need to be taken into account (see (14)). That adds additional orbital parameters to the set of timing parameters which have to be solved in the timing process (see below). In the simplest case, five Keplerian parameters need to be determined (see Fig. 9), i.e. orbital period, P_b; the projected semi-major axis of the orbit, $x \equiv a \sin i$ where i is the (usually unknown) orbital inclination angle; the orbital eccentricity, e; the longitude of periastron, ω; and and the time of periastron passage, T_0. For a number of binary systems this Keplerian description of the orbit is not sufficient and corrections need to be applied. These can be either time derivatives of Keplerian parameters or parameters describing completely new effects (e.g. those of a Shapiro delay due to curved spacetime near the companion). In any case, it is important to note that we do not have to assume a particular theory of gravity when measuring such relativistic corrections, called "post-Keplerian" (PK) parameters [20, 21]. Instead, we can take the observational values and compare them with predictions made within a framework of specific theories of gravity [22].

In GR, the five most important PK parameters are given by the expressions below [23, 24, 21, 25].

$$\dot{\omega} = 3T_\odot^{2/3} \left(\frac{P_b}{2\pi}\right)^{-5/3} \frac{1}{1-e^2} (M_p + M_c)^{2/3}, \tag{17}$$

$$\gamma = T_\odot^{2/3} \left(\frac{P_b}{2\pi}\right)^{1/3} e \frac{M_c(M_p + 2M_c)}{(M_p + M_c)^{4/3}}, \tag{18}$$

$$r = T_\odot M_c, \tag{19}$$

$$s = \sin i = T_\odot^{-1/3} \left(\frac{P_b}{2\pi}\right)^{-2/3} x \frac{(M_p + M_c)^{2/3}}{M_c}, \tag{20}$$

$$\dot{P_b} = -\frac{192\pi}{5} T_\odot^{5/3} \left(\frac{P_b}{2\pi}\right)^{-5/3} f(e) \frac{M_p M_c}{(M_p + M_c)^{1/3}}, \tag{21}$$

where all masses are expressed in solar units, G is Newton's gravitational constant, c the speed of light and

$$f(e) = \frac{\left(1 + (73/24)e^2 + (37/96)e^4\right)}{(1-e^2)^{7/2}}. \tag{22}$$

P_b is the period and e the eccentricity of the binary orbit. The masses M_p and M_c of pulsar and companion, respectively, are expressed in solar masses ($M_\odot$) where we define the constant $T_\odot = GM_\odot/c^3 = 4.925490947\mu s$. G denotes the Newtonian constant of gravity and c the speed of light.

The first PK parameter, $\dot{\omega}$, describes the relativistic advance of periastron in rad s^{-1}. It is the easiest to measure for orbits with non-zero eccentricities (note that ω is only poorly defined for $e \approx 0$ and so is $\dot{\omega}$). From a measurement of $\dot{\omega}$, we obtain from Eq. (17) the total mass of the system, $(M_p + M_c)$.

The orbital decay due to gravitational-wave damping is expressed by the (dimensionless) change in orbital period, $\dot{P}_b$. Any metric theory of gravity that embodies Lorentz-invariance in its field equations predicts gravitational radiation and, hence, $\dot{P}_b$. If a theory satisfies the strong equivalence principle (SEP, see next section), like GR, gravitational *dipole* radiation is not expected, but *quadrupole* emission will be the lowest multipole term. In alternative theories, while the *inertial* dipole moment may remain uniform, the *gravitational wave* dipole moment may not, and dipole radiation may be predicted. The magnitude of this effect depends on the difference in gravitational binding energies, expressed by the difference in coupling constants to a scalar gravitational field. We will discuss this further below.

The other two parameters, r and s, are related to the Shapiro delay caused by the curvature of spacetime due to the gravitational field of the companion. They are measurable, depending on timing precision, if the orbit is seen nearly edge-on.

5.5 Obtaining a Timing Solution

Eqs. (10) to (14) contain the set of timing parameters that need to be determined. Many of them are not known a priori (or only with limited precision after the discovery of a pulsar) and need to be determined precisely in a least squares fit analysis of the measured TOAs. The parameters can be categorized into three groups: (a) *astrometric parameters* (i.e. position, proper motion, parallax contained in the Römer and Shapiro delay, respectively); (b) *spin parameters* (i.e. rotation frequency, ν, and higher derivatives); (c) *binary parameters*.

Given a minimal set of starting parameters, a least squares fit is needed to match the measured arrival times to pulse numbers according to Eq. (10). We minimize the expression

$$\chi^2 = \sum_i \left(\frac{N(t_i) - n_i}{\sigma_i} \right)^2 \tag{23}$$

where n_i is the nearest integer to $N(t_i)$ and σ_i is the TOA uncertainty in units of pulse period (turns).

In order to obtain a phase-coherent solution that accounts for every single rotation of the pulsar between two observations, one starts off with a small set of TOAs that were obtained sufficiently close in time so that the accumulated uncertainties in the starting parameters do not exceed one pulse period. Gradually, the data set is expanded, maintaining coherence in phase. When successful, post-fit residuals expressed in pulse phase show a Gaussian distribution around zero with a root mean square that is comparable to the TOA uncertainties (see Fig 10).

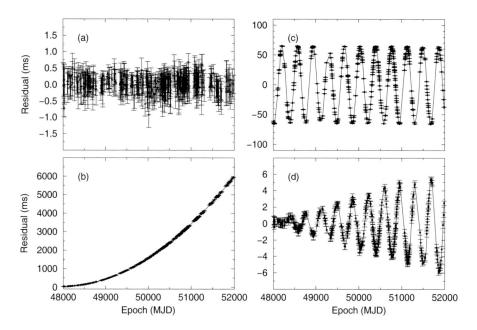

Fig. 10. (a) Timing residuals for the 1.19 s pulsar B1133+16. A fit of a perfect timing model should result in randomly distributed residuals. (b) A parabolic increase in the residuals is obtained if $\dot{P}$ is underestimated, here by 4 per cent. (c) An offset in position (in this case a declination error of 1 arcmin) produces sinusoidal residuals with a period of 1 yr. (d) The effect of neglecting the pulsar's proper motion, in this case $\mu_T = 380$ mas yr^{-1}. In all plots we have set the reference epoch for period and position to the first TOA at MJD 48000 to show the development of the amplitude of the various effects. Note the different scales on each of the vertical axes. For details see [17].

After starting with fits for only period and pulse reference phase over some hours and days, longer time spans slowly require fits for parameters like spin frequency derivative(s) and position. Incorrect or incomplete timing models cause systematic structures in the post-fit residuals identifying the parameter that needs to be included or adjusted (see Fig. 10). The precision of the parameters improves not only with the length of the data span and the frequency of observation, but also with orbital coverage in the case of binary pulsars.

6 Tests of Relativistic Gravity

In the following we will summarize tests of relativistic gravity using pulsars as accurate cosmic clocks. Some of these tests involve studies of the parameters in the so-called *Parameterized Post-Newtonian* (PPN) formalism. Other tests make use of PK parameters. For a detailed description of the underlying theoretical framework and its use in interpreting the observations, see the contribution by Thibault Damour in this volume.

6.1 Tests of Equivalence Principles

Metric theories of gravity assume *(i)* the existence of a symmetric metric, *(ii)* that all test bodies follow geodesics of the metric and *(iii)* that in local Lorentz frames the non-gravitational laws of physics are those of special relativity. Under these conditions we can study metric theories with the PPN formalism by describing deviations from simple Newtonian physics by a set of 10 real-valued PPN-Parameters [26]. Each of the parameters can be associated with a specific physical effect, like the violation of conservation of momentum or equivalence principles. Comparing measured PPN parameters to their theoretical values can tests theories in a purely experimental way. A complete description of the PPN formalism is presented by Will (2001) [26], whilst specific pulsar tests are detailed by Stairs (2003) [27].

Many tests of GR and alternative theories are related to tests of the *Strong Equivalence Principle* (SEP) which is completely embodied into GR. Alternative theories of gravity may predict a violation of some or all aspects of SEP. The SEP is, according to its name, stronger than both the *Weak Equivalence Principle* (WEP) and the *Einstein Equivalence Principle* (EEP). The WEP, included in all metric theories, states that all test bodies in an external gravitational field experience the same acceleration regardless of the mass and composition. The EEP also postulates Lorentz- and positional invariance, whilst the SEP includes both the WEP and the EEP, but postulates them also for gravitational experiments. As a consequence, both Lorentz- and positional invariance should be independent of the gravitational self-energy of the bodies in the experiment. Since all terrestrial bodies possess only a negligible fraction of gravitational self-energy, tests require the involvement of astronomical objects. A violation of SEP means that there is a difference between gravitational mass, M_g, and inertial mass, M_i. The difference can be written as

$$\frac{M_g}{M_i} \equiv 1 + \delta(\epsilon) = 1 + \eta\epsilon + \mathcal{O}(\epsilon^2), \tag{24}$$

where the parameter η characterizes the violation of SEP and ϵ is again the dimensionless self-energy. A possible difference can probed in the Earth–Moon system. Due to their different self-energies, the Earth and Moon would fall differently in the external gravitational field of the Sun, leading to a polarization of the Earth–Moon orbit ("Nordvedt-effect"). Lunar-laser-ranging experiments can be used to put tight limits on η which is a linear combination of PPN parameters representing effects due to preferred locations, preferred frames and the violation of the conservation of momentum [2]. However, as discussed earlier, solar system experiments cannot make tests in the strong-field regime where deviations to higher order terms of ϵ could be present. We can probe this possibility using pulsar–white dwarf systems. since for neutron stars $\epsilon \sim -0.2$ is much larger than for white dwarfs, $\epsilon \sim -10^{-4}$. A violation of the SEP would mean a acceleration that is different for the pulsar and the white dwarf in the external gravitational field of the Galaxy, and similar to the Nordvedt effect, this should lead to a polarization of the orbit [28]. Since the orbit is also slowly precessing, a careful analysis of all relevant low-eccentricity pulsar–white dwarf systems in a statistical manner is needed but leads to limits that are not only comparable to those obtained in the gravitational weak-field but which also probe the qualitatively different strong-field regime of gravity [29, 30].

Some metric theories of gravity violate SEP specifically by predicting preferred-frame and preferred-location effects. A preferred universal rest frame, presumably equivalent with that of the Cosmic Microwave Background, may exist if gravity is mediated in part by a long-range vector field. A violation of local Lorentz-invariance would cause a binary system moving with respect to a preferred universal rest frame to suffer a long-term change in its orbital eccentricity [31]. A statistical analysis of pulsar–white dwarf systems yields a limit of the PPN parameter $|\alpha_1| < 1.2 \times 10^{-4}$ (95% C.L.) [29] which is expected to be zero in GR.

Essentially any metric theory of gravity that embodies Lorentz-invariance in its field equations predicts gravitational radiation to be emitted by binary systems. In GR the *quadrupole* term is the lowest non-zero multipole component. In a theory that violates SEP gravitational *dipole* radiation can occur whereas the amplitude depends on the difference in gravitational binding energies, expressed by the difference in coupling constants to a scalar field, $(\hat{\alpha}_p - \hat{\alpha}_c)$, which are both zero in GR. For a white dwarf companion the coupling constant is much smaller than for a pulsar, $|\alpha_c| \ll |\alpha_p|$, so that the strongest emission should occur for short-orbital period pulsar–white dwarf systems. The currently best limit is provided by the binary pulsar PSR J1141−6545, consisting of a young pulsar and a heavy white dwarf companion ($M = 0.99 \pm 0.02 M_\odot$) in a compact eccentric 4.5-hr orbit [32, 33]. The derived tight upper limit of $\alpha_p \leq 0.004$, is an order of magnitude better than the previous limits [34].

6.2 Tests of GR with Double-Neutron-Stars

If we study the compact orbits of double neutron stars, we enter a regime where we expect very much larger effects than previously discussed. In this case a PPN approximation is no longer valid. Instead, we can use an existing theory of gravity and check if the observations are consistently described by the measured Keplerian and PK parameters. In each theory, for negligible spin-contributions the PK parameters should only be functions of the a priori unknown pulsar and companion mass, M_p and M_c, and the easily measurable Keplerian parameters. With the two masses as the only free parameters during the test, an observation of two PK parameters will already determine the masses uniquely in the framework of the given theory. The measurement of a third or more PK parameters then provides a consistency check.

In the past, only two binary pulsars had more than two PK parameters determined, the 59-ms pulsar B1913+16 and the 38-ms B1534+12. For PSR B1913+16 with an eccentric ($e = 0.61$) 7.8-hr orbit, the PK parameters $\dot{\omega}$, γ and $\dot{P_b}$ are measured very precisely. Correcting the observed $\dot{P_b}$ value for the Shklovskii effect (see Sect. 5.3), the measured value is in excellent agreement with the prediction of GR for quadrupole emission (see Fig. 11). It demonstrates impressively that GR provides a self-consistent and accurate description of the system as orbiting point masses, i.e. the structure of the neutron stars does not influence their orbital motion as expected from SEP. The precision of this test is limited by our knowledge of the galactic gravitational potential and the corresponding correction to $\dot{P_b}$. The timing results for PSR B1913+16 provide us with the currently most precise measurements of neutron star masses ever, i.e. $M_p = (1.4408 \pm 0.0003)M_\odot$ and $M_c = (1.3873 \pm 0.0003)M_\odot$ [35].

The 10-hr orbit of the second DNS B1534+12 ($e = 0.27$) is observed nearly edge-on. Thereby, in addition to the three PK parameters observed for PSR B1913+16, the Shapiro-delay parameters r and s can be measured, enabling non-radiative aspects of gravitational theories to be tested, as $\dot{P_b}$ is not needed. In fact, the observed value of $\dot{P_b}$ seems to be heavily influenced by Shklovskii-terms, i.e. it is larger than the value expected from GR by a term $\propto v_t^2/d$ (see Eq. (16)). Assuming that GR is the correct theory of gravitation, the measured proper motion and deviation from the predicted value can be used to derive the distance to the pulsar, $d = 1.02 \pm 0.05$ kpc [36].

6.3 Tests with Pulsar Structure Data

In addition to the use of pulsars as clocks, strong gravity effects can also be tested using pulse structure data, namely the effects of *geodetic precession* seen for PSRs B1913+16 [39, 37] and B1534+12 [40]. In both cases, the pulsar spin axis appears to be misaligned with the orbital angular momentum vector, so that general relativity predicts a relativistic spin–orbit coupling. The pulsar spin precesses about the total angular momentum, changing the relative orientation of the pulsar towards Earth. As a result, the angle between the pulsar spin axis and our line-of-sight changes with time, so that different portions of the emission beam are observed [41] leading to changes in the measured pulse profile (Fig. 11). In extreme cases, the precession may even move the beam out of our line-of-sight and the pulsar may disappear as predicted for PSR B1913+16 for the year 2025 [37].

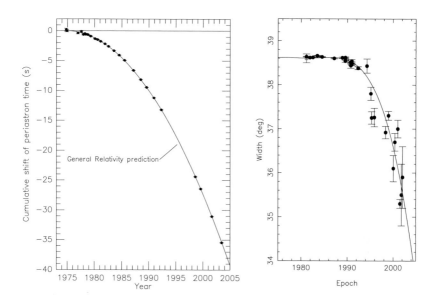

Fig. 11. (left) Shift in the periastron passage of the DNS PSR B1913+16 plotted as a function of time, resulting from orbital energy loss due to the emission of gravitational radiation. The agreement between the data, now spanning almost 30 yr, and the predicted curve due to gravitational quadrupole wave emission is now better than 0.5%. Figure provided by Joel Weisberg and Joe Taylor. (right) Component separation as measured for PSR B1913+16 [37, 38]. The component separation is shrinking as the line-of-sight changes due to geodetic precession. Extrapolating the model leads to the prediction that the pulsar should disappear in the year 2025, reappearing again around the year 2245.

7 The Double-Pulsar

The most remarkable discovery to date was made in 2003 with the DNS J0737−3039 [42, 8]. This system is not only the most relativistic one ever discovered, but it is also unique because both neutron stars are visible as radio pulsars with periods of 22.8 ms (PSR J0737−3039A, simply called "A" hereafter) and 2.8 s (PSR J0737−3039B, simply called "B" hereafter), respectively. It was discovered and is studied by a large collaboration involving colleagues from Australia, Canada, India, Italy and USA. The double pulsar's short and compact (orbital period of $P_b = 144$ min), slightly eccentric ($e = 0.09$) orbit makes the double pulsar the most extreme relativistic binary system ever discovered, demonstrated by the system's remarkably high value of periastron advance ($\dot{\omega} = 16.8995 \pm 0.0007 \, \mathrm{deg} \, \mathrm{yr}^{-1}$, i.e. four times larger than for the Hulse–Taylor pulsar!). Only four years after the discovery of the system, most of its timing parameters are determined with a precision that took several decades to achieve in the previously known best relativistic binary pulsars [43].

The measured decrease in orbital period of $\dot{P}_b = (1.25 \pm 0.2) \times 10^{-12}$ seconds per second means that the orbit is shrinking every day by 7.42 ± 0.09 mm, which agrees with GR's prediction of an orbital decay due to the emission of gravitational quadrupole waves within an uncertainty of 1%. Ultimately, the shrinkage leads to a coalescence of the two pulsars in only ~ 85 Myr. This boosts the hopes for detecting a merger of two neutron stars with first-generation ground-based gravitational wave detectors by a factor of several compared to previous estimates [42, 44]. Moreover, the detection of a young companion B around an old millisecond pulsar A confirms the evolution scenario proposed for the creation of recycled millisecond pulsars that was outlined earlier.

The two further PK parameters, r and s, related to a Shapiro delay are also measured in this system (see Fig. 12). Their measurement is possible, since – quite amazingly! – we observe the system almost completely edge-on. Hence, at superior conjunction the pulses of A pass the surface of B in only 30,000 km distance, needing to travel an extra length of curved spacetime and adding about 100 microseconds to the travel time to Earth. Within GR, we can interpret s as the sine of the orbital inclination angle. With a measurement of $\sin i \equiv s = 0.99974(-0.00039, +0.00016)$, this is indeed very close to an edge-on geometry of $i = 90°$.

When trying to see whether these PK parameter measurements are in agreement with the predictions of GR, we construct the test in a very elegant way [22]: The unique relationship between the two masses of the system predicted by any theory for each PK parameter can be drawn in a diagram showing the mass of A on one axis and that of B on the other. We expect all curves to intersect in a single point if the chosen theory is a valid description of the nature of this system (see Fig. 13).

Most importantly, the possibility of measuring the orbit of both A and B provides a new, qualitatively different constraint in such an analysis. Indeed, with a measurement of the projected semi-major axes of the orbits of both A and B, we obtain a precise measurement of the mass ratio simply from Kepler's third law, via $R \equiv M_A/M_B = x_B/x_A$ where M_A and M_B are the masses and x_A and x_B are the (projected) semi-major axes of the orbits of both pulsars, respectively. We can expect the mass ratio, R, to follow this simple relationship to at least the first Post-Newtonian (1PN or $(v/c)^2$ order) level. In particular, the R value is not only theory-independent, but also independent of strong-field (self-field) effects which is not the case for PK-parameters. Therefore, any combination of masses derived from the PK-parameters *must* be consistent with the mass ratio derived from Kepler's 3rd law. With five PK parameters already available, this additional constraint makes the double pulsar the most overdetermined system to date where the most relativistic effects can be studied in the strong-field limit. The theory of GR passes this new test at the record-breaking level of 0.05% [43].

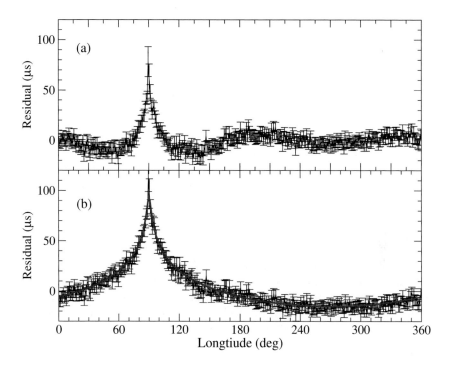

Fig. 12. The effect of the Shapiro delay as seen in the Double Pulsar caused by the gravitational potential of B seen in the timing residuals of A. (a) Observed timing residuals after a fit of all timing model parameters *except* the Shapiro-delay terms r and s which were set to zero. (b) Residuals illustrating the full Shapiro delay, obtained by holding all parameters to their fitted values, except the Shapiro delay terms which were set to zero. The line shows the predicted delay at the centre of the data span. In both cases, residuals were averaged in $1°$ bins of longitude [43].

The precision of the measured timing system parameters increases continuously with time as further and better observations are made. Soon, we expect the measurement of additional PK parameters, allowing more and new tests of theories of gravity. Some of these parameters arise from a relativistic deformation of the pulsar orbit and those which find their origin in aberration effects and their interplay with geodetic precession. In a few years, we will measure the decay of the orbit so accurately, that we can put limits on alternative theories of gravity which should even surpass the precision achieved in the solar system. On somewhat longer time scales, we will even achieve a precision that will require us to consider post-Newtonian terms that go beyond the currently used description of the PK parameters. Indeed, we already achieve a level of precision in the $\dot{\omega}$ measurement where we expect corrections and contributions at the 2PN level. One such effect involves the prediction by GR that, in contrast to Newtonian physics, the neutron stars' spins affect their orbital motion via spin–orbit coupling. This effect modifies the observed $\dot{\omega}$ by an amount that depends on the pulsars' moment of inertia, so that a potential measurement of this effect would allow the moment of inertia of a neutron star to be determined for the very first time [45, 8]. We do not expect this measurement to be easy, but we will certainly try!

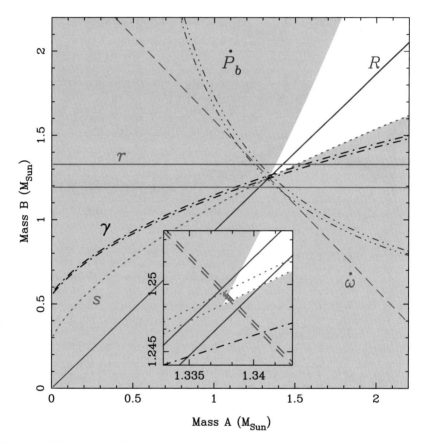

Fig. 13. 'Mass–mass' diagram showing the observational constraints on the masses of the neutron stars in the double pulsar system J0737–3039. The shaded regions are those that are excluded by the Keplerian mass functions of the two pulsars. Further constraints are shown as pairs of lines enclosing permitted regions as given by the observed mass ratio, R, and the PK parameters shown here as predicted by general relativity (see text). Inset is an enlarged view of the small square encompassing the intersection of these constraints. See [43] for details.

With the measurement of already five PK parameters and the unique information about the mass ratio, the double pulsar indeed provides a truly unique test-bed for relativistic theories of gravity. Again, GR has passed these new tests with flying colours. The precision of these tests and the nature of the resulting constraints go beyond what has been possible with other systems in the past. However, we have only just started to study and exploit the relativistic phenomena that can be investigated in great detail in this wonderful cosmic laboratory.

8 Future Tests Using Pulsars and Black Holes

Whilst the currently possible tests of GR are exciting, they are only the prelude to what will be possible once the Square-Kilometre-Array (SKA) comes online. The SKA project is a global effort to built a radio telescope interferometer with a total collecting area of 10^6 m^2. It will be about 100 times more sensitive than the VLA, GBT or Effelsberg and about 200 times more sensitive than the Lovell telescope. Pulsar surveys with the SKA will essentially discover all active pulsars in the Galaxy that are beamed toward us. In addition to this complete Galactic Census, pulsars will be discovered in external galaxies as far away as the Virgo cluster. Most importantly for probing relativistic gravity is the prospect that the SKA will almost certainly discover the first pulsar orbiting a black hole (BH). Strong-field tests using such unprecedented probes of gravity have been identified as one of the key science projects for the SKA [46]. The SKA will enable us to measure both the BH spin and the quadrupole moment using the effects of classical and relativistic spin–orbit coupling – impossible with the timing precision affordable with present-day telescopes [47]. Having extracted the dimensionless spin and quadrupole parameters, χ and q,

$$\chi \equiv \frac{c}{G}\frac{S}{M^2} \quad \text{and} \quad q = \frac{c^4}{G^2}\frac{Q}{M^3}, \tag{25}$$

where S is the angular momentum and Q the quadrupole moment, we can use these measured properties of a BH to confront the predictions of GR [47, 46] such as the "Cosmic Censorship Conjecture" and the "No-hair theorem".

In GR the curvature of spacetime diverges at the centre of a BH, producing a singularity, which physical behaviour is unknown. The Cosmic Censorship Conjecture was invoked by Penrose in 1969 [48] to resolve the fundamental concern that if singularities could be seen from the rest of spacetime, the resulting physics may be unpredictable. The Cosmic Censorship Conjecture proposes that singularities are always hidden within the event horizons of BHs, so that they cannot be seen by a distant observer. Whether the Cosmic Censor Conjecture is correct remains an unresolved key issue in the theory of gravitational collapse. If correct, we would always expect $\chi \leq 1$, so that the complete gravitational collapse of a body always results in a BH rather than a naked singularity [49]. In contrast, a value of $\chi > 1$ would imply that the event horizon has vanished, exposing the singularity to the outside world. Here, the discovered object would not be a BH as described by GR but would represent an unacceptable naked singularity and hence a violation of the Cosmic Censorship Conjecture [50].

One may expect a complicated relationship between the spin of the BH, χ, and its quadrupole moment, q. However, for a rotating Kerr BH in GR, both properties share a simple, fundamental relationship [51, 52],

$$q = -\chi^2. \tag{26}$$

This equation reflects the "no-hair" theorem of GR which implies that the external gravitational field of an astrophysical (uncharged) BH is fully determined by its mass and spin [12]. Therefore, by determining q and χ from timing measurements with the SKA, we can confront this fundamental prediction of GR for the very first time.

Fig. 14. Artistic impression of the SKA as conceived in its reference design. Parabolic dishes shown in the background are complemented by aperture arrays monitoring the whole sky at once visible in the foreground. (Figure provided by the International SKA Project Office.)

Finally, about 1000 millisecond pulsars to be discovered with the SKA can also be used to directly detect gravitational radiation in contrast to the indirect measurements from orbital decay in binaries. Pulsars discovered and timed with the SKA act effectively as the endpoints of arms of a huge, cosmic gravitational wave detector which can measure a stochastic background spectrum of gravitational waves predicted from energetic processes in the early Universe. This "device" with the SKA at its heart promises to detect such a background, at frequencies that are below the band accessible even to LISA [46].

9 Conclusions

Pulsars provide some of the most stringent and in many cases the only constraints for theories of relativistic gravity in the strong-field limit. Being precise clocks, moving in deep gravitational potentials, they are a physicist's dream-come-true. With the discoveries of the first pulsar, the first binary pulsar, the first millisecond pulsar, and now recently also the first double pulsar, a wide range of parameter space can be probed. The SKA will provide yet another leap in our understanding of relativistic gravity and hence in the quest for quantum gravity.

10 Further Reading

Naturally, the above can only touch upon many of the involved topics and issues. Many details are left for the interested reader to follow up with the help of the literature suggested below:

- *Damour T. & Taylor J.H., 1992, Phys.Rev.D, 45, 1840*: The classical complete description of the use of binary pulsars for strong-field tests of gravity.
- *Lorimer D. & Kramer M., 2005, Handbook of Pulsar Astronomy, Cambridge University Press*: A recent hands-on description of the classical and modern observation and data analysis techniques, published in the Cambridge "Observer Handbook" series.
- *Lyne A. & Smith F.G., 2005, Pulsar Astronomy, Cambridge University Press*: A completely revised version of this classical textbook on pulsars, covering all aspects of pulsar astronomy.
- *Stairs I., 2003, Testing General Relativity with Pulsar Timing, Living Reviews, 6, 5*: This excellent review provides a recent account of the use of binary pulsars in the study of gravitational physics.
- *Will C., 1993, Theory and Experiment in Gravitational Physics, Cambridge University Press*: The classical textbook on the experimental tests of gravitational theories. A recent update on particular aspects covered in the book can be found on-line as part of the Living Reviews in Relativity.

References

1. A. Hewish, S. J. Bell, J. D. H. Pilkington, P. F. Scott, R. A. Collins, *Nature* **217**, 709 (1968).
2. K. Nordtvedt, *Phys. Rev.* **170**, 1186 (1968).
3. I. I. Shapiro, *Phys. Rev. Lett.* **13**, 789 (1964).
4. B. Bertotti, L. Iess, P. Tortora, *Nature* **425**, 374 (2003).
5. T. Damour, G. Esposito-Farèse, *Phys. Rev. D* **53**, 5541 (1996).
6. T. Damour, G. Esposito-Farèse, *Phys. Rev. D* **58**, 1 (1998).
7. I. H. Stairs, S. E. Thorsett, Z. Arzoumanian, *Phys. Rev. Lett.* **93**, 141101 (2004).
8. A. G. Lyne, *et al.*, *Science* **303**, 1153 (2004).
9. J. M. Lattimer, M. Prakash, *ApJ* **550**, 426 (2001).
10. J. R. Oppenheimer, G. Volkoff, *Phys. Rev.* **55**, 374 (1939).
11. V. E. Zavlin, G. G. Pavlov, *A&A* **329**, 583 (1998).
12. S. L. Shapiro, S. A. Teukolsky, *Black Holes, White Dwarfs and Neutron Stars. The Physics of Compact Objects* (Wiley–Interscience, New York, 1983).
13. M. A. McLaughlin, *et al.*, *ApJ* **591**, L135 (2003).
14. M. Kramer, A. G. Lyne, J. T. O'Brien, C. A. Jordan, D. R. Lorimer, *Science* **312**, 549 (2006).
15. M. D. Young, R. N. Manchester, S. Johnston, *Nature* **400**, 848 (1999).
16. M. A. Alpar, A. F. Cheng, M. A. Ruderman, J. Shaham, *Nature* **300**, 728 (1982).
17. Lorimer, D. R. and Kramer, M., *Handbook of Pulsar Astronomy* (Cambridge University Press, 2005).
18. J. H. Taylor, *Philos. Trans. Roy. Soc. London A* **341**, 117 (1992).
19. D. C. Backer, R. W. Hellings, *Ann. Rev. Astr. Ap.* **24**, 537 (1986).
20. T. Damour, N. Deruelle, *Ann. Inst. H. Poincaré (Physique Théorique)* **43**, 107 (1985).
21. T. Damour, N. Deruelle, *Ann. Inst. H. Poincaré (Physique Théorique)* **44**, 263 (1986).
22. T. Damour, J. H. Taylor, *Phys. Rev. D* **45**, 1840 (1992).
23. H. P. Robertson, *Ann. Math.* **38**, 101 (1938).
24. R. Blandford, S. A. Teukolsky, *ApJ* **205**, 580 (1976).
25. P. C. Peters, *Phys. Rev.* **136**, 1224 (1964).
26. C. Will, *Living Reviews in Relativity* **4**, 4 (2001).
27. I. H. Stairs, *Living Reviews in Relativity* **6**, 5 (2003).
28. T. Damour, G. Schäfer, *Phys. Rev. Lett.* **66**, 2549 (1991).
29. N. Wex, Kramer *et al.* [53], pp. 113–116.
30. I. H. Stairs, *et al.*, *ApJ* **632**, 1060 (2005).
31. T. Damour, G. Esposito-Farèse, *Phys. Rev. D* **46**, 4128 (1992).
32. V. M. Kaspi, *et al.*, *ApJ* **543**, 321 (2000).
33. M. Bailes, S. M. Ord, H. S. Knight, A. W. Hotan, *ApJ* **595**, L49 (2003).
34. G. Esposito-Farese (2004). Contribution to 10th Marcel Grossmann meeting, gr-qc/0402007.
35. J. M. Weisberg, J. H. Taylor, Bailes *et al.* [54], pp. 93–98.
36. I. H. Stairs, S. E. Thorsett, J. H. Taylor, A. Wolszczan, *ApJ* **581**, 501 (2002).
37. M. Kramer, *ApJ* **509**, 856 (1998).
38. M. Kramer, O. Löhmer, A. Karastergiou, Bailes *et al.* [54], pp. 99–102.
39. J. M. Weisberg, R. W. Romani, J. H. Taylor, *ApJ* **347**, 1030 (1989).
40. I. H. Stairs, S. E. Thorsett, J. H. Taylor, Z. Arzoumanian, Kramer *et al.* [53], pp. 121–124.

41. T. Damour, R. Ruffini, *Academie des Sciences Paris Comptes Rendus Ser. Scie. Math.* **279**, 971 (1974).
42. M. Burgay, *et al.*, *Nature* **426**, 531 (2003).
43. M. Kramer, *et al.*, *Science* **314**, 97 (2006).
44. V. Kalogera, *et al.*, *ApJ* **601**, L179 (2004).
45. T. Damour, G. Schäfer, *Nuovo Cim.* **101**, 127 (1988).
46. M. Kramer, D. C. Backer, T. J. W. Lazio, B. W. Stappers, S. Johnston, *New Astronomy Reviews* **48**, 993 (2004).
47. N. Wex, S. Kopeikin, *ApJ* **513**, 388 (1999).
48. S. W. Hawking, R. Penrose, *Royal Society of London Proceedings Series A* **314**, 529 (1970).
49. R. M. Wald, *General relativity* (Chicago: University of Chicago Press, 1984, 1984).
50. S. W. Hawkings, G. F. R. Ellis, *The Large Scale Structure of spacetime* (Cambridge University Press, Cambridge, 1973).
51. K. S. Thorne, *Reviews of Modern Physics* **52**, 299 (1980).
52. K. S. Thorne, R. H. Price, D. A. Macdonald, *Black Holes: The Membrane Paradigm* (New Haven: Yale University Press, 1986).
53. M. Kramer, N. Wex, R. Wielebinski, eds., *Pulsar Astronomy - 2000 and Beyond, IAU Colloquium 177* (Astronomical Society of the Pacific, San Francisco, 2000).
54. M. Bailes, D. J. Nice, S. Thorsett, eds., *Radio Pulsars* (Astronomical Society of the Pacific, San Francisco, 2003).

Perspective in the Search for Relativistic Pulsars

Nichi D'Amico and Marta Burgay

INAF- Osservatorio Astronomico di Cagliari, Loc. Poggio dei Pini, Strada 54, 09012 Capoterra, Italy damico@ca.astro.it

Pulsar research makes major contributions to the field of fundamental physics, ranging from ultra-dense matter physics to relativistic gravity, cosmology and stellar evolution. A striking example is the confirmation of the existence of gravitational radiation, as predicted by Einstein's theory of General Relativity. Large scale surveys and searches carried out at radio wavelengths with large radio telescopes have greatly increased our understanding of the pulsar population, and have detected a number of peculiar objects. In the last few years, with a series of successful pulsar search experiments carried out using the Parkes 64m radio telescope in NSW (Australia), we have produced an unprecedented boom of radio pulsar discoveries, including the first ever known double-pulsar. The main virtue of the Parkes experiments has been the discovery of many pulsars which are intrinsically rare in the population, but very interesting for their physical applications. We have identified several young energetic pulsars, relativistic binary systems, binary pulsars with a massive star companion and millisecond pulsars in exotic binary systems. We have also used millisecond pulsar properties to probe the dynamical status of Globular Clusters, and we have found evidence of a high density of unseen dark remnants in a globular cluster. Future deep searches with similar equipment and long term timing programmes should produce many more exotic binary systems, and might represent the first real attempt to find a binary pulsar with a black hole companion. In this chapter we review the pulsar search techniques, with particular attention to the detection and precise timing of relativistic binaries, and their impact in our understanding of Gravitation.

1 Introduction

Since the discovery of pulsars 40 years ago [1], many different searches for these objects have contributed to the 700 or so pulsars known prior to mid-1997, when new generation pulsar search experiments commenced at Parkes. Some efforts with a relatively narrow focus have resulted in the discovery of extremely important objects, for example the Crab pulsar [2] or the first millisecond pulsar [3]. However, the vast majority of known pulsars have been found in larger-scale searches. These searches generally have well-defined selection criteria and hence provide samples of the galactic population which can be modeled to determine the properties of the parent population. Most of our knowledge about the galactic distribution and the evolution of pulsars has come from such studies [4, 5, 6, 7, 8].

Of particular significance are young pulsars. With their higher spin-down luminosities, young pulsars are likely to be detected at X-ray and γ-ray energies, providing an important observational diagnostic for the yet mysterious rotation-powered neutron-star energetics. Young pulsars often exhibit glitch behavior [9, 10], useful for diagnosing the physics of neutron-star interiors. Young pulsars also typically show interesting random rotational irregularities [11], and their rapid spin-down often allows the measurement of deterministic rotational properties that constrain electromagnetic braking [12, 13]. Young pulsars are often associated with supernova remnants [14]. Pulsar–supernova associations are potentially valuable for understanding the fate of massive stars and the birth and evolution of neutron stars and supernova remnants (SNRs) in the Galaxy. Establishing that a neutron star is part of an SNR is important for classifying the remnant in terms of one type of explosion (Type II or Type Ib, Ic). Furthermore, pulsar associations provide independent estimates of crucial remnant properties such as age and distance. The detection of a pulsar can also clarify unusual morphology in a remnant, as in the proposed association between PSR B1757−24 and G5.4−1.2 [15, 16], where the morphology might otherwise lead to a misclassification [17].

Of comparable importance though, are millisecond pulsars, as they allow us to address many interesting physical and astrophysical issues. Due to the high stability of their periodic signal and thanks to the precision with which we can measure the times of arrival of their pulses, millisecond pulsars are remarkably stable clocks [18, 19], and provide many potential applications of long-term precision timing, such as testing gravitational theories, detecting gravitational waves and establishing an improved standard of terrestrial time [20]. Furthermore, finding pulsars with periods shorter than the shortest-known period of 1.56 ms [3] would be crucial for constraining the neutron star equations of state (EoS) [21]. The smallest possible spin period reachable by a rotating neutron star can be roughly estimated imposing that the gravitational force is equal to the centrifugal force at the breakup limit. This simple assumption gives a limit spin period $P_{min} \propto R^{9/4} M^{-1/2}$. Different EoS give a different relation between masses and radii, hence, through detailed calculations, to a different value of P_{min}. Millisecond pulsars are relatively abundant in Globular Clusters. These stellar systems contain a relative wealth of low-mass X-ray binaries compared with the galactic disk, so they are prone to host the resulting spun-up or 'recycled' pulsars [22]; also, the high stellar densities in the cluster cores facilitate encounters in which a neutron star can capture an ordinary star to form an X-ray binary system [23, 24]. In turn, precision timing observations of millisecond pulsars in globular clusters can be used to determine a variety of physical properties of the host cluster. In particular, the period derivatives measured for pulsars in globular clusters can be dominated by the dynamical effects of the cluster gravitational field, and can be used to constrain the surface mass density of the cluster. This information, coupled with the accurate positioning of the pulsars in the cluster projected potential well, gives precious information on the cluster dynamical status.

Also of great importance are the serendipitous discoveries of unusual and often unique objects by large-scale surveys; for instance, the first binary pulsar, PSR B1913+16 [25], the first star with planetary-mass companions [26], the first pulsar with a massive stellar companion [27], and the first eclipsing pulsar [28]. Pulsars show an amazingly diverse range of properties and most major surveys turn up at least one object with new and unexpected characteristics. Some of these are of great significance. The prime example is of course PSR B1913+16, which has provided the first observational evidence for gravitational waves and the first evidence that general relativity is an accurate description of gravity in the strong-field regime [18].

Pulsars are relatively weak radio sources. Successful pulsar surveys therefore require a large radio telescope, low-noise receivers, a relatively wide bandwidth and long observation times. Pulsars have steep spectra, typically $S(\nu) \propto \nu^{-1.7}$, and low-frequency cut-off occurs usually below 200–300 MHz. So, in principle, we would better observe them at a relatively low frequency of around 400 MHz. However, pulsar signals suffer dispersion due to the presence of charged particles in the interstellar medium. The dispersion delay across a bandwidth of $\Delta\nu$ centred at a frequency ν is

$$\tau_{\mathrm{DM}} \simeq 8.30 \times 10^3 \, \mathrm{DM} \, \Delta\nu \, \nu^{-3} \ \mathrm{s}, \tag{1}$$

where the dispersion measure, DM, is in units of $\mathrm{cm}^{-3}\mathrm{pc}$ and the frequencies are in MHz. To retain sensitivity, especially for short-period, high-dispersion pulsars, the observing bandwidth must be sub-divided into many channels. In most pulsar searches, this has been achieved using a filterbank system.

The sensitivity of pulsar searches is also limited by the galactic radio continuum background and by interstellar scattering, especially for low radio frequencies and at low galactic latitudes. Interstellar scattering results in a one-sided broadening of the observed pulse profile with a frequency dependence $\sim \nu^{-4.4}$ [29] which cannot be removed by using narrow bandwidths.

While relatively young pulsars tend to be located at low galactic latitudes, old pulsars and in particular recycled binary pulsars and millisecond pulsars, can be found at high galactic latitudes where dispersion is rather low, and multipath scattering is negligible. It has been proved that large scale surveys carried out at low frequency are prone to discover many millisecond pulsars. While the larger telescope beams available at low frequency make such surveys relatively fast, they are not very efficient to probe the inner regions of the galactic disk, because of the high background temperature and strong dispersion and scattering reduce significantly the sensitivity. Recently, the availability of multibeam receivers has significantly reduced the limitations of high frequency large scale surveys and has allowed us to fully exploit the capabilities of high frequency observations to take into account dispersion and interstellar scattering and probe deeper regions of the Galaxy. In general, there is no ideal radio frequency to be adopted for pulsar observations, and a given frequency choice, coupled with the other system parameters, simply results in a better sampling of a given volume of the pulsar parameters space. This means that observations at low frequency are ideal for observing the local sample of millisecond pulsar, at high galactic latitude, while relatively high frequencies, around 1400 MHz, are more efficient to probe the inner part of the Galaxy, at low galactic latitude.

Most pulsar searches along the galactic plane therefore have been carried out at higher radio frequencies, often around 1400 MHz [30, 31]. One of the main limitations of searches at relatively high frequencies is the narrow size of radiotelescope beams, which make large scale surveys relatively slow. This limitation has been solved only recently with the availability of multibeam receivers.

The Parkes multibeam receiver was conceived with the aim of undertaking large-scale and sensitive searches for relatively nearby galaxies ($z \leq 0.04$) by detection of their emission in the 21-cm line of neutral hydrogen. The receiver has 13 feeds with a central feed surrounded by two rings, each of six feeds, arranged in a hexagonal pattern [32]. This arrangement permits the simultaneous observation of 13 regions of sky, increasing the speed of surveys by approximately the same factor. It was quickly realized that this system would make a powerful instrument for pulsar surveys, provided that the bandwidth was increased above the original specification and that the necessary large filterbank system could be constructed. A new data acquisition system capable of handling multibeam data sets was also a fundamental component of the system. These requirements were met as a result of a joint effort among the Italian pulsar group (http://pulsar.ca.astro.it/pulsar/), now based in Cagliari, the Jodrell Bank pulsar group (http://www.jb.man.ac.uk/pulsar/) based at the University of Manchester in UK, and the Australia Telescope National Facility pulsar group (http://www.atnf.csiro.au/research/pulsar/) in Sydney.

The first Parkes multibeam pulsar survey commenced in August 1997. This survey covered a strip with $|b| < 5^o$ along the galactic plane between galactic longitudes of 260° and 50°. The Parkes pulsar group continued the effort with a complementary high-latitude pulsar survey, covering a strip of the southern sky perpendicular to the galactic plane up to high latitudes ($b = \pm 60^o$), and with a targeted search of the Globular Cluster system for millisecond pulsars. In the same period, a group at the University of Swinbourne, in Australia, using the same equipment has carried out an intermediate latitude pulsar survey.

1.1 Focus of the Chapter

Most of the observations related to the survey programmes mentioned above have been completed and most of the data have been processed. This chapter describes the characteristics of such experiments and reviews the discoveries whose number and nature matches, and in one case goes beyond, the expectations according to which the surveys had been undertaken.

Many of the investigations mentioned in the previous section require collecting observations of the new discoveries for at least one year and in some cases several years (pulsar ages, population statistics, pulsar velocities, millisecond pulsar timing array) and hence are still works in progress at the time of the delivery of this Chapter.

In Sect. 2 a review of the basic knowledge of pulsars is presented, briefly treating their discovery, emission mechanism, evolution and observational properties. Sect. 3 describes the technical characteristics of the Parkes equipment, the data acquisition and analysis and the follow-up data treatment (confirmations and timing of new discoveries). In Sect. 4 the main results of the Parkes surveys are reported, with particular attention to some of the most exotic objects. Finally, Sect. 5 focuses on the most important result: the discovery of the first double pulsar system, PSR J0737$-$3039A/B, and on the implications that this discovery has in many fields of physics and astrophysics. In particular Sect. 5.3 shows the importance of the timing of the two pulsars in this binary system for testing General Relativity and relativistic gravity, Sect. 5.3 describes the impact that this discovery has on the determination of the Double Neutron Stars (DNSs) merger rate and, by implication, on the possibility to detect gravitational wave bursts produced during the merging of the two stars in such systems.

2 Pulsars: Overview

2.1 A Brief History of Radio Pulsars

The first pulsar [1], was discovered in 1967 by Jocelyn Bell, and Anthony Hewish, during an experiment on the interplanetary scintillation of extragalactic radio sources. The astronomical origin of the recorded signal, that strongly resembled terrestrial radio frequency interference (RFI), became clear when it was noticed that it reappeared four minutes earlier each day, as expected from a celestial source observed with a transit telescope.

This object showed extremely regular pulsations with a period of 1.337 s and lasting about 20 ms. On the basis of the the so-called causality principle, stating that a signal of duration t must be emitted from a source with a diameter $d \leq t \times c$ (where c is the speed of light), it was argued that the object responsible for the radio pulsations should have a very small radius. At first, because of the small inferred size and of the regularity of the pulses, the hypothesis that this could be a signal from an extraterrestrial intelligence was taken into account and the first pulsar was nicknamed LGM1 (for Little Green Man 1). Very soon Professor Hewish and Miss Bell understood that this couldn't be an artificial signal, firstly because three more similar recordings were found in different positions of the sky and secondly because there was no evidence of an orbital motion of the emitting object which would have been detected if it was a message coming from an extraterrestrial planet orbiting its sun.

The attention then moved to small sized stellar objects: white dwarfs, whose radius is about a few thousand kilometers, or neutron stars, whose existence was postulated in 1934 [33] but never observationally verified. The ultimate evidence that pulsars were indeed neutron stars came from the discovery, a year later, of the pulsar in the Crab Nebula [34], a plerionic supernova remnant that Franco Pacini [35], before the discovery of the first radio pulsar theorised that could be powered by a rotating magnetised neutron star. In fact, the short period of the Crab pulsar (only 33 milliseconds) ruled out the white dwarf hypothesis: at this rotational speed such an object would be disrupted by the centrifugal force overwhelming the gravity. The observation of spin-down in this object also confirmed that the source of periodicity was indeed rotation and not vibration, as also theorised in the first years, for which slow down is not expected.

2.2 Radio Emission

Pulsars are highly magnetised rapidly rotating neutron stars (NSs). The origin of both the fast spinning and high magnetisation is not fully understood yet, but it is probably related to angular momentum and magnetic flux conservation during the supernova explosion that produces the neutron star. Since the magnetic axis is not aligned with the rotational axis the neutron star's dipole moment is time-varying and thus the star radiates a significant amount of energy in the form of electromagnetic waves at the rotation frequency Ω:

$$\dot{E}_{dipole} = -\frac{2B^2 R^6 \Omega^4 \sin^2 \alpha}{3c^3} \tag{2}$$

here B is the surface magnetic field, R is the radius of the NS (typically ~ 10 km) and α is the angle between the magnetic and the rotational axis. The energy emitted via magnetodipole radiation is supplied by the loss of rotational energy:

$$\dot{E}_{rot} = I\Omega\dot{\Omega} \tag{3}$$

where I is the moment of inertia of the star ($\sim 10^{45}$ g cm^2) and $\dot{\Omega}$ is the derivative of the spin frequency.

Equating the two expressions above we can obtain, assuming $\alpha = 90°$, an estimate of the surface magnetic field of the rotating neutron star in terms of its period $P = 2\pi/\Omega$ and the period derivative $\dot{P}$:

$$B = \sqrt{\frac{3c^3 I}{8\pi^2 R^6} P\dot{P}} \simeq 3.2 \times 10^{19} \sqrt{P\dot{P}} \quad G \qquad (4)$$

The slow-down of pulsars is in general described by a relation $\Omega \propto \dot{\Omega}^n$, where n is the braking index (equal to 3 in the magneto-dipole model described above):

$$n = \frac{\Omega\ddot{\Omega}}{\dot{\Omega}^2} \qquad (5)$$

Integrating the Eq. (5), if $n \neq 1$, we obtain an estimate of the age of the pulsar:

$$t = \frac{P}{(n-1)\dot{P}} \left[1 - \left(\frac{P_0}{P}\right)^{n-1} \right] \qquad (6)$$

For $n = 3$ (dipole emission) and assuming that the initial spin period $P_0 \ll P$ we obtain the characteristic age τ of the pulsar:

$$\tau = \frac{P}{2\dot{P}} \qquad (7)$$

According to the picture seen above, we can follow the evolution of an isolated pulsar (for pulsars in binary systems see Sect. 2.3) using the P–$\dot{P}$ diagram (Fig. 1).

After the supernova explosion the pulsar is rapidly spinning and has a high magnetic field strength, hence, according to Eq. (4), it has a high value of $\dot{P}$, placing it in the top-left side of the P–$\dot{P}$ diagram. As time passes the pulsar spins down and it moves right in the diagram, following the constant B-field lines (or slightly turning downwards if the magnetic field decays). When the pulsar crosses the so-called *death-line*, the emission mechanism is no more efficient and the pulsar switches off. The exact location of this theoretical line in the P–$\dot{P}$ diagram depends on the details of the electrodynamical model adopted to explain the coherent radio emission.

One of the most commonly accepted models for the radio emission mechanism of pulsars, the so-called polar-cap model, predicts that the presence of the high magnetic fields ($B \sim 10^8 - 10^{13}G$) in a fast spinning object produces an electric field so intense that it extracts charged particles from the neutron star surface. The particles are constrained to move along the magnetic field lines co-rotating with the neutron star. The co-rotation is possible only within a radius R_{lc} , the light cylinder radius (see Fig. 2) given by:

$$R_{lc} = c\frac{P}{2\pi} \qquad (8)$$

Beyond this radius the speed of the field lines would be greater than the speed of light: the field lines must hence be open. Using the equation that describes the magnetic field lines of a dipolar field ($\sin^2 \theta/r = const$) we can define the angle θ_p formed at the magnetic poles of the star by the last open field line:

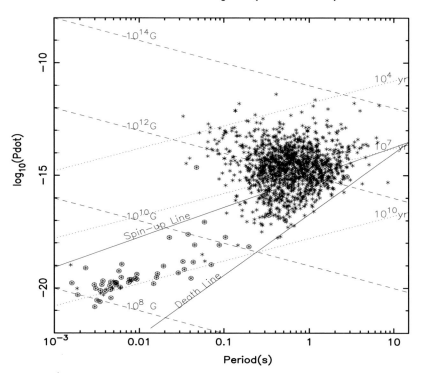

Fig. 1. P–$\dot{P}$ diagram. The dashed lines denote constant magnetic fields, the dotted ones constant age, the upper solid one is the spin-up line (see text) for an Eddington mass transfer rate and the lower solid line is the death line (see text). Circles denote the binary pulsars.

$$\frac{\sin^2 \theta_p}{R} = \frac{\sin^2 90 \deg}{R_{lc}} \qquad \rightarrow \qquad \sin \theta_p = \sqrt{\frac{2\pi R}{cP}} \qquad (9)$$

This angle naturally delimits a conal-like set of field lines. The particles moving along these open field lines give rise to the radio emission (probably curvature radiation) confined in a beam of semi-amplitude θ_p. Because of the misalignment of the magnetic and rotational axes, the collimated emission appears pulsating with a period equal to the spin period of the neutron star to any observer whose line of sight crosses the radio beam.

2.3 Binary and Millisecond Pulsars

When a pulsar crosses the death line the emission mechanism fails and the neutron star ceases to behave as a radio pulsator. If the neutron star belongs to a binary system, during the evolution of the companion star matter and angular momentum can be transferred via wind or more efficiently via Roche Lobe Overflow (RLO) on the neutron star. During this accretion phase the

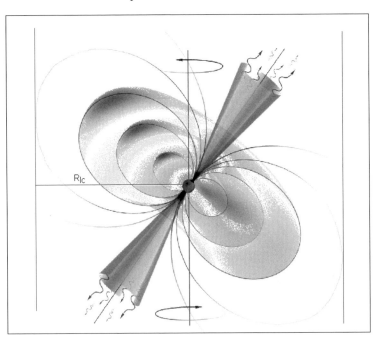

Fig. 2. Polar-cap model for the radio pulsar emission mechanism.

neutron star is visible as an X-ray source, its period decreases (the star is spun-up by angular momentum transfer) and its magnetic field decays, probably because of the accretion itself (see [36, 37] for alternative explanations of the magnetic field decay). Depending on the mass of the companion and hence on the duration of the mass transfer phase, at the end of the accretion the spin period and magnetic field can be such that the neutron star is again above the death line, in the bottom-left part of the P–$\dot{P}$ diagram. These reborn pulsars are called millisecond pulsars (because of their short rotational periods) or *recycled* pulsars. The model described above for the formation of this class of objects is the so-called *recycling model*, developed in the early 80s by Ali Alpar and collaborators [38].

Let's see in more detail the evolution followed by a binary system giving rise to a millisecond pulsar (Fig. 3).

We start with a system composed of a primary star having mass $M > 8$ $M_\odot$ and a secondary with mass $M_2 < 1$–2 $M_\odot$. The more massive star evolves becoming a red giant, filling its Roche Lobe and losing mass until only a helium core is left. The first stage of the evolution ends with the helium core implosion, the supernova explosion and the creation of a compact remnant. If the system doesn't remain gravitationally bound, the newly born neutron star evolves as an isolated pulsar as described in Sect. 2.2, otherwise we obtain a binary system containing a neutron star with a typical mass $M_{NS} \sim 1.4$ $M_\odot$

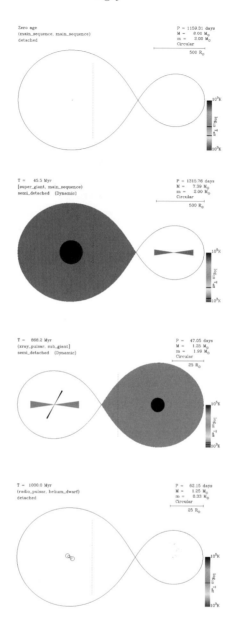

Fig. 3. Evolutionary path for a binary system containing a 9.0 $M_\odot$ and a 2.0 $M_\odot$ stars (from http://www.manybody.org/cgi-bin/starlab/binary_demo.pl). From top to bottom: main sequence - main sequence; first Roche Lobe overflow (followed by a supernova explosion that leaves a NS); second RLO (the NS shines as X-ray source); end of the mass transfer on the NS: the NS shines as radio pulsar and secondary star is a WD.

and a low mass main sequence star. At that point the neutron star can be seen as a young radio pulsar in a binary system. As long as the (possible) wind from the companion star is swept by the pulsar radiation pressure up to the light cylinder radius and/or the gravitational radius, given by:

$$R_G = \frac{2GM}{v_w^2} \qquad (10)$$

(where v_w is the velocity of the wind from the companion), the pulsar steadily spins down following Eq. (4) (*ejector* phase). Because of the slowing down, the pulsar wind pressure drops off and the matter carried in the companion wind reaches the magnetospheric radius R_m. This quantity, defined as a fraction $\phi < 1$ of the the Alfven radius, at which the magnetic field pressure $P_{mag} = B^2/8\pi$ equals the pressure of a spherically infalling matter $P_{gas} = \rho_{gas}/v_w^2$ (with ρ_{gas} the density of the gas), can be written as:

$$R_{mag} = \phi 9.8 \times 10^5 B_8^{4/7} M_{NS}^{-1/7} R_6^{10/7} \dot{M}_E^{-2/7} \qquad \text{cm} \qquad (11)$$

where B_8 is the pulsar magnetic field in units of 10^8 Gauss, M_{NS} is the mass of the neutron star in solar masses, R_6 is the radius of the neutron star in units of 10^6 cm and $\dot{M}_E$ is the mass transfer rate in Eddington units (i.e. $1.5 \times 10^{-8} R_6$ $M_\odot$ yr^{-1}, [39]). At this point, if at R_{mag} the velocity of the magnetic field lines is greater than the keplerian velocity, the *propeller* phase begins in which the centrifugal force exerted by the magnetosphere expels the infalling matter from the system. In this phase the neutron star loses angular momentum and the pulsar spins down. When the velocity of the magnetic field lines at the magnetospheric radius becomes smaller than the keplerian velocity, the centrifugal barrier is open and the matter carried by the wind of the companion star starts to be accreted on the neutron star surface. At this point the neutron star's spin period slowly decreases. The key phase for the recycling model is the final phase, when the companion star fills its Roche Lobe and starts to transfer matter onto the neutron star surface at much higher rates than before (up to 10^{-8} M$_\odot$/yr). For an accretion rate $\dot{M}_{acc}$, the matter transferred can spin-up the neutron star to a minimum period $P_{min} \propto B^{6/7} \dot{M}_{acc}^{-3/7}$ at which the rotational velocity equals the keplerian velocity of the inner rim of the accretion disk: the accretion of angular momentum stops and the neutron star is placed on the P–$\dot{P}$ diagram on the so-called spin-up line (in Fig. 1 traced for $\dot{M}_{acc} = \dot{M}_E$). In low mass binary systems, as the one described here, the RLO phase (or *accretor* phase) lasts $10^7 - 5 \times 10^8$ yr at the end of which the neutron star reaches periods of few milliseconds and magnetic fields in the range 10^8–10^9 G. When the stellar surroundings are free of matter (at least up to the light cylinder radius), the neutron star, being above the death line, starts shining again as radio pulsar.

2.4 Basic Observational Properties

Following what discussed in Sect. 2.2 and Sect. 2.3, pulsars can be subdivided in two families: the long period pulsars, with periods ranging from ~ 0.1 to 8 seconds and magnetic fields $B \sim 10^{11}$–10^{13} G, and the millisecond (or recycled) pulsars, with periods $P < 50$ ms and magnetic fields in the range $10^8 - 10^9$ G.

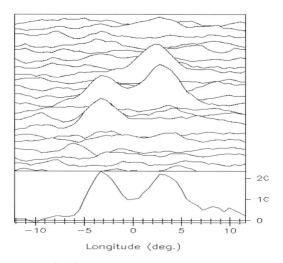

Fig. 4. Single pulses (top) and averaged profile (bottom) for PSR B0834+06.

Duty cycles (i.e. the duration of the pulse over the period) are typically 1–5% for long period pulsars and from $\sim 5\%$ up to $> 30\%$ for millisecond pulsars [40]. Pulse shapes are highly variable from pulse to pulse (Fig. 4) but the integrated profile (averaged over few hundreds of cycles) is extremely stable and represents a sort of "fingerprint" for a pulsar.

The observed pulse shapes, because of the interaction between the broad band radio signal and the interstellar medium (ISM), usually appear different from the emitted pulses. The principal effects of the ISM on the radio emission are the dispersion and the scattering of the pulsed signal both producing a broadening of the pulse profile.

Dispersion

The dispersion of a broad band pulsed signal is due to the fact that the group velocity v_g of an electromagnetic wave in a plasma depends on the emission frequency ν of the signal. This quantity can be written as

$$v_g = c\sqrt{1 - \left(\frac{\nu_p}{\nu}\right)^2} \tag{12}$$

where the term under square root represents the refractive index of the medium and $\nu_p = \sqrt{e^2 n_e / \pi m_e}$ (with e, n_e and m_e, the charge, the number density and the mass of the electron respectively) is the plasma frequency, below which a signal is totally absorbed. For a frequency $\nu \gg \nu_p$ we can write the group velocity as:

$$v_g = c\left[1 - \frac{1}{2}\left(\frac{\nu_p}{\nu}\right)^2\right] \tag{13}$$

The time taken by a signal with a frequency ν to reach the observer is then given by:

$$t = \int_0^d \frac{dl}{v_g} \sim \int_0^d \frac{1}{c}\left[1 + \frac{1}{2}\left(\frac{\nu_p}{\nu}\right)^2\right] dl = \frac{d}{c} + \frac{1}{2}\frac{e^2}{\pi m_e c}\frac{1}{\nu^2}DM \tag{14}$$

where $DM = \int_0^d n_e dl$ is the *dispersion measure* and d is the distance to the observer. Since the radio pulsar signal is broad band, the dependence of t on the frequency produces a broadening Δt_{DM} of the pulse profile:

$$\Delta t_{DM} = \frac{e^2}{2\pi m_e c}\left(\frac{1}{\nu_1^2} - \frac{1}{\nu_2^2}\right) DM \sim 8.3 \cdot 10^3 \frac{\Delta\nu_{MHz}}{\nu_{MHz}^3}DM \quad \text{s} \tag{15}$$

where ν_{MHz} and $\Delta\nu_{MHz}$ are the central observing frequency and the total bandwidth in MHz respectively and DM is given in pc cm^{-3}.

One method for minimizing the effects of the dispersion is to split the total bandwidth into several channels having a frequency width $\delta\nu$. In this way, in each frequency channel, the pulse is poorly affected by the dispersion in the ISM and the signal in the different channels appears as in Fig. 5, with the pulses arriving in each channel at a different time. Knowing the dispersion measure allows us to correct for this time shift integrating the signal along the line connecting the edges of the pulses in each channel (*dedispersion* of the signal) and obtaining a sharp and high signal-to-noise profile.

Better results can be obtained applying the so-called *coherent de-dispersion* method, in which the incoming signals are de-dispersed over the whole bandwidth using a filter which has the inverse transfer function to that of the interstellar medium [41]. As a result, the pulse profile is perfectly aligned in frequency, without any residual dispersive smearing caused by finite channel bandwidths (Fig. 6).

Interstellar Scattering

The inhomogeneities in the ISM act like center of scattering onto the signal emitted from a pulsar. Because of this interaction with clumps of matter, the

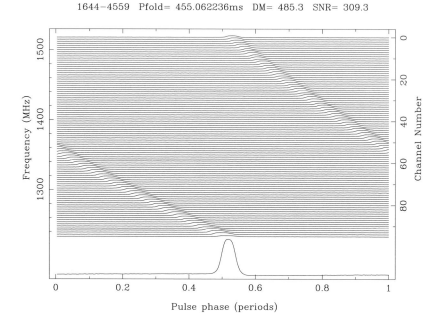

Fig. 5. Effect of the dispersion on the pulse of PSR J1644-4559. In this observation we used a total bandwidth $\Delta\nu = 288$ MHz, split into 96×3 MHz channels, and a central frequency $\nu = 1372.5$ MHz. The bottom integrated profile is obtained by dedispersing the signal (see text).

radio waves follow different paths and arrive at the observer at different times (Fig. 7) thus producing a broadening δt_{scatt} of the pulse. According to the simple "thin screen" model δt_{scatt} scales as:

$$\delta t_{scatt} \propto \frac{DM^2}{\nu^4} \qquad (16)$$

Unlike the case of the dispersion, no technical skills can be adopted to mitigate the effects of interstellar scattering. The only way to reduce its effect is to observe at higher frequencies.

Using a high frequency reduces the broadening of the pulses caused by the interaction of the radio signal with the interstellar medium (both dispersion and scattering) and also reduces the sky background noise ($\propto \nu^{-2.7}$). On the other hand the typical pulsar spectrum at radio wavelength is a steep power-law $S(\nu) \propto \nu^{-\alpha}$ with spectral index $\alpha \sim 1.6$. The choice of the observing frequency, hence, is a crucial point in any observational campaign.

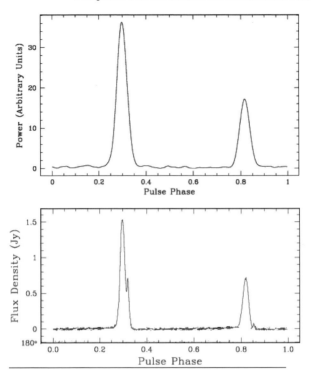

Fig. 6. Profiles of PSR B1937+21 obtained with filterbanks (top) and using coherent dedispersion (bottom).

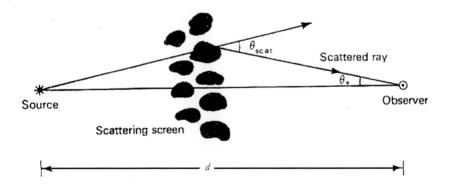

Fig. 7. Effect of the interstellar scattering on the photons emitted by a celestial source.

3 The Multibeam Pulsar Equipments at Parkes: Data Acquisition and Data Reduction Systems

3.1 Hardware System

The new receiver system installed in 1997 on the Parkes 64m radio telescope (NSW, Australia) operates at a central frequency of 1347 MHz and consists of 13 separate beams (Fig. 8), with a central feed surrounded by two rings, each of six feeds, arranged in a hexagonal pattern [32]. This arrangement permits the simultaneous observation of 13 regions of sky, increasing the speed of surveys by approximately the same factor. The 13 beams of the multibeam receiver are spaced by approximately two beamwidths in the sky so that a given area is entirely covered by four pointings placed as shown in the right panel of Fig. 8. This cluster of pointings covers a region about 1.5° across, with adjacent beams overlapping at the half power points. An interactive programme, hexaview can be used to display the status of each pointing ("selected", "observed" or "processed", see as an example Fig. 9) during survey observations, and to select pointings for observation. The total 288 MHz bandwidth of each feed is subdivided into 2 × 96 3 MHz channels in order to minimize the pulse broadening due to dispersion in the interstellar medium (see Sect. 2.4). The signals of the individual channel of each beam are one-bit sampled, typically every 125 or 250 μs, and recorded on Digital Linear Tapes for off-line analysis.

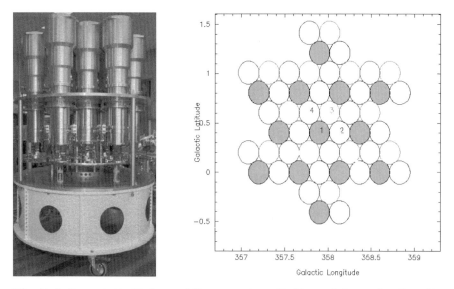

Fig. 8. Left panel: the Parkes multibeam receiver. Right panel: beams locations for a cluster of four pointings.

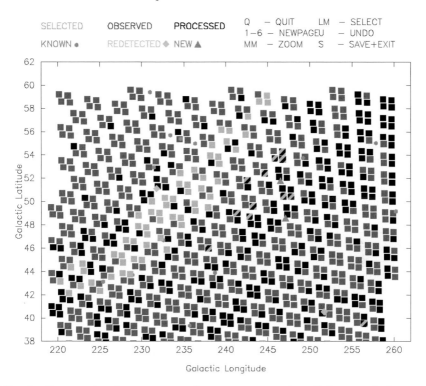

Fig. 9. Output plot for `hexaview`: the gray squares, from lighter to darker, are respectively selected, observed and processed pointings. The striped squares are the not (yet) observed pointings.

The minimum detectable flux density for a pulsar of period P can be described by the following equation [42]:

$$S_{min} = \alpha \times \epsilon n_\sigma \frac{T_{sys} + T_{sky}}{G\sqrt{N_p \Delta t \Delta \nu_{MHz}}} \sqrt{\frac{W_e}{P - W_e}} \qquad \text{mJy} \qquad (17)$$

where n_σ is the minimum signal-to-noise ratio S/N considered (here 8.0), T_{sys} and T_{sky} the system noise temperature and the sky temperature in K respectively, G the gain of the radio telescope (in K/Jy), Δt the integration time in seconds, N_p the number of polarizations and $\Delta \nu_{MHz}$ the bandwidth in MHz. ϵ is a factor ~ 1.5 accounting for sensitivity reduction due to digitization and other losses. α is a normalization factor ~ 2 that takes into account the difference between the reported Eq. (17) and the more precise limit sensitivity formula obtained by [43]. This factor has been calculated in such a way to obtain the reported sensitivity limit (0.14 mJy) for the center of the central beam of the Parkes Multibeam survey [43]. Finally, W_e is the effective width of the pulse:

$$W_e = \sqrt{W^2 + \delta t^2 + \delta t_{DM}^2 + \delta t_{scatt}^2} \qquad (18)$$

where its value depends on the intrinsic pulse width W, on the time resolution δt of the detection apparatus and on the broadening of the pulse introduced both by the dispersion of the signal in each channel (δt_{DM}) and by the scattering induced by inhomogeneities in the ISM (δt_{scatt}).

Table 1. Feed and receiver parameters of the Parkes multibeam system.

Number of Beams	13		
Polarizations/Beam	2		
Frequency Channels/Polarization	96 × 3 MHz		
System Temperature (K)	23		
Beam	Central	Inner Rim	Outer Rim
Telescope Gain (K/Jy)	0.735	0.690	0.581
Half Power Beamwidth (arcmin)	14.0	14.1	14.5

Given the parameters of the multibeam receiver (Table 1) and the other survey parameters, we obtain, depending on the pulsar period and dispersion measure DM, a sensitivity limit typically in the range $0.14. - 0.3$ mJy for an integration time of 35min (as adopted in the survey of the galactic plane), and in the range $0.43 - 1$ mJy, for an integration time of 265 s (as adopted in the high latitude pulsar survey).

Fig. 10 shows as an example the sensitivity limit of the high latitude pulsar survey, as a function of P for a range of dispersion measures and for the center of the central beam of the multibeam receiver. In this case, the contribution of the sky temperature, low at high galactic latitudes, is neglected.

Using an average value of the telescope gain in the different beams, and adopting a further degrading factor [43] which takes into account the non-uniform beam response at the position of the pulsar relatively to the beam center, the realistic flux limit for the high latitude pulsar survey is about 0.68 mJy.

3.2 Data Analysis

For off-line processing we used a network of 10 Alpha workstations at the Bologna Astronomical Observatory and, more recently, a Linux cluster consisting of 196 CPUs at the Jodrell Bank Observatory (Manchester, UK) and a Linux cluster consisting of 40 CPUs at Cagliari Observatory.

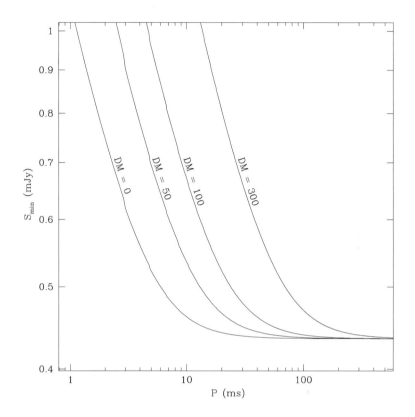

Fig. 10. Flux limits as a function of the pulsar period P and for different values of DM for an integration time of 265 s and a sampling time of 125 μs, as adopted in the high latitude pulsar survey, calculated for the centre beam of the multibeam receiver. The plotted curves refer to the center of the central beam of the multibeam receiver and do not take into account the (negligible) contribution of the sky temperature. The value adopted here for the pulse width W is 5% of the spin period.

Processing

In the typical pipeline, the first stage of processing consists of identifying spurious signals in the power spectrum of non de-dispersed (zero-DM) data set. Radio Frequency Interferences (RFIs), being produced mainly on Earth, are indeed often undispersed. Known signals as, for instance, the power line at 50 Hz and its harmonics, are first identified and their bandwidth determined. The remaining spectrum is searched for signals that appear many times in different positions in the sky: if a frequency is seen in more than four beams of a pointing or if it appears three times in the same beam of different pointings or if it appears more than seven times in the whole tape, it is flagged as an

RFI. These signals and their harmonics are then deleted in the single pointing where they have been found or, for the last case, in the entire tape.

The subsequent step of the data analysis consists of de-dispersing the signal at different trial DMs. The 'Tree' algorithm [44] is typically used for processing survey observations, while a set of individual de-dispersed time series is created for the targeted search of the Globular Cluster system. In the survey case, the raw data are de-dispersed using DM values ranging from zero to the value corresponding to a broadening of the signal in a single frequency channel equal to the sampling time (i.e. from 0 to 17 pc cm^{-3}, for a sampling time of 125 μs as in the case of the high latitude survey). Then the data samples are summed in pairs so that the new effective sampling time is double the original one. The rebinned time series is again de-dispersed as described above (hence with DMs ranging from 17 to 35) and the process is reiterated five more times. One then ends with seven .dedisp files containing time series de-dispersed with DM values ranging from 0 to a maximum value which is determined by the sampling time adopted during the data collection. In the case of the high latitude survey, the nominal maximum DM value is 1088 (corresponding to a 4 ms smearing), but the actual maximum valued adopted is chosen as the minimum between 1088 and $42/\sin|b|$ pc cm^{-3} (the higher the galactic latitude b the smaller the amount of interstellar medium).

The de-dispersed time series are then Fourier transformed using a Fast Fourier Transform (FFT) routine. The resulting power spectra are harmonically summed in such a way that at each fundamental is added its first harmonic (Fig. 11). The harmonic summing process is particularly efficient for signals - as the radio pulsars ones - having narrow peaks: a pulse having width W will have approximately $P/2W$ harmonics with amplitude comparable to that of the fundamental, while a sinusoidal wave will have a large fundamental peak and small power in the harmonics.

The original power spectra and those obtained summing 2, 4, 8 and 16 harmonics are searched for significant peaks to give a set of 50 candidate periods (10 from the fundamental spectrum and 10 from each harmonic sum) for each DM. For each candidate a pulse profile in the time domain is created by inverse transformation of the complex Fourier components for the fundamental and its harmonics. For all the candidates having signal-to-noise ratios greater than four (to a maximum of 265 candidates per beam, sorted by S/N), the appropriate de-dispersed time series are summed in 4 sub-bands and folded in 32 to 64 sub-integrations (depending on the candidate period), using the nominal period and dispersion measure. Sub-bands and sub-integrations are then summed with a range of delays in frequency and time, up to one sample per sub-band and per sub-integration, to search for the maximum S/N over a range of parameters around the nominal P and DM. The resulting profiles, the new maximized S/N ($P - DM$ signal-to-noise ratio) and related parameters are then stored for later inspection.

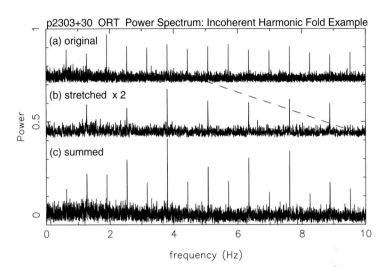

Fig. 11. The first half of the power spectrum is stretched in such a way that at each fundamental is summed its first harmonic.

Results Inspection

In the last step of the survey data reduction, the candidates of an entire tape are compared and searched for common periods. If a signal appears more than six times in a tape, it is flagged as an interference. All the remaining candidates, having a S/N greater than 8.0 (usually) are visually inspected using an interactive display (Fig. 12) and classified as candidates or rejected as interferences.

Although the classification of candidates is somehow subjective, there are several features characterizing a pulsar signal that one can search for in the display plots: a high signal-to-noise ratio, together with a well defined peak in the DM versus S/N plot and continuity of the pulse signal across subbands and sub-integrations are the most important criteria to discriminate between real pulsars and an RFIs. The signal should also be linear or parabolic (indicating a constant acceleration) in the phase-time plot and linear in the phase-frequency one (see Fig. 12 caption).

Candidate Confirmation: Gridding and Binary Search

Once a candidate is selected, it is necessary to confirm its pulsar nature. To do so the suspect is observed at 5 grid positions, one at the nominal coordinates and four with an offset in latitude and longitude of 9 arcminutes, using the central beam of the multibeam receiver. These observations are searched in period and dispersion measure around the nominal values and, if two or three

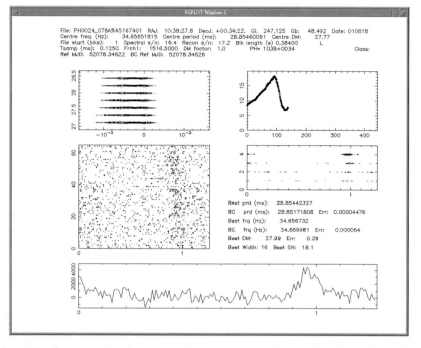

Fig. 12. Output plot for a candidate subsequently confirmed as the pulsar PSR J1038+0034. Clockwise from the top left the sub-plots show a gray-scale of the dependence of S/N on DM and offset (in ms) from the nominal period, the dependence of S/N on DM trial number, a gray-scale plot of S/N versus pulse phase in four sub-bands, the integrated pulse profile and a gray-scale plot of S/N versus pulse phase for 64 sub-integrations.

detections are obtained, an improved position is computed from the relative signal-to-noise ratios. This technique is known as "gridding" of a suspect.

For millisecond candidates, if there is no detection, the central pointing data are searched for parabolic shifts in phase. If the pulsar belongs to a tight binary system, indeed, the simple linear search approach doesn't work because *(i)* the period can be significantly different from the nominal one (hence not searched) and *(ii)* the integrated pulse can be completely smeared out summing in phase sub-integrations where the phase of the signal is not constant (Doppler shifted). To perform such a search, a programme called `binary-confirm` is used. After the de-dispersion of the data, `binary-confirm` launches, with different values of the input period, a programme that searches, over 128 sub-integrations, for the best linear or parabolic shift (i.e. the phase shift that gives the highest signal-to-noise ratio). The period range over which the search programme is launched (i.e. the maximum variation of the period from the nominal value that we consider) is chosen in such a way that $pr/P_i =$

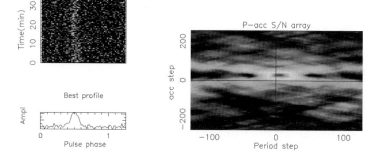

Tape: PM0048 File: 21   Start Block: 1 Date: 980706   UT: 14:21:36.0000
LST: 19:12:03.93   R.A.(J2000): 18:01:56.3172   Dec(J2000): −21:30:16.847
Source: J1801−2130   Freq1(MHz): 1516.50000   96 x −3.0000 MHz Tsamp(ms): 0.250000

Zero−Acc − Best DM

Ndat: 8388608   MJD: 51000.616902   Nsub: 128
DM(cm^{-3} pc): 149.40

Best s/n: 21.8
P(ms):   12.643271 Pb(ms):   12.642983
a (m/s^2):   3.585
da/dt (cm/s^3):   0.000

Best Acc − Best DM

P−acc S/N array

Best profile

Fig. 13. Binary-confirm output plot for the binary pulsar PSR J1801−2130. On the left: top panel shows the phase versus sub-integrations plot before the phase shift correction, below the same plot after the correction and at the bottom the pulse profile after correction. On the right: period step (centered on the input period in ms) versus acceleration step (centered on 0.0 m/s^2); the grayscale indicates increasing S/N.

0.001 (where P_i is the ith input period). An example of a binary-confirm output plot is shown in Fig. 13.

3.3 Timing

When a new pulsar is discovered and confirmed the only known parameters are its approximate position (within few arcminutes), rotational period P and dispersion measure DM, that gives a rough estimate of the pulsar's distance, once a model for the electrons' distribution in the Galaxy [45] is assumed.

A follow-up observational campaign then starts in order to better assess new pulsar's position and rotational parameters (P and $\dot{P}$). These quantities allow us to estimate the pulsar age and magnetic field and to cross-correlate its position with that of possibly related sources, such as X-ray or γ-ray objects or supernova remnants. The regular monitoring ('timing') of new discoveries also allows us to determine whether a pulsar is isolated or belongs to a binary system and, in the latter case, to derive orbital parameters.

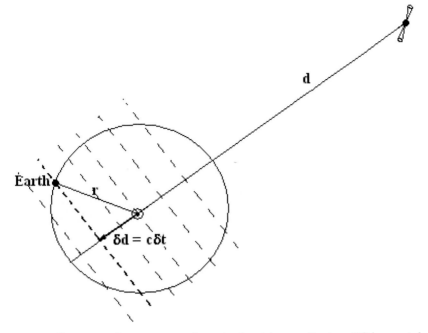

Fig. 14. Correction from topocentric to baricentric coordinates: TOAs must be corrected for δt, the time necessary to the wavefront to cover the distance δd.

Timing a pulsar means measuring and phase connecting the pulses times of arrival (TOAs). To do so, a standard profile (obtained summing in phase an adequate number $->1000$ – of single pulses) is convolved with the integrated profiles of each observation: the topocentric TOA is then calculated adding at the starting time of the observation the fraction of period, τ, at which the χ^2 of the convolution is minimized.

The measured times are then compared with TOAs predicted by a given pulsar model. The best fit positional, rotational and orbital parameters are then obtained minimizing the differences between measured and predicted times of arrivals (the timing residuals) with a multiparametric fit.

The first step of a timing analysis is the transformation of the topocentric TOAs to the solar system baricenter, in first approximation an inertial reference frame.

The time correction takes into account the Roemer delay $\Delta R_\odot$ (due to the motion of the Earth), the delays due to the dispersion in the ISM and the corrections due to the relativistic effects known as Einstein delay $\Delta E_\odot$ and Shapiro delay $\Delta S_\odot$). In summary the baricentrised TOA t_b is obtained applying the formula:

$$t_b = t + \Delta R_\odot - \frac{D}{\nu^2} + \Delta E_\odot - \Delta S_\odot \tag{19}$$

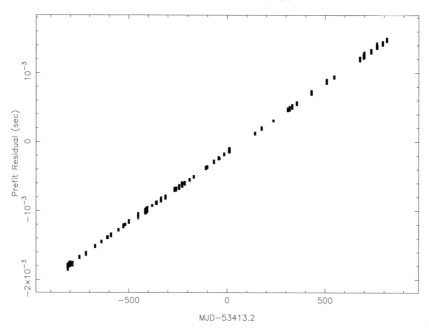

Fig. 15. Timing residuals with an error in the estimate of the spin period.

where t is the topocentric TOA. The Roemer delay can be written as

$$\Delta R_\odot = \frac{\mathbf{r} \cdot \mathbf{n}}{c} + \frac{\mathbf{r} \cdot \mathbf{n} - |r|}{2cd} \qquad (20)$$

where $\mathbf{n}$ is the versor on the line going from the Sun to the pulsar and $\mathbf{r}$ is the vector connecting the Sun and the Earth. The second term in Eq. (19) takes into account the delay introduced by the dispersion (see Sect. 2.4). Here D is the so-called dispersion constant defined as:

$$D = \frac{t_2 - t_1}{\nu_2^{-2} - \nu_1^{-2}} = \frac{e^2}{2\pi m_e c} \int_0^d n_e dl \qquad (21)$$

Einstein's delay is a combination of gravitational redshift and relativistic time delay due to the motion of the Earth and the other solar system's planets. It's derivative is given by:

$$\frac{d\Delta E_\odot}{dt} = \sum_i \frac{Gm_i}{c^2 r_i} + \frac{v_\oplus^2}{2c^2} \qquad (22)$$

where G is the gravitational constant, m_i are the masses of the planets, r_i their distances to the Sun and $v_\oplus$ the velocity of the Earth with respect

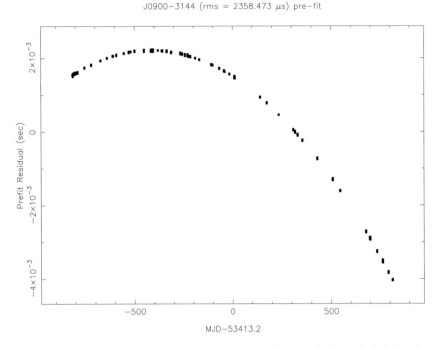

Fig. 16. Timing residuals with an error in the estimate of the period derivative.

to the baricenter of the solar system. The Shapiro delay [46] measures the propagation time of an electromagnetic wave in the gravitational field of the Sun. This quantity is a function of the angle θ from the pulsar to the Earth and the Sun:

$$\Delta S_\odot = -\frac{2GM_\odot}{c^3} \log\left(1 + \cos\theta\right) \qquad (23)$$

If the pulsar is orbiting a companion star, the TOAs vary along the orbit anticipating when the pulsar is in front of the companion and delaying when it's behind it (orbital Doppler variations). Fitting for the orbital modulations, one can hence derive five keplerian parameters describing the binary system: the orbital period P_b, the eccentricity e, the projection of the semi-major axis $x = a\sin i$, the periastron longitude ω and the epoch of periastron T_0. Combining these quantities one can derive the *mass function*:

$$f(M) = \frac{(M_c \sin i)^3}{(M_{NS} + M_c)^2} = \frac{4\pi^2 (a \sin i)^3}{G P_b^2} \qquad (24)$$

where M_c is the companion mass. Assuming a typical value for $M_{NS} = 1.4\,M_\odot$ and an edge on orbit ($i = 90°$) we can obtain a lower limit for the companion mass.

J0900−3144 (rms = 1403.749 μs) pre−fit

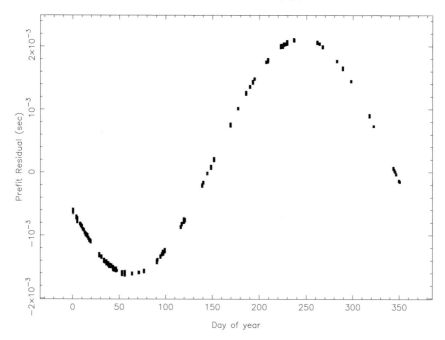

Fig. 17. Timing residuals with an error in the estimate of the position.

When a pulsar is in an eccentric orbit with a compact massive companion (producing a strong near gravitational field), up to five further parameters – the so-called post-keplerian parameters – can be measured (see Sect. 5).

From the practical point of view timing a pulsar means comparing the baricentrised TOAs with the ones predicted by a model describing rotational, positional and orbital parameters of the given pulsar and searching for systematic trends in the time residuals. A poor estimate of the spin period, for instance, will give a linear trend in the timing residuals (Fig. 15) and an error in the period derivative will give a parabolic trend (Fig. 16). Collecting TOAs for roughly a year allows to determine accurately the pulsar position. An error in the position, in fact, corresponds to an error in the baricentric TOAs correction and gives rise to a sinusoidal trend with a period of one year in the residuals (Fig. 17). Continuing the timing observations over a period of several years also allows us to measure the pulsar proper motion. In fact the changing position of the pulsar produces a sinusoidal modulation of the residuals with a period of a year and with increasing amplitude (Fig. 18).

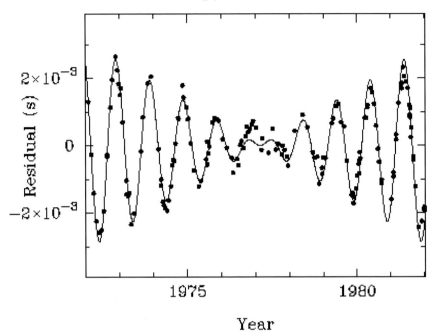

Fig. 18. Timing residuals with proper motion set to zero.

Timing Instabilities

After applying a correct timing model, one would expect a set of uncorrelated timing residuals randomly scattered about a zero mean value. This is not always the case: the residuals of many pulsars, mostly the youngest ones [47, 48], exhibit long term variations with time, known as "timing noise". Millisecond pulsars, on the contrary, are very stable clocks and are virtually not affected by timing instabilities. While the physical processes of this phenomenon are not well understood, it seems likely that it may be connected to superfluid processes and temperature changes in the interior of the neutron star [49], or processes in the magnetosphere [50, 51].

4 Results of the Recent Parkes Experiments

In this section we review the main results of the pulsar survey programmes carried out at Parkes in the last few years adopting the multibeam setup.

4.1 The PM Survey of the Galactic Plane

This survey covered a strip with $|b| < 5^o$ along the galactic plane between galactic longitudes of 260° and 50°. The data were sampled every 250 μs, and the observation time per pointing was 35 min, giving a very high sensitivity, about seven times better than previous similar surveys [30, 31], at least for pulsars not in short-period binary systems. The survey has been outstandingly successful, with over 700 pulsars discovered so far. Fig. 19 shows the locations of the new pulsars in the P–$\dot{P}$ diagram, showing that a substantial number of relatively young pulsars are now available in the sample.

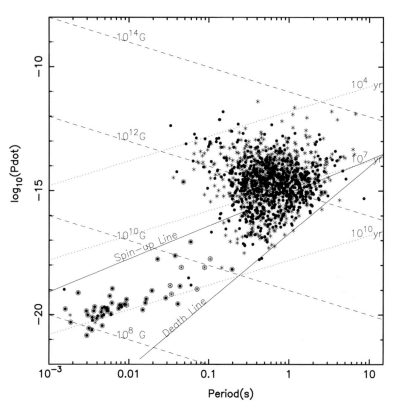

Fig. 19. P–$\dot{P}$ diagram containing the pulsars discovered by the PM survey (black dots),for which timing parameters have been already published, overlaid on the previous known population.

Several reports on the multibeam survey and its results are available [43, 52, 53, 54, 55, 56]. Also, papers on the discovery of several pulsars of particular interest have been published. PSR J1811−1736 [57], is a pulsar with a period of 104 ms in a highly eccentric orbit of period 18.8 d with a companion of minimum mass 0.7 $M_\odot$, most probably a neutron star. PSR J1119−6127 and J1814−1744, [58] are two young pulsars which have the highest surface dipole magnetic field strengths among known radio pulsars. PSR J1119−6127 has a characteristic age τ_c of only 1600 years, a measured braking index $n = 2.91 \pm 0.05$ and is associated with a previously unknown supernova remnant [59, 60]. PSR J1814−1744 has a much longer period, 3.975 s, and one of the highest inferred surface dipole field strengths among known radio pulsars, 5.5×10^{13} G, in the region of so-called "magnetars" [61]. PSR J1141−6545 is a relatively young pulsar ($\tau_c = 1.4$ Myr) in an eccentric 5-hour orbit for which the relativistic precession of periastron has been measured [62]. This implies that the total mass of the system is 2.30 $M_\odot$, indicating that the companion is probably a massive white dwarf formed before the neutron star we observe as the pulsar. PSR J1740−3052 [63] is a high-mass binary system in a highly eccentric 230-day orbit with a companion star of minimum mass 11 $M_\odot$. A possible companion is a late-type star identified on infrared images, but the absence of the expected eclipses and precession of periastron due to tidal interactions suggest that the actual companion may be a main-sequence B-star or a black hole hidden by the late-type star. We have also discovered [64] several circular-orbit binary systems with orbital periods in the range 1.3 − 15 days. Finally, we have discovered [65] several young pulsars which may be associated with EGRET γ-ray sources.

In terms of pulsar counting, this is the most successful pulsar survey carried out so far. Fig. 20 shows the distribution of the new pulsars in the Galaxy, and clearly indicates that much deeper regions of the Galaxy were probed by the PM survey

4.2 The Parkes High Latitude Survey

The analysis of the data of the Parkes High-Latitude Pulsar Survey (PH survey), collected in three years of observations starting from November 2000, resulted in the discovery of 17 new pulsars, four of which belonging to the class of millisecond, or recycled, pulsars. Of these, three are in a binary system, and one is an isolated mildly recycled object.

All the previously known pulsars in the survey area, having a flux density above our sensitivity limit have been re-detected. Three of these are millisecond pulsars; the PH survey has more than doubled the number of such sources in the selected region, fulfilling one of the main aims of this project, that is increasing the statistics on these objects. Fig. 21 shows the region of the sky covered by the PH survey: the filled dots are the pulsars discovered during this work, the open ones are the previously known re-detected pulsars and the starry ones are the undetected ones, below our flux limits. The spatial distribution of millisecond pulsars, in previous large area surveys, appeared isotropic; in the present work, as well as in the intermediate and high latitude Swinburne surveys [66, 67], that have the same observational parameters as the PH survey, the millisecond pulsar population seems to be more concentrated in a latitude range $|b| < 20°$. This suggests that the flux limit for these surveys is low enough to sample the real spatial distribution of these sources

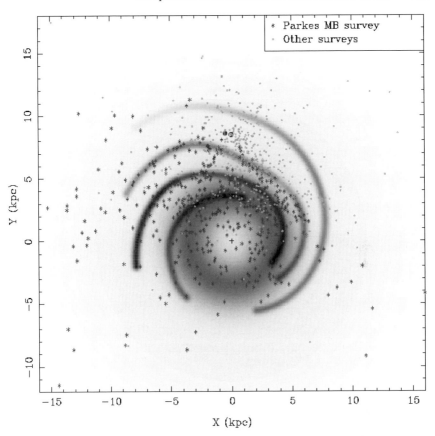

Fig. 20. Schematic diagram of the galactic coordinate system viewed from above, showing the distribution of the new discovered pulsars overlaid on the previous known sample.

and that the previously observed isotropy was essentially due to selection effects.

As expected from a survey looking at high latitudes (where the effect of the ISM in broadening the pulses is less severe) and using a fast sampling time (125 μs), the ratio between millisecond and long period pulsars is much higher for the population of the PH survey, devoted to the discovery of fast spinning objects indeed, than for the rest of the population (most of which were discovered in galactic plane surveys using longer sampling times).

PH SURVEY

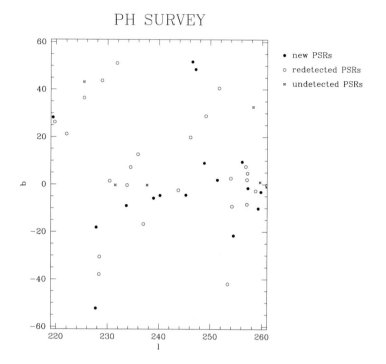

Fig. 21. Survey area of the PH survey in galactic coordinates. Filled dots are the new discovered pulsars, empty ones are the previously known objects re-detected by the present survey and starry points are the undetected ones.

5 The Double Pulsar J0737−3039

5.1 The Discovery

PSR J0737−3039A, a millisecond pulsar with a spin period of 22.7 ms, was discovered in April 2003 [68]. The original detection plot is shown in the top panel of Fig. 22. From the very pronounced curvature in the phase versus sub-integrations box (see Sect. 3.2 for a full description of output plots), de-noting a time dependent change in the phase of arrival of the pulses, it was immediately clear that the pulsar signal was affected by a significant Doppler effect in only 4 minutes of integration. That in turn suggested that the pulsar was experiencing the gravitational pull of a nearby and relatively massive companion. The bottom panel of Fig. 22 shows the result of the analysis of the same 4-minutes data processed with the binary-confirm code (Sect. 3.2). This programme found the best correction to the Doppler phase shift for an acceleration of 99 m/s^2 suggesting a binary period of just few hours for a

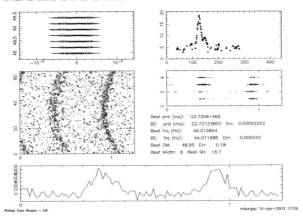

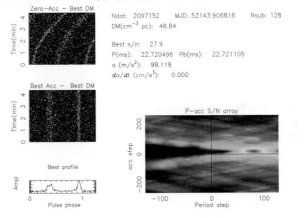

Fig. 22. Top panel shows the original detection plot for PSR J0737−3039A. Bottom panel shows the same 4-minutes data analysed with `binary-confirm`.

companion of ~ 1 M$_\odot$ (as a comparison, the typical observed binary pulsars having a $0.2 - 0.3$ M$_\odot$ white dwarf companion display acceleration < 50 m/s^2). After correction, the signal-to-noise ratio increases by $\sim 30\%$ with respect to the detection value.

Follow-up observations performed in May 2004, consisting of three ~ 5-hour integrations, confirmed that the orbit is indeed very tight and far from being circular: the binary period P_b is only 2.4 hr and the eccentricity $e \sim 0.09$. That makes J0737−3039A's orbit the tightest among those of all known binary pulsars in eccentric systems. Fig. 23 shows the radial velocity curve obtained plotting the (barycentric) spin period of the pulsar, measured at different times, versus the binary phase: the fact that the orbit is eccentric can be easily argued from the asymmetric shape of that curve.

PSR 0738−3033

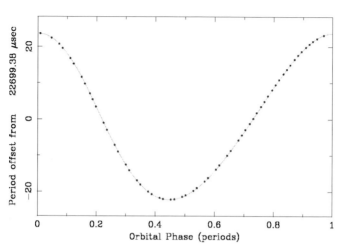

Fig. 23. Radial velocity curve for PSR J0737−3039A.

From the orbital parameters available at that time ($P_b \sim 2.4$ hr, $a \sin i \sim$ 1.4 lt-s; for the current best estimates of these parameters see Table 2), the pulsar mass function was calculated and resulted $M_f = 0.29$ M$_\odot$, implying a minimum companion mass of about 1.24 M$_\odot$, assuming $M_{NS} = 1.35$ M$_\odot$. Given such a high value for the minimum mass three possible hypotheses for the nature of the companion star were examined: a non-degenerate object, a massive Carbon–Oxygen white dwarf (CO-WD) or a second neutron star. The first hypothesis could be immediately ruled out since the radius of a non degenerate object of the required mass would almost completely fill the orbit of the system, probably producing deep and extended eclipses or, at least, strongly affecting the radio emission, which is not seen in the signal of PSR J0737−3039A. On the other hand, also the CO-WD scenario appeared unlikely considering the eccentricity of the orbit: a system containing a recycled neutron star and a white dwarf is indeed expected to be highly circular since it had the time to reach the minimum energy configuration (i.e. to circularise). If a second, recent supernova explosion had occurred forming another neutron star, the energy and momentum released would have distorted the system and this would explain the observed eccentricity.

A further strong constraint on the mass of PSR J0737−3039A companion came few days later from the first fit of a binary model to the times of arrival of the pulsations. By using a data span of only 6 days, a 10-σ determination of the advance of the periastron, $\dot{\omega}$, was possible. This parameter resulted in an extraordinarily high value: $\dot{\omega} \sim 17 \,°/\mathrm{yr}$ (note that the the previously highest observed value for this parameter was $5.33 \,°/\mathrm{yr}$, measured for PSR J1141$-$6545 [69]).

The periastron advance is one of the five post-keplerian parameters (see Sect. 3.3) that can be used to describe the corrections to the classical laws of orbital motion and signal propagation when the emitting source travels in a strong gravitational field.

In any given theory of relativistic gravity, the post-keplerian (PK) parameters can be written as function of the pulsar and companion masses and of the keplerian parameters. In the case of General Relativity, the equations describing the PK parameters assume the form [70, 77, 71]:

$$\dot{\omega} = 3 \left(\frac{P_b}{2\pi}\right)^{-5/3} (T_\odot M)^{2/3} \left(1 - e^2\right)^{-1} \tag{25}$$

$$\gamma = e \left(\frac{P_b}{2\pi}\right)^{1/3} T_\odot^{2/3} M^{-4/3} m_2 \left(m_1 + 2m_2\right) \tag{26}$$

$$\dot{P}_b = -\frac{192\pi}{5} \left(\frac{P_b}{2\pi T_\odot}\right)^{-5/3} \left(1 + \frac{73}{24}e^2 + \frac{37}{96}e^4\right) \left(1 - e^2\right)^{-7/2} \frac{m_1 \, m_2}{M^{1/3}} \tag{27}$$

$$r = T_\odot \, m_2 \tag{28}$$

$$s = x \left(\frac{P_b}{2\pi}\right)^{-2/3} T_\odot^{-1/3} M^{2/3} m_2^{-1} \tag{29}$$

where m_1 and m_2 are the two star masses, $M = m_1 + m_2$, $x = a \sin i$ and $T_\odot \equiv GM_\odot/c^3 = 4.925490947 \,\mu s$. The parameter $\dot{\omega}$, as just mentioned, measures the advance of the periastron, γ is a term taking into account gravitational redshift and time dilatation, $\dot{P}_b$ is the so-called orbital decay and measures the rate at which the orbital period decreases due to losses of binding energy via gravitational radiation. Finally, r and $s \equiv \sin i$ are respectively the rate and the shape of the Shapiro delay, a parameter measuring the time delays of the signal caused by the space-time deformations around the companion star.

The measured value of $\dot{\omega}$ implied a total mass for the system containing PSR J0737$-$3039A of about 2.58 M$_\odot$ giving a maximum mass for the pulsar of about 1.34 M$_\odot$ and a minimum mass for the companion ~ 1.24 M$_\odot$. While the maximum mass for the pulsar perfectly agrees with the other measurements of neutron star masses [72], the mass of the companion is a little lower than average. In absence of additional information the white dwarf hypothesis could not be completely rejected. It is important to point out, anyway, that the $\dot{\omega}$ value that one measures, in general, is given by the term of Eq. (25) plus two classical extra terms, arising (i) from tidal deformations of the companion star (relevant only if the companion is non-degenerate) and (ii) from rotationally induced quadrupole moment of the companion star, applicable to the case of a fast rotating white dwarf. For a neutron star both the additional contributions are negligible. If the companion to J0737$-$3039A was a white dwarf the relativistic $\dot{\omega}$, and by consequence the total system mass, would be smaller than the measured one (since $\dot{\omega}_{GR} = \dot{\omega}_{obs} - \dot{\omega}_{classical}$) implying an implausibly small (< 1 M$_\odot$) maximum allowed mass for J0737−3039A (that, being a pulsar, is certainly a neutron star).

All these pieces of evidence strongly suggested that the discovered binary was the sixth, and by far the most relativistic, Double Neutron Star system known (see Sect. 5.3 for the implications of its discovery for testing relativistic gravity).

The ultimate confirmation of the above picture came few months later when, analysing the follow-up observations of PSR J0737−3039A, a strong signal with a repetition period of ~ 2.8 seconds occasionally appeared [73]. The newly discovered pulsar, henceforth called PSR J0737−3039B (or simply 'B'), had the same dispersion measure as PSR J0737−3039A (or 'A'), and showed orbital Doppler variations that identified it, without any doubt, as the companion to the millisecond pulsar. The first ever Double Pulsar system had been eventually discovered.

The timing parameters and the observational characteristic of the two pulsars are given in Sect. 5.2. Pulsar B was not detected in the 4-minute survey integration because its signal is clearly visible for only $\sim 20\%$ of the orbit [73]. The discovery of such a unique system has important implications in many fields of astrophysics: in Sect. 5.3 the importance of the discovery of PSR J0737−3039A and B in testing relativistic gravity is shown while in Sect. 5.3 the repercussions on the possibility to detect gravitational waves are described.

5.2 PSR J0737−3039A and B: Observational Characteristics

In this section the measured parameters and main characteristics of PSR J0737−3039A and B are summarized. The interpretation and implications of the properties of the pulsars are presented in Sect. 5.3.

Timing Results

In order to obtain a coherent timing solution over the few initial months of observation (from which deriving the spin period derivative of PSR J0737-3039A), it has been necessary to perform an independent measurement of the pulsar's position. Inspecting NVSS catalogue and maps, a point-like source with $\sim 20\%$ of linearly polarized flux density was found in the 7′ gridding error circle. To confirm this likely identification and improve the positional accuracy, in June 2003 a ~ 5 hours observation was performed at the Australian Telescope Compact Array (ATCA) in Narrabri (NSW, Australia) yielding the pulsar's position with arcsecond precision. After collection of data spanning several months, a positional term was also fitted to the times of arrival, allowing further improvement of the accuracy of the pulsar location in the sky.

Table 2. Measured parameters for PSR J0737−3039A and B. The 2-σ errors on the last quoted digit(s) are given in parentheses.

	PSR J0737−3039A	PSR J0737−3039B
Right Ascension α	$07^{\mathrm{h}}37^{\mathrm{m}}51^{\mathrm{s}}.24927(3)$	−
Declination δ	$-30°39'40''.7195(5)$	−
Proper motion in the RA direction (mas yr^{-1})	$-3.3(4)$	−
Proper motion in Declination (mas yr^{-1})	$2.6(5)$	−
Parallax, π (mas)	$3(2)$	−
Spin frequency ν (Hz)	$44.054069392744(2)$	$0.36056035506(1)$
Spin frequency derivative $\dot{\nu}$ (s^{-2})	$-3.4156(1) \times 10^{-15}$	$-0.116(1) \times 10^{-15}$
Timing Epoch (MJD)	53156.0	53156.0
Dispersion measure DM (cm^{-3}pc)	$48.920(5)$	−
Orbital period P_b (day)	$0.10225156248(5)$	−
Eccentricity e	$0.0877775(9)$	−
Projected semi-major axis $x = (a/c)\sin i$ (s)	$1.415032(1)$	$1.5161(16)$
Longitude of periastron ω (deg)	$87.0331(8)$	$87.0331 + 180.0$
Epoch of periastron T_0 (MJD)	$53155.9074280(2)$	−
Advance of periastron $\dot{\omega}$ (deg/yr)	$16.89947(68)$	$[16.96(5)]$
Gravitational redshift parameter γ (ms)	$0.3856(26)$	−
Shapiro delay parameter s	$0.99974(-39, +16)$	−
Shapiro delay parameter r (μs)	$6.21(33)$	−
Orbital period derivative $\dot{P}_b$	$-1.252(17) \times 10^{-12}$	−

The parameters measured in three years of regular timing are listed in Table 2, while the derived parameters for both pulsars are given in Table 3 [74].

It is important to point out that spin period and magnetic field of the two pulsars nicely agrees with what is predicted by the standard evolutionary model for the formation of Double Neutron Stars [75]. According to this model, in fact, the first born neutron star (in this case pulsar A), having accreted matter from a massive companion, has been spun-up to millisecond periods and its magnetic field (according to the scenario presented in Sect. 2.3) has been somehow reduced by the accretion itself. When the companion star explodes in a supernova it becomes the second neutron star of the system (here B). Since this second-born neutron star is not subjected to any mass transfer on its surface, it is expected to be a 'normal' pulsar, with a long period and a high magnetic field. PSR J0737−3039B nicely fits this picture providing confirmation of the standard scenario.

To be fully consistent with the model presented above both pulsars should have the same age: the (re-)birth of the millisecond pulsar should be roughly coincident with the end of the mass transfer phase and with the supernova explosion, hence with the birth of the second pulsar. The characteristic ages of PSR J0737−3039A and B, on the contrary, differs by a factor ~ 4. This discrepancy can be easily reconciled by questioning one or more of the assumptions inherent in the calculation of characteristic ages: a negligible birth spin period and a magnetic dipole braking torque (see Eqs. (6) and (7)). At very least, the initial (post-accretion) spin period of PSR J0737−3039A cannot have been negligible with respect to the present value: at most, the accretion could have pushed it up to the spin-up line (as explained in Sect. 2.3). If we

Table 3. Derived parameters for PSR J0737−3039A and B. The 2-σ errors on the last quoted digit(s) are given in parentheses.

	PSR J0737−3039A	PSR J0737−3039B
Magnetic Field B (G)	6.3×10^9	1.2×10^{12}
Characteristic age τ (Myr)	210	50
Spin-down luminosity $\dot{E}$ (erg/s)	6000×10^{30}	2×10^{30}
Total proper motion (mas yr^{-1})	4.2(4)	
Distance d(DM) (pc)	~ 500	
Distance $d(\pi)$ (pc)	$200 - 1000$	
Transverse velocity ($d = 500$ pc) (km s^{-1})	10(1)	
Orbital inclination angle (deg)	88.69(-76,+50)	
Mass function ($M_\odot$)	0.29096571(87) 0.3579(11)	
Mass ratio, R	1.0714(11)	
Total system mass ($M_\odot$)	2.58708(16)	
Neutron star mass ($m_\odot$)	1.3381(7) 1.2489(7)	

simply assume a constant magnetic field and a constant dipole spin-down after the end of the accretion, we can calculate that the initial rotational period value should have been in the range 10–18 ms [73], enough to explain the characteristic age discrepancy.

The importance and uniqueness of the timing measurements obtained for these two objects are presented in Sect. 5.3.

5.3 Implications

In this section we discuss the interpretation and implications of the observed properties of the double pulsar PSR J0737−3039A/B

Testing Relativistic Gravity

Since the only unknowns in the left-hand side of Eqs. (25)–(29) (or their analogues in other relativistic gravity theories) are the masses of the two stars in the binary system, the measurement of two post-keplerian parameters yields the masses and the measurement of three or more PK-parameters over-determines the system. In other words, once the masses are obtained measuring any two PKs, their value can be placed in the equation of 3rd, 4th and 5th parameter and the values obtained can be compared with the measured ones, giving a self-consistency test of the theory on which the given equations are based.

Timing of PSR J0737−3039A over three years led to the measurement of all 5 post-keplerian parameters (see Table 2). For the two other pulsars belonging to a double neutron star system on which these kind of measurements have been successfully done, B1913+16 [25] and B1534+12 [76], timing observations yielded the measurement of three PK parameters in ~ 30 years [77, 78], and all five PK parameters in ~ 10 years [79] respectively. This comparison gives an idea of the possibilities that J0737−3039 system opens in this field. It is worth noting, for instance, that using only the information given by pulsar A and in just 36 months of follow-up observations, J0737−3039 system already tests General Relativity with very high accuracy: the measured value of the shape of the Shapiro delay, s_{obs}, agrees with Einstein's predictions, s_{GR}, at 0.05% level:

$$\frac{s_{obs}}{s_{GR}} = 0.99987(50) \tag{30}$$

This can be compared with the 0.2% agreement resulting from the measurement of the orbital decay in PSR B1913+16 [78].

Having also detected the pulsations from the second neutron star in the binary system allows us for the first time to perform even better and significant tests of relativistic gravity: the timing measurement of the projected semi-major axis of both pulsars, in fact, yields the measurement of the mass ratio R of the two neutron stars. This value gives a qualitatively different constraint to the masses of the stars, since the relation

$$R \equiv \frac{a_B}{a_A} = \frac{M_A}{M_B} \tag{31}$$

is largely independent on the adopted theory of gravity. In fact, Eq. (31) is valid for all 'fully conservative' theories [80] and in particular for all Lagrangian-based theories [71]. Hence, for any given set of equations describing the post-keplerian parameters, the lines formed on the mass-mass diagram must cross on the line indicating the mass ratio.

In Fig. 24 all the constraints on the masses of PSR J0737−3039A and PSR J0737−3039B are plotted: the light grey regions of the diagram is ruled out by A and B's mass functions, solid lines enclose the region permitted by the calculated range for the mass ratio, dashed curves indicate the constraint imposed by the advance of periastron measurement, dot-dashed lines are from the measured gravitational redshift – time dilatation parameter γ, triple dot-dashed lines and dotted curves enclose respectively the region permitted by the measured values of the range r and shape s of the Shapiro delay. The inset in Fig. 24 is an enlarged view of the small square which encompasses the intersection of the three tightest constraints (with the scales increased by a factor of 16). Note that all the plotted curves derived from the General Relativity equations (25)–(29), intersect each other on the mass ratio R line.

Another important and extreme characteristic of PSR J0737−3039A and B is the predicted geodetic spin-precession due to curvature in space-time around the pulsars: according to Einstein's theory, if the rotational and orbital angular momenta are misaligned, the spin axis of pulsar A should be moving around the axis perpendicular to the orbital plane at a rate of 4.79 °/yr, four time larger that the largest measured to date (1.2 °/yr for B1913+16). This implies a period for the geodetic precession of only 75 years for pulsar A. The same calculations applied to pulsar B give a geodetic precession period of 71 yr. This property should allow detailed mapping of the pulsar emission beam [81] in a few years, since the observer's line of sight through it should rapidly changing. On the other hand geodetic precession may also have the undesirable effect

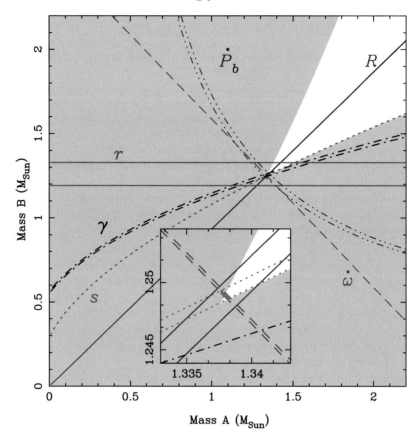

Fig. 24. The observational constraints upon masses M_A and M_B [74]. For an explanation of the different curves see text.

of making the pulsar undetectable in a short time (PSR B1913+16, because of this, is expected to disappear in 2025 [82]).

In the next few years, several more relativistic effects are expected to be measured, some dependent on higher-order terms in the post-Newtonian expansion (i.e. $\ddot{\omega}$). These will provide the tightest constraints yet on theories of gravity in the strong-field regime.

Gravitational Waves Detection

Double neutron star systems lose orbital energy because of emission of gravitational waves [77]. As a consequence, the two stars undergo an orbital spiral-in at the end of which they merge producing a burst of gravitational waves.

Coalescence of DNS are one of the prime target for ground based gravitational wave detectors such as LIGO [83], GEO [84] and VIRGO [85] and the expected rate of these events is clearly an important factor for the development of these instruments [86].

Of the five double neutron star system known [87] before the discovery of PSR J0737−3039 system, only three have orbits tight enough that the two neutron stars will coalesce within a Hubble time. Two of them (PSR B1913+16 and PSR B1534+12) are located in the galactic field, while the third (PSR B2127+11C) is found on the outskirts of a globular cluster. The contribution to the galactic DNS coalescence rate $\mathcal{R}$ of globular cluster's pulsars is estimated to be negligible [88]; also, recent studies [90] have demonstrated that the current estimate of $\mathcal{R}$ relies mostly on PSR B1913+16. Hence, a rough assessment of the contribution of the discovery of J0737−3039 system to the estimates of double neutron star merger rate can be done by comparing its properties with those of PSR B1913+16.

PSR J0737−3039A and B will coalesce due to the emission of gravitational waves in a merger time $\tau_m \sim 85$ Myr, a timescale that is a factor 3.5 shorter than that for PSR B1913+16 [89]. In addition, the estimated distance for J0737−3039 system (~ 0.5 kpc), is an order of magnitude less than that of PSR B1913+16. These properties have a substantial effect on the prediction of the rate of merging events in the Galaxy.

For a given class k of binary pulsars in the Galaxy, in fact, apart from a beaming correction factor, the merger rate $\mathcal{R}_k$ is calculated as $\mathcal{R}_k \propto N_k/\tau_k$ [90]). Here τ_k is the binary pulsar lifetime defined as the sum of the time since birth, τ_b, and the remaining time before coalescence, τ_m whereas N_k is the scaling factor defined as the number of binaries in the Galaxy belonging to the given class. The value of τ_b for PSR J0737−3039A can be computed as the time since the pulsar left the spin-up line [91], and it results ~ 150 Myr. Alternatively one can assume that the characteristic age of pulsar B is a reliable estimate of the true age of both pulsars. In this case $\tau_b \sim 50$ Myr. In either case the life time of PSR J0737−3039 is much shorter than that of PSR B1913+16 ($\tau_{1913}/\tau_{0737} = (365 \text{ Myr})/(235 - 135 \text{ Myr}) \sim 1.6 - 2.7$, where the subscript numbers refer to the pulsars), implies roughly a doubling of the ratio $\mathcal{R}_{0737}/\mathcal{R}_{1913}$. A much more substantial increase results from the computation of the ratio of the scaling factors N_{0737}/N_{1913}.

Since the available numbers for the coalescence rate of double neutron star systems in the Galaxy, to which the computed scaling factors apply, are derived using a population synthesis code whose parameters are based mostly on the results of surveys done at 400 MHz [92, 91, 93, 90], we adopt here 400 MHz as reference frequency for comparing the luminosity of the two pulsars in question.

The flux at 400 MHz for PSR J0737−3039A has been obtained by observations with the Parkes radio telescope resulting ~ 100 mJy [68]. The pulsed luminosity at 400 MHz, considering a distance of 0.5 kpc, for PSR J0737−3039A hence results $L_{0737} \sim 25$ mJy kpc^2, much lower than that of PSR B1913+16 (~ 200 mJy kpc^2).

For a planar homogeneous distribution of pulsars in the Galaxy, the ratio N_{0737}/N_{1913} scales as $L_{1913}/L_{0737} \sim 8$. Therefore we obtain $\mathcal{R}_{0737}/\mathcal{R}_{1913} \sim 16$. Including the moderate contribution of the longer-lived PSR B1534+12 system to the total rate we obtain an increase factor for the total merger rate $(\mathcal{R}_{0737} + \mathcal{R}_{1913} + \mathcal{R}_{1534})/(\mathcal{R}_{1913} + \mathcal{R}_{1534})$ of about an order of magnitude. A better estimate of this factor and its uncertainty can be obtained using a bayesian statistical approach [90].

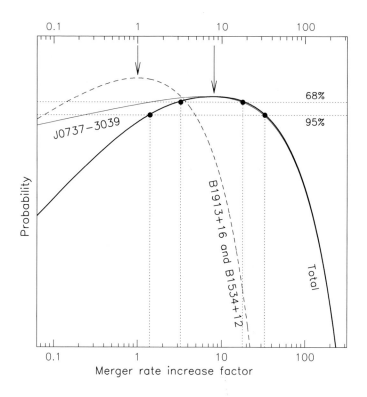

Fig. 25. Probability density function for the increase in the double neutron star merger rate $(\mathcal{R}_{0737} + \mathcal{R}_{1913} + \mathcal{R}_{1534})/(\mathcal{R}_{1913} + \mathcal{R}_{1534})$ resulting from the discovery of J0737−3039 system. The dashed curve represents the reference probability density function corresponding to the merger rate calculated on the basis of PSR B1913+16 and B1534+12 only, the thick solid curve is the new probability density function accounting also for J0737−3039 system. The dotted vertical lines delimits the 68% and 95% confidence levels on the determination of the increase factor. The dominant role of J0737−3039 system in shaping the new statistics of DNS coalescence rate is evident (solid thin curve).

Fig. 25 displays such an estimate, reporting the probability density function for the merger rate increase factor. It shows a peak value of ~ 8 and an upper limit of ~ 30 at a 95% confidence level. For a given class k of binary pulsars, the probability density function $P(\mathcal{R}_k)$ for the corresponding merger rate $\mathcal{R}_k$ is obtained from the relation $P(\mathcal{R}_k) = A^2 \mathcal{R}_k e^{-A\mathcal{R}_k}$ where $A = \tau_k/(N_k f_k)$ and f_k is the correction factor due to the beamed nature of the radio pulsar emission [90]. The beaming factor for PSR J0737−3039A is chosen as the average of the beaming factors of the two other coalescing DNSs.

The conservative result obtained here, in the simple assumption of a fixed pulsar luminosity and a uniform disk distribution of pulsar binaries, can be refined by including the parameters of the Parkes High-Latitude Pulsar Survey into a simulation programme modelling survey selection effects (and hence detection probability) and the galactic population of pulsars. Extensive simulations [94] produce results consistent with that derived in Fig. 25, and show that the peak of the merger rate increase factor, resulting from the discovery of J0737−3039 system, lies in the range 6 to 7 and is largely independent on the adopted pulsar population model.

In summary, the discovery of J0737−3039 system sets a lower limit to the merger rate increase factor of about six. The actual predicted value of the galactic merger rate, and hence the detection rate by gravity wave detectors, depends on the shape of the pulsar luminosity function. For the most favorable distribution model available, the updated cosmic detection rate for first-generation gravity wave detectors such as VIRGO, LIGO and GEO can be as high as 1 every 1–2 years at 95% confidence level. It is the first time that the estimated double neutron star detection rate enters within astrophysically relevant regimes.

After a few years of operation of the gravity wave detectors, it should be possible to test these predictions directly and thus place better constraints on the cosmic population of double neutron star binaries.

6 Conclusion

In the last few years, using new generation equipment, we have doubled the number of radio pulsars available and we have discovered a significant number of exotic objects, particularly interesting for their applications in a wide range of fundamental research. We have discovered the most interesting binary system ever found, PSR J0737−3039. This is the most relativistic Double Neutron Star known (with an orbital velocity of 0.1% the speed of light), providing the best ever laboratory for testing General Relativity and implying an increase of almost an order of magnitude in the estimates of the DNSs coalescence rate (Sect. 5.3), and it is also the first known Double Pulsar. The presence of a second detectable pulsation in such a system provides the measurement of the mass ratio of the two stars, allowing new and more precise tests for the theories of relativistic gravity (Sect. 5.3). Sensitive pulsar searches are continuing at Parkes, and new sensitive experiments are in progress at Green Bank and Arecibo. We expect that, in a couple of years, more than 2000 pulsars will be available. We expect a growing number of millisecond pulsars, and we hope in the discovery of additional double pulsars. As the total number of pulsars available increases, we might be approaching the discovery of a pulsar-black hole binary system. In a few years, systematic long term high precision timing observations of millisecond pulsar could be used to provide a precise astronomical time base standard – the so-called "Pulsar Timing Array", which might represent a new challenge for the direct detection of gravitational waves.

References

1. Hewish, A., Bell, S. J., Pilkington, J. D. H., Scott, P. F. and Collins, R. A. 1968, Nature, 217, 709
2. Staelin, D. H.and Reifenstein, E C. 1968, Science, 162, 1481
3. Backer, D. C., Kulkarni, S. R., Heiles, C., Davis, M. M. and Goss, W. M. 1982, Nature, 300, 615
4. Lyne, A. G., Manchester, R. N. and Taylor, J. H. 1985, MNRAS, 213, 613
5. Lorimer, D. R., Bailes, M., Dewey, R. J. and Harrison, P. A. 1993, MNRAS, 263, 403
6. Hartman, J. W., Bhattacharya D., Wijers R. and Verbunt F. 1997a, A&A, 322, 477
7. Cordes, J. M. and Chernoff, D. F. 1998, ApJ, 505, 315
8. Lyne, A. G., Manchester, R. N., Lorimer, D. R., Bailes, M., D'Amico, N., Tauris, T. M., Johnston, S., Bell, J. F. and Nicastro, L. 1998, MNRAS, 295, 743
9. McKenna, J. and Lyne, A. G. 1990, Nature, 343, 349
10. Kaspi, V. M., Lyne, A. G., Manchester, R. N., Johnston, S., D'Amico, N. and Shemar, S. L. 1993, ApJ, 409, L57
11. Johnston, S., Manchester, R. N., Lyne, A. G., Kaspi, V. M. and D'Amico, N. 1995, A&A, 293, 795
12. Lyne, A. G., Pritchard, R. S. and Smith, F. G. 1988, MNRAS,233, 667
13. Kaspi, V. M., Manchester, R. N., Siegman, B. and Johnston, S. and Lyne, A. G. 1994, ApJ, 422, L83
14. Kaspi, V. M. 2000, ASP Conference Series, 202, 485
15. Frail, D. A. and Kulkarni, S. R. 1991, Nature, 352, 785
16. Manchester, R. N., Kaspi, V. M., Johnston, S., Lyne, A. G. and D'Amico, N. 1991, MNRAS, 253, 7P
17. Becker, R. H. and Helfand D. J. 1985, Nature, 313, 115
18. Taylor, J. H., Wolszczan, A., Damour, T. and Weisberg, J. M. 1992, Nature, 355, 132
19. Kaspi, V. M., Taylor, J. H. and Ryba, M. 1994, ApJ, 428, 713
20. Taylor, J. H. 1991, Basic Physics and Cosmology from Pulsar Timing Data, National Academy Press
21. Cook, G. B., Shapiro, S. L. and Teukolsky, S. A. 1994, ApJ, 424, 823
22. Hamilton, T. T., Helfand, D. J. and Becker, R. H. 1985, AJ, 90, 606
23. Fabian, A. C., Pringle, J. E. and Rees, M. J. 1975, MNRAS, 172, 15P
24. Hills, J. G. 1976, MNRAS, 175, 1P

25. Hulse, R. A. and Taylor, J. H. 1974, ApJ, 191, L59
26. Wolszczan, A. and Frail, D. A. 1992, Nature, 355, 145
27. Johnston, S., Manchester, R. N., Lyne, A. G., Bailes, M., Kaspi, V. M., Guojun Quiao and D'Amico, N. 1992, ApJ, 387, L37
28. Fruchter, A. S., Stinebring, D. R. and Taylor, J. H. 1988, Nature, 333, 237
29. Ricket, B. J. 1977, ARA&A, 15, 479
30. Clifton, T. R., Lyne, A. G., Jones, A. W., McKenna, J. and Ashworth, M. 1992, MNRAS, 254, 177
31. Johnston, S., Lyne, A. G., Manchester, R. N., Kniffen, D. A., D'Amico, N., Lim, J. and Ashworth, M. 1992, MNRAS, 255, 401
32. Staveley-Smith, L., Wilson, W. E., Bird, T. S., Disney, M. J., Ekers, R. D., Freeman, K. C., Haynes, R. F., Sinclair, M. W., Vaile, R. A., Webster, R. L. and Wright, A. E. 1996, PASA, 13, 243
33. Baade, W. and Zwicky, F. 1934, Phys. Rev., 45, 138
34. Cocke, W. J., Disney, M. J. and Taylor, D. J. 1969, Nature, 221, 525
35. Pacini, F. 1967, Nature, 216, 567
36. Miri M. J. and Bhattacharya D. 1994, MNRAS, 269, 455
37. Urpin, V., Geppert, U. and Konekov, D. 1998, A&A, 331, 244
38. Alpar, M. A., Cheng, A. F., Ruderman, M. A. and Shaham, J. 1982, Nature, 300, 728
39. Burderi, L., King, A. R. and Wynn, G. A. 1996, ApJ, 457, 351
40. Kramer, M., Xilouris, K. M., Lorimer, D. R., Doroshenko, O., Jessner, A., Wielebinski, R., Wolszczan, A. and Camilo, F. 1998, ApJ, 501, 270
41. Hankins, T. H. and Rickett, B. J. 1975, Meth. Comp. Phys., 14, 55
42. Manchester, R. N., Lyne, A. G., D'Amico, N., Bailes, M., Johnston, S., Lorimer, D. R., Harrison, P. A., Nicastro, L. and Bell, J. F. 1996, MNRAS, 279, 1235
43. Manchester, R. N., Lyne, A. G., Camilo, F., Bell, J. F., Kaspi, V. M., D'Amico, N., McKay, N. P. F., Crawford, F., Stairs, I. H., Possenti, A., Kramer, M. and Sheppard, D. C. 2001, MNRAS, 328, 17
44. Taylor, J. H. 1974, A&A, 15, 367
45. Taylor, J. H. and Cordes, J. M. 1993, ApJ, 411, 674
46. Shapiro, I. I. 1964, Phys. Rev. Lett., 13, 789
47. Manchester, R. N. and Taylor, J. H. 1974, ApJ, 191, L63
48. Cordes, J. M. and Helfand, D. J. 1980, ApJ, 239, 640
49. Alpar, M. A., Nandkumar, R. and Pines, D. 1986 , ApJ, 311, 197
50. Cheng, K. S. 1987, ApJ, 321, 799
51. Cheng, K. S. 1987, ApJ, 321, 825
52. Morris, D. J., Hobbs, G., Lyne, A. G., Stairs, I. H., Camilo, F., Manchester, R. N., Possenti, A., Bell, J. F., Kaspi, V. M., D'Amico, N., McKay, N. P. F., Crawford, F. and Kramer, M. 2002, MNRAS, 335, 275
53. Kramer, M., Bell, J. F., Manchester, R. N., Lyne, A. G., Camilo, F., Stairs, I. H., D'Amico, N., Kaspi, V. M., Hobbs, G., Morris, D. J., Crawford, F., Possenti, A., Joshi, B. C., McLaughlin, M. A., Lorimer, D. R., Faulkner, A.J. 2003, MNRAS, 342, 1299
54. Hobbs, G., Faulkner, A., Stairs, I. H., Camilo, F., Manchester, R. N., Lyne, A. G., Kramer, M., D'Amico, N., Kaspi, V. M., Possenti, A., McLaughlin, M. A., Lorimer, D. R., Burgay, M., Joshi, B. C. and Crawford, F. 2004, accepted by MNRAS

55. Faulkner, A. J., Stairs, I. H., Kramer, M., Lyne, A. G., Hobbs, G., Possenti, A., Lorimer, D. R., Manchester, R. N., McLaughlin, M. A., D'Amico, N., Camilo, F., Burgay, M. 2006, MNRAS, 355, 147

56. Lorimer, D. R., Faulkner, A. J., Lyne, A. G., Manchester, R. N., Kramer, M., McLaughlin, M. A., Hobbs, G., Possenti, A., Stairs, I. H., Camilo, F., Burgay, M., D'Amico, N., Corongiu, A., Crawford, F. 2006, MNRAS, 372, 777

57. Lyne, A. G., Camilo, F., Manchester, R. N., Bell, J. F., Kaspi, V. M., D'Amico, N., McKay, N. P. F., Crawford, F., Morris, D. J., Sheppard, D. C. and Stairs, I. H. 2000, MNRAS, 312, 698

58. Camilo, F., Kaspi, V. M., Lyne, A. G., Manchester, Bell, J. F., D'Amico, N., McKay, N. P. F.and Crawford F.2000, ApJ, 541, 367

59. Crawford, F., Gaensler, B. M., Kaspi, V. M., Manchester, R. N., Camilo, F., Lyne, A. G. and Pivovaroff, M. J. 1991, ApJ, 554, 152

60. Pivovaroff, M. J., Kaspi, V. M., Camilo, F., Gaensler, B. M., Crawford, F. 2001, ApJ, 554, 161

61. Pivovaroff, M. J., Kaspi and V. M., Camilo, F. 2000, ApJ, 535, 379

62. Kaspi V. M., Lackey J. R., Mattox J., Manchester R. N., Bailes M. and Pace R. 2000, ApJ, 528, 445

63. Stairs, I. H., Manchester, R. N., Lyne, A. G., Kaspi, V. M., Camilo, F., Bell, J. F., D'Amico, N., Kramer, M., Crawford, F., Morris, D. J., Possenti, A., McKay, N. P. F., Lumsden, S. L., Tacconi-Garman, L. E., Cannon, R. D., Hambly, N. C., Wood, P. R. 2001, MNRAS, 325, 979

64. Camilo, F., Lyne, A. G., Manchester, R. N., Bell, J. F., Stairs, I. H., D'Amico, N., Kaspi, V. M., Possenti, A., Crawford, F., McKay, N. P. F. 2001, ApJ, 548, L187

65. D'Amico, N., Kaspi, V. M., Manchester, R. N., Camilo, F., Lyne, A. G., Possenti, A., Stairs, I. H., Kramer, M., Crawford, F., Bell, J. F., McKay, N. P. F., Gaensler, B. M. and Roberts, M. S. E. 2001, ApJ, 552, L45

66. Edwards, R. T., Bailes, M., van Straten, W., Britton, M. C. 2001, MNRAS, 326, 358

67. Jacoby, B. A. 2003, Radio Pulsars, ASP Conference Series, 302, 133

68. Burgay, M., D'Amico, N., Possenti, A., Manchester, R. N., Lyne, A. G., Joshi, B. C., McLaughlin, M. A., Kramer, M., Sarkissian, J. M., Camilo, F., Kalogera, V., Kim, C. and Lorimer, D. R. 2003, Nature, 426, 531

69. Kaspi, V. M., Lyne, A. G., Manchester, R. N., Crawford, F., Camilo, F., Bell, J. F., D'Amico, N., Stairs, I. H., McKay, N. P. F., Morris, D. J., Possenti, A. 2000, ApJ, 543, 321

70. Damour, T. and Deruelle, N. 1986, Ann. Inst. H. Poincaré (Physique Théorique), 44, 263

71. Damour, T. and Taylor, J. H. 1992, Phys. Rev. D, 45, 1840

72. Thorsett, S. E. and Chakrabarty, D. 1999, ApJ, 512, 288

73. Lyne, A. G., Burgay, M., Kramer, M., Possenti, A., Manchester, R. N., Camilo, F., McLaughlin, M. A., Lorimer, D. R., D'Amico, N., Reynolds, J., Freire, P. C. C. 2004, Science, 303, 1153

74. Kramer, M., Stairs, I.H., Manchester, R. N., McLaughlin, M. A., Lyne, A. G., Ferdman, R. D., Burgay, M., Lorimer, D. R., Possenti, A., D'Amico, N., Sarkissian, J. M., Hobbs, G. B., Reynolds, J. E., Freire, P.C.C., Camilo, F. 2006, Science 314, 97

75. Bhattacharya, D., van den Heuvel, E. P. J. 1991, Phis. Rev., 203, 1

76. Wolszczan, A. 1991, Nature, 350, 688

77. Taylor, J. H. and Weisberg, J. M. 1989, ApJ, 345, 434

78. Weisberg, J. M. and Taylor, J. H. 2003, Radio Pulsars, ASP Conference Series, 302, 93

79. Stairs, I. H., Thorsett, S. E., Taylor, J. H. and Wolszczan 2002, ApJ, 581, 501

80. Will, C. 1992, Nature, 355, 111

81. Weisberg, J. M. and Taylor, J. H. 2002, ApJ, 576, 942

82. Kramer, M. 1998, ApJ, 509, 856

83. Abramovici, A., Althouse, W. E., Drever, R. W. P., Gursel, Y., Kawamura, S., Raab, F. J., Shoemaker, D., Sievers, L., Spero, R. E., Thorne, K. S. 1992, Science, 256, 325

84. Danzmann, K. et al. 1995, First Edoardo Amaldi Conference on Gravitational Wave Experiments, World Scientific, 100

85. Caron, B. et al. 1997, Nucl. Phys., B54, 167

86. Thorne, K. S. and Cutler, C. 2002, General Relativity and Gravitation, World Scientific, 72

87. Taylor, J. H. 1992, Phil. Trans. Royal Soc. of London, 341, 117

88. Phinney, E. S. 1991, ApJ, 380, L17

89. Taylor, J. H., Fowler, L. A. and McCulloch, P. M. 1979, Nature 277, 437

90. Kim, C., Kalogera, V. and Lorimer, D. R. 2003, ApJ, 584, 985

91. Arzoumanian, Z., Cordes, J. M. and Wasserman, I. 1999, ApJ, 520, 696

92. van den Heuvel, E. P. J. and Lorimer. D. R. 1996, MNRAS, 283, L37

93. Kalogera, V., Narayan, R., Spergel, D. N. and Taylor, J. H. 2001, ApJ, 556, 340

94. Kalogera, V., Kim, C., Lorimer, D. R., Burgay, M., D'Amico, N., Possenti, A., Manchester, R. N., Lyne, A. G., Joshi, B. C., McLaughlin, M. A., Kramer, M., Sarkissian, J. M. and Camilo, F. 2004, ApJ, 601, L179

The Formation and Evolution of Relativistic Binaries

E.P.J. van den Heuvel[1]

Astronomical Institute "Anton Pannekoek" and Center for High Energy Astrophysics, University of Amsterdam, Kruislaan 403, 1098SJ Amsterdam, The Netherlands edvdh@science.uva.nl

1 Summary and Introduction

In 1971, only four years after the discovery of the first radio pulsar, the first neutron star in a close binary was discovered: the 4.84 s X-ray pulsar Centaurus X-3, which is moving in a 2.087 day orbit around an O-star with a mass $> 16 M_\odot$ [135]. Several more of these High Mass X-ray Binaries (HMXBs) were discovered soon after and it was found that, contrary to what is observed in radio pulsars, the pulse periods of several of these X-ray pulsars are steadily decreasing in the course of time, moving to shorter and shorter values on timescales of order 10^4 years. It was soon realized that the same accretion process of matter flowing over from the massive companion star that is the cause of the X-ray emission, also causes this "spin-up". The matter flow in the binary system has angular momentum – derived from the system's orbital motion – and this angular momentum is fed to the neutron star, causing its rotation rate to increase. A few years later, the suggestion was made by [6] that these pulsating X-ray sources in binaries may later in life, when their massive companion stars have exploded as a supernova, become observable as radio pulsars. Such pulsars, which had a history of accretion and spin-up in binaries were later given the name "recycled pulsars" [119]. In 1973 it was calculated [176] that before the second supernova explosion in a HMXB takes place, the orbit of the system will have become very narrow, as a consequence of extensive mass transfer to the neutron star and loss of mass with high angular momentum from the system, leading to final orbital periods of only a few hours. The resulting close system then consists of a helium star (the helium core of the massive companion) plus the neutron star. In 1974 the Hulse–Taylor binary radio pulsar PSRB 1913+16 was discovered, which in addition to its very narrow and eccentric orbit ($P_{\rm orb} = 7.75$ h, $e = 0.615$) appeared to have very abnormal characteristics as a radio pulsar: its magnetic field strength is only 2×10^{10} G, some two orders of magnitude lower than that of the other pulsars then known, and its spin period is abnormally short (0.059 s), which at the time made it the second fastest radio pulsar known, after the Crab pulsar ($P = 0.033$ s). Its orbital period and eccentricity were almost exactly what one would obtain if the helium star in the 4 hour orbit binary (resulting from a HMXB like Centaurus X-3, as calculated in 1973) would explode as a supernova and itself would leave a neutron star. This model for the origin of the Hulse–Taylor binary pulsar was therefore proposed immediately after its discovery [44, 31]. It was thought in these days that the magnetic fields of neutron stars decay on a relatively short timescale, of order 5 million years. The abnormally weak magnetic field of PSRB1913 +16 therefore led [139] to the suggestion that the observed pulsar is the oldest of the two neutron stars in the system, which after a long period of field decay had been spun up by accretion in an X-ray binary system, before the second star exploded. It was subsequently shown [142] that this spin-up idea is the only explanation possible for this peculiar combination of rapid spin and weak magnetic field observed in PSRB 1913+16. This then immediately implies that the companion of this pulsar must also be a neutron star. The reason for this is that during the phases of accretion, orbital shrinking and spin up, the orbit of the system will have become completely circularized by tidal and frictional forces. The only way to then subsequently obtain the large observed orbital eccentricity of the system is: if a second supernova explosion took place. This then implies that the companion of PSRB 1913+16 must itself also be a neutron star: the younger one of the two. As the last-born neutron star did not undergo any accretion, and after the second explosion the system was free of gas, the second neutron star is expected to be a normal newborn "garden variety" radio pulsar with a normal strong magnetic field of order 10^{12} G [142]. Such pulsars rapidly spin down on a timescale of order a few million years, after which they become unobservable. On the other hand, due to its weak magnetic field, the spin-down timescale of PSRB1913+16 is longer than 10^8 years. This much longer lifetime of the old spun-up pulsar is the simple explanation why in PSRB 1913 +16 (and in 5 of the 7 other double neutron stars known in the galactic disk, see Sect. 3) we only observe the old rapidly spinning weak magnetic field neutron star and not its strong-field younger companion [142]. The discovery of the double pulsar PSRJ 0737-3039 [85] has fully confirmed this prediction: the companion of the rapidly spinning weak-magnetic-field pulsar PSRJ 0737-3039A is a normal "garden-variety" radio pulsar with a magnetic field of 10^{12} G and a spin period of about 2 s. Although the above-described model for the formation of double neutron stars seems very straightforward, there are many detailed parts of the stellar evolution and the physics that remain to be filled in. For example, the model only works if the magnetic fields of non-accreting neutron stars do not undergo substantial spontaneous decay on a relatively short ($\sim 5 \times 10^6$ yrs) timescale, as originally was thought to be the case for radio pulsars (e.g. see [87]). Indeed, in the past decades strong evidence has been found that magnetic fields of non-accreting neutron stars remain stable on very long timescales ($\sim 10^9$ years or longer; e.g. see [75, 178, 9]). On the other hand, ample observational evidence has been found indicating that accretion of matter onto the neutron star surface weakens the magnetic field, and this weakening somehow increases in step with the amount of matter accreted [147]. Several models have been proposed to explain this effect (see [202, 26, 27, 127]; see references in [9]). After the accretion has terminated, the field decay stops and the field strength remains stable again. This chapter describes our present understanding of the formation and evolution of neutron stars in binary systems, with a particular focus on the formation of double neutron stars. It will appear that the orbital and spin characteristics of the double neutron stars give unique new information on the formation processes of neutron stars. As will be explained in Sect. 4 it is indeed thanks to the double neutron stars that we now, for the first time, have confirmation of the theoretical prediction [93] that there are two different formation processes of neutron stars: (i) by the electron-capture collapse of a degenerate O–Ne–Mg core in stars of intermediate mass, presumably in the mass range 8–$12\,M_\odot$, and (ii) by iron-core collapse in stars that started out with masses $\geq 12\,M_\odot$ [171, 172, 173, 174, 110]. This chapter is organized as follows: Sect. 2 summarizes the observed characteristics of the different types of binaries with compact objects: the X-ray binaries and their descendents, the binary radio pulsars. In this section also a brief account is given of the spin evolution of an accreting magnetized neutron star in a binary system. Sect. 3 describes the formation and evolution of binary systems that contain one compact star (X-ray binaries) and describes how X-ray binaries may evolve into binary radio pulsars and other binary systems consisting of two compact objects. Sect. 4 specifically describes in more detail the formation of double neutron stars. From the orbital and other characteristics of the double neutron stars arguments are derived indicating that there are two different formation mechanisms of neutron stars: one which does not impart a sizeable kick velocity to the neutron star, and occurs only in intermediate-mass binary systems, while the second mechanism does impart a large kick velocity and occurs in single stars and in binaries of higher mass. Sect. 5 describes the formation of Low-Mass X-ray binaries and of millisecond pulsars. Sect. 6 describes in more detail the concept of "recycling" of neutron stars and discusses magnetic field decay. Sect. 7 briefly describes the formation of close binaries consisting of a neutron star and a black hole or of two black holes, and Sect. 8 addresses the galactic formation rate of the various types of relativistic binary systems.

2 Observed Properties of Binaries with Compact Objects

2.1 Introduction

The binaries with compact objects can be divided into the categories and types listed in Table 1. Basically they fall into the categories "compact star plus ordinary star" and "two compact stars". The first category consists of the X-ray binaries and systems like the Cataclysmic Variables (abbreviated as CVs; including Symbiotic binaries, pre-CVs, etc.), the second category consists of the binary radio pulsars and the double white dwarfs. Each of these categories can be divided into a few main types, which can be further divided into sub-types, as indicated in the table, where also examples of the different sub-types are given. A few binaries with compact objects do not fit into the above categories, notably the so-called "anti-deluvian" binary radio pulsars, which are young pulsars in an eccentric orbit around an unevolved star, which presumably are the progenitors of HMXBs. Examples of such systems are PSR 1259-63 (P_{orb} = 3.4 yr), PSR 1820-11 (P_{orb} = 357.8 d), PSR J1740-3052 (P_{orb} = 231 d) and PSR J0045-7319(P_{orb} = 51 d), the latter one in the Large Magellanic Cloud. In the first system and the last-mentioned two systems, the companion of the pulsar is a B-type main sequence star (in two cases very rapidly rotating and with emission lines), in PSR 1820-11 the companion is probably a low-mass star. Furthermore, there are the two peculiar X-ray binaries with relativistic jets, SS433 and Cyg X-3. In SS433, the companion of the jet-producing compact object is a very luminous object which may be a Wolf–Rayet star [165, 12]. In Cyg X-3, it is a Wolf–Rayet star (helium star, see [185], which confirmed the prediction about the nature of this system made by [176]. In recent years two more WR X-ray binaries like Cyg X-3 were discovered in external galaxies: IC 10 X-1 (P_{orb} = 34.8 h; [115]) and NGC 300 X-1 (P_{orb} = 32.8 h; [19]); in the latter two systems the compact companions of the very massive WR stars (about 40 $M_{\odot}$) are most probably black holes (see Sect. 7). I now briefly describe the main characteristics of each of the various types of neutron-star and black-hole binary systems listed in the table. In this chapter I will concentrate mainly on these systems as these are the ones that are relevant for the formation of relativistic binary systems. Only where it is relevant I will also mention CV-like systems and double white dwarfs.

Table 1. Main categories and types of binaries with compact objects (after [157]).

X-RAY BINARIES

main type	sub-type	obs. example
high-mass donor ($M_{\mathrm{donor}} \geq 10\,M_\odot$)	'standard' HMXB	Cen X–3, $P_{\mathrm{orb}} = 2.087^{\mathrm{d}}$ (NS) Cyg X–1, $P_{\mathrm{orb}} = 5.60^{\mathrm{d}}$ (BH)
	Be-star HMXB	A0535+26, $P_{\mathrm{orb}} = 104^{\mathrm{d}}$ (NS)
low-mass donor ($M_{\mathrm{donor}} \leq 1\,M_\odot$)	galactic disk LMXB soft X-ray transient globular cluster	Sco X–1, $P_{\mathrm{orb}} = 0.86^{\mathrm{d}}$ (NS) A0620–00, $P_{\mathrm{orb}} = 7.75^{\mathrm{hr}}$ (BH) X 1820–30, $P_{\mathrm{orb}} = 11^{\mathrm{min}}$ (NS)
	millisecond X-ray pulsar $P_{\mathrm{orb}} = 2.0^{\mathrm{d}}$ (NS)	SAX J1808.4–36,
intermediate-mass donor ($1 < M_{\mathrm{donor}}/M_\odot < 10$)		Her X–1, $P_{\mathrm{orb}} = 1.7^{\mathrm{d}}$ (NS) Cyg X–2, $P_{\mathrm{orb}} = 9.8^{\mathrm{d}}$ (NS) V 404 Cyg, $P_{\mathrm{orb}} = 6.5^{\mathrm{d}}$ (BH)

BINARY RADIO PULSARS

main type	sub-type	obs. example
'high-mass' companion ($0.5 \leq M_{\mathrm{c}}/M_\odot \leq 1.4$)	NS + NS (double) NS + (ONeMg) WD NS + (CO) WD	PSR 1913+16, $P_{\mathrm{orb}} = 7.75^{\mathrm{hr}}$ PSR 1435–6100, $P_{\mathrm{orb}} = 1.35^{\mathrm{d}}$ PSR 2145–0750, $P_{\mathrm{orb}} = 6.84^{\mathrm{d}}$
'low-mass' companion ($M_{\mathrm{c}} < 0.45\,M_\odot$)	NS + (He) WD	PSR 0437–4715, $P_{\mathrm{orb}} = 5.74^{\mathrm{d}}$ PSR 1640+2224, $P_{\mathrm{orb}} = 175^{\mathrm{d}}$
non-recycled pulsar	(CO) WD + NS	PSR 2303+46, $P_{\mathrm{orb}} = 12.3^{\mathrm{d}}$
unevolved companion	B-type companion low-mass companion	PSR 1259–63, $P_{\mathrm{orb}} = 3.4^{\mathrm{yr}}$ PSR 1820–11, $P_{\mathrm{orb}} = 357^{\mathrm{d}}$

CV-LIKE BINARIES

main type	sub-type	obs. example
novae-like systems	($M_{\mathrm{donor}} \leq M_{\mathrm{WD}}$)	DQ Her, $P_{\mathrm{orb}} = 4.7^{\mathrm{hr}}$ SS Cyg, $P_{\mathrm{orb}} = 6.6^{\mathrm{hr}}$
super soft X-ray sources	($M_{\mathrm{donor}} > M_{\mathrm{WD}}$)	CAL 83, $P_{\mathrm{orb}} = 1.04^{\mathrm{d}}$ CAL 87, $P_{\mathrm{orb}} = 10.6^{\mathrm{hr}}$
AM CVn systems (RLO)	(CO) WD + (He) WD	AM CVn, $P_{\mathrm{orb}} = 22^{\mathrm{min}}$
double WD (no RLO)	(CO) WD + (CO) WD	WD1204+450, $P_{\mathrm{orb}} = 1.6^{\mathrm{d}}$
sdB-star systems	(sdB) He-star + WD	KPD 0422+5421, $P_{\mathrm{orb}} = 2.16^{\mathrm{hr}}$

2.2 X-Ray Binaries

a) High- and low-mass X-ray binaries

The X-ray binaries can, broadly speaking, be divided into two main groups, the HMXBs and the LMXBs, which differ in a number of important characteristics, listed in Table 2 and graphically represented in Fig. 1. (The characteristics listed and depicted are for systems containing neutron stars, but most of them also hold for the X-ray binaries that contain black holes; also these can be divided into HMXBs and LMXBs, see [89]). For further details I refer the reader to the reviews of these two types of systems in the book by [79]. In the HMXBs the companion of the X-ray source is a luminous early-type star of spectral type O or B, like in the Cen X-3 system, with a mass typically between 10 and 40 $M_\odot$. In the LMXBs it is a faint star of mass $\leq 1\,M_\odot$; in most LMXBs the stellar spectrum is not even visible as the light of the system is dominated by that of the accretion disk around the compact star. The orbital periods of the LMXBs are generally short (mostly < 0.5 d), though on average somewhat longer than those of the Cataclysmic Variables.

Table 2. The two main classes of strong galactic X-ray sources.

	HMXB	LMXB
X-ray spectra:	$kT \geq 15$ keV (hard)	$kT \leq 10$ keV (soft)
Type of time variability:	regular X-ray pulsations no X-ray bursts	only a very few pulsars often X-ray bursts
Accretion process:	wind (or atmos. RLO)	Roche-lobe overflow
Timescale of accretion:	10^5 yr	$10^7 - 10^9$ yr
Accreting compact star:	high **B**-field NS (or BH)	low **B**-field NS (or BH)
Spatial distribution:	galactic plane	galactic center and spread around the plane
Stellar population:	young, age $< 10^7$ yr	old, age $> 10^9$ yr
Companion stars:	luminous, $L_{opt}/L_x > 1$ early-type O(B)-stars $> 10\,M_\odot$ (Pop. I)	faint, $L_{opt}/L_x \ll 0.1$ blue optical counterparts $\leq 1\,M_\odot$ (Pop. I and II)

b) Intermediate-mass X-ray binaries

During the last decade it was realized that apart from these two main groups of X-ray binaries, which each contain some 100 systems in our Galaxy (see [187]), there are a few X-ray binaries in which the companion is or has been a star of "intermediate mass", i.e. between 1 $M_\odot$ and 10 $M_\odot$. In fact Her X-1, in which the donor star has a mass of 2 $M_\odot$, is a system of this type and Cyg X-2

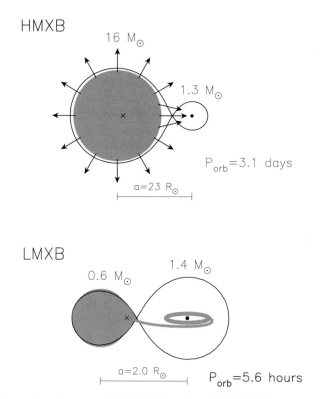

Fig. 1. Examples of a typical HMXB (top) and LMXB (bottom). The neutron star in the HMXB is fed by a strong high-velocity stellar wind and/or by beginning atmospheric Roche-lobe overflow. The neutron star in an LMXB is surrounded by an accretion disk which is fed by Roche-lobe overflow. There is also observational evidence for HMXBs and LMXBs harbouring black holes.

as well. In the latter system the companion star has presently a mass $\leq 1\,M_\odot$, but it is highly overluminous for its mass, indicating that it is an evolved star, and it was suggested already long ago [166] that it started out with a mass similar to that of the companion of Her X-1; the present idea is that its original mass may even have been as high as 3 to $4\,M_\odot$ [107, 70, 154, 74].

c) Spin and magnetic field in neutron-star X-ray binaries

From the characteristics of the HMXBs and LMXBs listed in Table 2 and graphically depicted in Fig. 1 it will be clear that the HMXBs belong to a very young stellar population, as O- and early B-stars do not live longer than about 10^7 years. This implies that the neutron stars in these systems in general have ages of at most a few million years. On the other hand, the characteristics of the LMXBs show that they tend to belong to a much older stellar population, with ages ranging from a few hundred million years to over 10^{10} years. (The last-mentioned age is typical for the about a dozen known strong globular cluster X-ray sources.) Hence, the neutron stars in these systems tend to be old. This age difference is strikingly clear from their different galactic distributions, the HMXBs being strongly concentrated in a very thin layer around the galactic plane, and not towards the galactic center, while the LMXBs are strongly concentrated towards the galactic center and have a much broader distribution around the plane (see [187]). A further difference is the presence of strong magnetic fields ($B \geq 10^{12}$ G) in the neutron stars in the HMXBs, as evidenced by their regular X-ray pulsations, and the absence of such regular pulsations and thus of strong magnetic fields in the bulk of the LMXBs. The occurrence of thermonuclear X-ray bursts in many LMXBs confirms that their magnetic fields must be weak, as fields stronger than 10^{10}–10^{11} G suppress such bursts [78]. In recent years a number of the LMXBs was found to be millisecond X-ray pulsars (e.g. see [116, 184]). Such short rotation periods of the accreting neutron stars are possible only if their magnetic fields are not stronger than 10^8 to 10^9 G (see Sect. 3 and Sect. 6). This difference in magnetic field strength between the HMXBs and the LMXBs is most probably due to field decay related to evolution in a binary system [147]. The neutron stars in the LMXBs have been living next to their companions and accreting and spinning up on average for a much longer period (10^8 to 10^9 years) than those in the HMXBs, which in most cases have been accreting less than a few times 10^5 years (see below). The various ways in which this large difference in accretion timescales may affect the surface magnetic field strength are discussed in Sect. 5 and 6 (see [9, 127, 27]). With the long-lasting accretion through a disk, as occurs in the LMXBs (see Sect. 5), the neutron star will be spun up to millisecond periods, as mentioned in Sect. 1. The relation between X-ray binaries and binary and millisecond radio pulsars will be considered in Sect. 4, 5 and 6.

d) The black hole X-ray binaries

Here a great breakthrough came with the discovery by [89] that the K5V dwarf companion of the "X-ray NOVA Monoceros 1975" – the source A0620-00 – is a spectroscopic binary with an orbital period of 7.75 hours and a radial velocity amplitude > 470 km s^{-1}. This large orbital velocity indicates that even if the K-dwarf would have zero mass, the compact object still has a mass > $3\,M_{\odot}$, and therefore must be a black hole [96, 64]. Since then about twenty such systems have been discovered consisting of a low-mass star together with a black hole (see [90]). Apart from these systems with low-mass donor stars, there is, of course the HMXB Cygnus X-1 which consists of a black hole and a O9.7Iab supergiant star, and the system of LMC X-3, in which the donor is an "intermediate mass" B3V star of about $6\,M_{\odot}$.

e) The high-mass X-ray binaries in more detail

It appears that the HMXBs fall into three main sub-types which differ in a number of important characteristics: (i) The "standard" HMXBs in which the massive OB-type companion of the X-ray source is close to filling its Roche lobe, as is evidenced from its double-wave optical lightcurve, that shows that the star is tidally deformed ("pear-shaped"; see [187]). These are the systems depicted in the top part of Fig. 1, resembling Cen X-3 and Cyg X-1. They are persistent sources and, except for one, have short orbital periods, < 11 days. (ii) The B-emission X-ray binaries, in which the B-type companion is a rapidly rotating main-sequence star which is deep inside its Roche lobe. These systems have orbital periods between 15 days and several years, and in most cases are "transient" X-ray sources that may occasionally turn on for a few weeks to several months, with long "off" periods in between. B-emission stars are known with such time intervals to exhibit "emission" phases, during which they eject mass from their equatorial regions. This circumstellar matter, in the form of a rapidly rotating equatorial disk, causes their hydrogen lines to go in emission. At the same time, the accretion of this ejected matter causes the neutron star companion to temporarily become an X-ray source, as was first pointed out by [88]. When the equatorial disk is present, the X-ray source shows periodic outbursts, every time the neutron star in its inclined eccentric orbit passes through the disk (see Fig. 2). (iii) The new class of supergiant fast X-ray transient HMXBs recently discovered with the gamma-ray satellite INTEGRAL. These sources are very heavily absorbed at the traditional X-ray energies of 2–10 keV, and appear to be located in the spiral arms of our Galaxy. They have been discovered only in very hard X-rays: 20–60 keV. Their transient outbursts observed in this energy range last typically of the order of thousands of seconds only. The fact that these sources do not emit at energies below about 20 keV is the reason why they were not detected with earlier X-ray satellites. The absorption of the softer X-rays is very heavy, suggesting hydrogen column densities of order 10^{24} cm^{-2}. The steady X-ray luminosity of these sources is generally low, of order 10^{33} erg s^{-1}. The companion stars appear to be O-type supergiants which, contrary to the case of the "standard" HMXBs, are deep inside their Roche lobes. Some of the sources are slow X-ray pulsars, others fast ones [193]. So far only one orbital period has been measured, for the system of IGRJ 11215-5952 which appears to be very long (330 days, [137]), and the feeling is that the orbital periods are generally much longer than those of the "standard" HMXBs . The idea is that the steady X-ray emission is powered by wind accretion, and that the steady accretion rate is quite low due to the large orbital separation. The winds of early supergiants are thought to be clumpy, and the flares are thought to be due to the accretion of high-density clumps from the wind [193]. In the top part of Table 1 a few characteristic examples of the first two types of HMXBs are listed (see also [123, 179]). The B-emission X-ray binaries are by far the largest group of HMXBs: already some 50 of such systems are known in our Galaxy and the

Magellanic Clouds (see [81]). Judging from these numbers, the total number of B-emission X-ray binaries in our Galaxy may easily be several thousands. On the other hand, not more than about a few dozen "standard" HMXBs are known in our sector of the Galaxy and it seems likely that there are not more than some 50 to 100 such binaries in the entire Galaxy. Good estimates of the number of supergiant fast X-ray transient HMXBs in our Galaxy are not yet available.

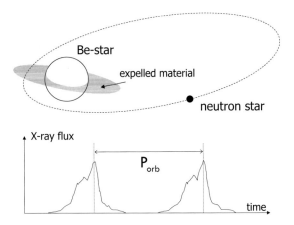

Fig. 2. Schematic model of a Be-star X-ray binary system. The neutron star moves in an eccentric orbit around the Be-star which is not filling its Roche-lobe. However, near the periastron passage the neutron star accretes circumstellar matter, ejected from the rotating Be-star, resulting in an X-ray outburst lasting several days.

2.3 The Binary and Millisecond Radio Pulsars

a) Spin and magnetic field

The binary radio pulsars are characterized by in general much shorter spin periods than the ordinary single radio pulsars. This can be clearly observed in the $\dot{P}$ versus P diagram of radio pulsars, where P is the pulse period and $\dot{P}$ is the period derivative. Fig. 3 shows this diagram for the about 1600 pulsars that were known in the galactic disk in 2007 [18]. Dots are single pulsars, circles are binary pulsars. The surface dipole magnetic field strength of pulsars B is related to P and $\dot{P}$ by the equation (see [87, 86]):

$$B = \sqrt{\frac{3c^3 I}{8\pi^2 R^6} P\dot{P}} \simeq 3 \times 10^{19} \sqrt{P\dot{P}} \quad \text{Gauss} \tag{1}$$

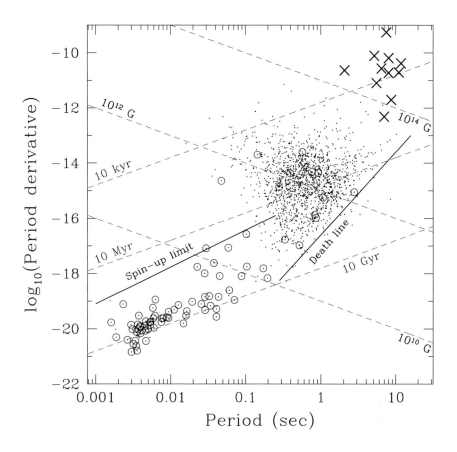

Fig. 3. $(P, \dot{P})$-diagram of ~ 1600 observed radio pulsars (courtesy F.Camilo 2007). Binary pulsars are marked by a circle. Soft gamma-ray repeaters (SGR) and anomalous X-ray (AXP) pulsars are marked by crosses. Also shown are lines of constant surface dipole magnetic field strength (dashed) and characteristic ages (dotted). The "death line"(pair-creation limit for generating radio pulses) and the "spin-up limit"("spin-up line") are indicated (explanation in text and Sect. 6).

where I and R are the moment of inertia ($\sim 10^{45}$ g cm^2) and radius of the neutron star, respectively (this follows from equating the rotational energy loss to the electromagnetic energy loss from the rotating magnetized neutron star). In Fig. 3 lines of constant B are indicated. The figure shows that a large fraction of the binary pulsars is millisecond pulsars, with spin periods shorter than 0.01 seconds. In the figure also lines are drawn of constant spindown age, as defined by [87]:

$$\tau \equiv P/2\dot{P} \qquad (2)$$

The figure shows that the binary pulsars typically have magnetic field strengths in the range $10^8 - 10^{10}$ G and "ages" of order $(1 - 10) \times 10^9$ years, whereas the "garden variety" pulsars, that make up the cloud of ordinary single pulsars typically have magnetic field strengths of 10^{11} to 10^{13} G and "ages" $< 10^7$ years. The nine double neutron stars in our galaxy, of which one is in a globular cluster and eight are in the galactic disk, typically have magnetic field strengths between 10^9 and 10^{10} G. Table 1 lists some vital data of representative examples of these systems and of some characteristic representatives of the other types of binary radio pulsars which have a white dwarf as a companion star. Many pulsars are known in globular clusters. Most of these are millisecond pulsars and some 40 per cent − and in some clusters, like 47 Tuc, over 60 per cent − of them are in binaries, against some 5 per cent of the total galactic pulsar population.

b) Classes of binary pulsars

With the exception of the "anti-deluvian" systems mentioned in Table 1 (consisting of a young pulsar and an unevolved main-sequence companion star), and a few eclipsing systems in globular clusters, the companion stars in radio pulsar binaries are themselves also dead stars: neutron stars or white dwarfs. They can be divided into the following categories, of which examples are depicted in Fig. 4: (i) Systems in which the pulsar has a relatively weak magnetic field (10^9 to 10^{10} G) and rapid spin, and the companion star is a neutron star. These systems all have eccentric orbits, which tend to be relatively narrow, with periods between 0.2 and 18 days. (In only one out of the 8 known double neutron stars in the galactic disk (listed in Table 3) the system of PSR J 1906+0746, the observed pulsar is a normal strong-magnetic-field pulsar, and in only one of the other systems, PSRJ 0737-3039 the companion of the weak-magnetic field pulsar is observed, and happens to have a normal strong magnetic field). (ii) Systems consisting of a neutron star and a massive ($\sim 1\,M_\odot$) white dwarf in an eccentric orbit. In these systems, of which PSR 2303+46 and PSRJ 1141-6545 are examples (see [186, 2]), the pulsar is a normal strong-magnetic-field neutron star (10^{11} to 10^{13} G). The orbits cover a range from very narrow (0.2 d) to very wide. (iii) Systems in which the pulsar has a relatively weak magnetic field (10^9 to 10^{10} G) and rapid spin, and the companion is a "massive" ($> 0.6\,M_\odot$) white dwarf. These systems all have perfectly circular orbits, with relatively short periods, in the same range as that of the double neutron stars. (iv) Systems consisting of a neutron star and a low-mass white dwarf (mass typically between 0.2 and 0.4 $M_\odot$) in relatively wide and circular orbits. A characteristic example of this class is the first-discovered binary millisecond pulsar PSR 1953+29 [13], with an orbital period of 117.35 days. In most cases the neutron stars in these systems are spinning very rapidly, with periods of less than 10 milliseconds, and have magnetic field strengths typically between 10^8 and 10^9 G.

Examples of the orbits of the binary pulsars with "massive" compact companions (categories (i),(ii) and (iii)) are depicted in the left-hand part of Fig. 4 and examples of the orbits of category (iv) in the right-hand part.

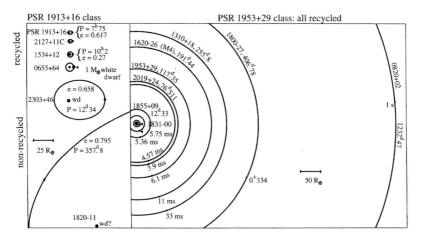

Fig. 4. The main classes of binary radio pulsars (orbits drawn to scale). Left: The class of systems "resembling" PSR 1913+16, which tend to have narrow and eccentric orbits; the companion of the pulsar is itself a neutron star or a massive white dwarf. When, in the case of the white dwarf companions, the orbit is circular, the neutron star is recycled; if the orbit is eccentric, the neutron star is not recycled - this is the case in 2303+46 (see text). Right: the PSR 1953+29 class systems tend to have wide and circular orbits; here the companion stars have a low mass, in the range 0.2–$0.4 M_\odot$, or even smaller, and most probably are helium white dwarfs (see text).

2.4 Spin Evolution of Accreting Neutron Stars; Alfven Radius, Co-Rotation Radius, Equilibrium Spin, Spin-Up and -Down

a) Observed spin behavior of accreting neutron stars

Figure 5 shows the pulse-period versus orbital period relation of the normal strong-magnetic field accreting X-ray pulsars [80]. (The accreting millisecond X-ray pulsars have not been included). Also the binary radio pulsars with B-type companion stars are indicated (the open squares). One observes in this figure that many of the X-ray pulsars have very long pulse periods, between 10 and 1000 seconds. In persistent sources spin periods of the order of a few seconds are found only in systems where from UV and optical observations we have clear evidence for the existence of an accretion disk. These are systems where (a large part of) the mass transfer is due to Roche-lobe overflow: the low-mass X-ray binary Her X-1 and the high-mass systems in which the

Table 3. Double neutron star binaries and the eccentric-orbit white-dwarf neutron star system J1145-6545.

Pulsar Name	Spin Per. (ms)	P_{orb} (d)	e	Compan. Mass ($M_\odot$)	Pulsar Mass ($M_\odot$)	Sum of masses ($M_\odot$)	B_s (10^{10} G)	Ref
J0737-3039A	22.7	0.10	0.088	1.250(5)	1.337(5)	2.588(3)	0.7	(1)
J0737-3039B	2770	0.10	0.088	1.337(5)	1.250(5)	2.588(3)	$1, 2.10^2$	(1)
J1518+4904	40.9	8.63	0.249	1.05 (+0.45) (-0.11)	1.56 (+0.13) (-0.45)	2.62(7)	0.1	(2)
B1534+12	37.9	0.42	0.274	1.3452(10)	1.3332(10)	2.678(1)	1	(3)
J1756-2251	28.5	0.32	0.18	1.18(3)	1.40(3)	2.574(3)	0.54	(4)
J1811-1736	104	18.8	0.828	1.11 (+0.53) (-0.15)	1.62 (+0.22) (-0.55)	2.60(10)	1.3	(3)
J1829+2456	41.0	1.18	0.139	1.27 (+0.11) (-0.07)	1.30 (+0.05) (-0.05)	2.53(10)	~ 1	(5)
J1906+0746	144.1	0.165	0.085	–	–	2.61(2)	$1, 7.10^2$	(7)
B1913+16	59	0.33	0.617	1.3873(3)	1.4408(3)	2.8281(1)	2	(3)
J1145-6545	394	0.20	0.172	1.00(2)	1.28(2)	2.288(3)	$\sim 10^2$	(6)

References: (1) [85]; (2) Nice et al. (1996); (3) [143];
(4) [42]; (5) Champion et al. (2004); (6) Bailes (2004);
(7) Lorimer et al. (2006)

supergiant donors are just beginning to overflow their Roche lobes: Cen X-3, SMC X-1 and LMC X-4. These are indicated by the crosses in the diagram. These sources are expected to be powered by a combination of stellar wind and beginning Roche-lobe overflow [129, 130]. In most of these systems the X-ray pulsars show a secular decrease of the spin period ("spin-up") on a relatively short timescale: a few thousand years in the massive systems and of order 10^5 years in Her X-1 see [5, 43]. Fig. 6 for example shows the observed spin evolution of Cen X-3 as measured by GRO-BATSE. Although on short timescales episodes of spin-up and spin-down alternate, the average trend is that of spin-up. On the other hand, in persistent systems that are purely wind-fed, such as the HMXBs with blue supergiant companions that do not yet fully fill their Roche lobes, the pulse periods are very long, and they vary erratically in time showing no clear secular trends. This can be explained by the fact that the amount of angular momentum carried by the supersonic winds is negligible, and eddies form in the wind downstream of the neutron star, which alternating may feed co-rotating and counter-rotating angular momentum to it [148, 47]. The B-emission X-ray binaries show an again different spin behavior: these sources are transients, in most cases recurrent. During a transient outburst they in general show rapid spin-up, indicating that a disk has been formed. However, between outbursts they spin down, as at the beginning of the next outburst the spin period is generally observed to be much longer than at end of the previous one. An important clue to what drives the spin-up and down in the B-e X-ray systems is given by the correlation between spin periods and orbital periods in these systems (Fig. 5), discovered by [22]. In order to understand this we first have to consider the concept of equilibrium spin period, which is the topic of the next paragraph.

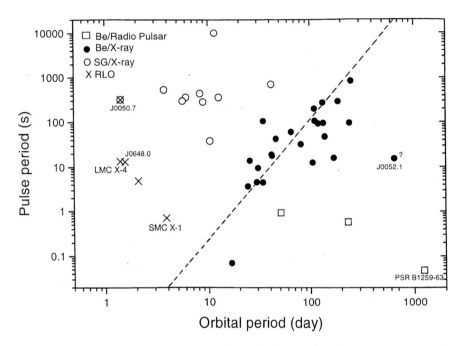

Fig. 5. The pulse period versus orbital period diagram for the accretion-powered pulsars with known orbital periods. Crosses indicate persistent sources with accretion disks (Roche-lobe overflow), open circles are persistent sources with blue supergiant companions (wind accretors), and filled circles are the B-emission X-ray binaries, which are mostly transient sources. Squares are radio pulsars with unevolved B-type companions. The dashed line indicates Corbet's [22] relation between pulse period and orbital period for the Be-X-ray systems (after [80]).

b) The concepts of Alfven radius, magnetosphere and equilibrium spin

These concepts were introduced by [29, 76]. The Alfven radius R_A is the distance from the neutron star where the kinetic energy density $0.5\,\rho\,v^2$ of the inflowing matter equals the magnetic energy density $0.5\,\mu\,B^2$ of the neutron star's magnetic dipole field, so:

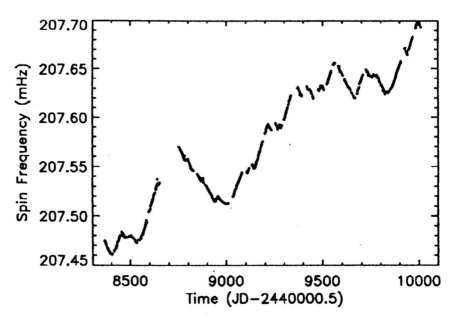

Fig. 6. About 5 years of Cen X-3 spin frequency measurements from GRO-BATSE (after [5, 43]).

$$\rho\, v^2 = \mu\, B^2 \qquad (3)$$

where

$$B(r) = B_s\, (R/r)^3 \qquad (4)$$

in which B_s is the magnetic field strength at the neutron-star radius R and μ is the permeability of the vacuum. Here $v(r)$ is either the free-fall velocity or the Kepler velocity in the disk (which at a given r are the same, except for a factor $\sqrt{2}$), and $\rho(r)$ can be expressed in terms of the accretion rate $\dot{M}$, $v(r)$ and r (see for example [8, 169]). This leads to:

$$R_A = \left(\frac{B_s{}^2\, R_s{}^2}{\dot{M}\, \sqrt{2\, G\, M}} \right)^{2/7} \qquad (5)$$

For $r > R_A$ the flow of matter is not influenced by the neutron star's magnetic field. On the other hand, for distances smaller than the Alfven radius the magnetic field forces the matter to flow in along the field lines. The region around the neutron star closer than the Alfven radius is called the *magnetosphere*. For $r < R_A$ the inflowing matter is forced to co-rotate with the magnetosphere of the neutron star, for $r > R_A$ matter can freely orbit the star.

A second important radius is the co-rotation radius R_{co}. This is the distance where the Kepler velocity of the matter orbiting around the neutron star just equals the rotational velocity of the magnetosphere $\omega\, r$. The value of R_{co} is given by:

$$\omega^2 r = G M r^{-2} \tag{6}$$

where $\omega = 2\pi/P$, P being the rotation period of the star. Thus:

$$R_{\text{co}} = \left(G M P^2/4\pi^2\right)^{1/3} \tag{7}$$

There are now two possibilities for the accretion, illustrated in Fig. 7:

(i) If $R_A \geq R_{co}$, matter at the magnetospheric boundary cannot flow in: as soon as the matter enters the magnetosphere it is forced to co-rotate; however, it then rotates faster than the Keplerian velocity and is centrifuged out of the magnetosphere again. In this situation the accretion therefore is shut off.

(ii) If $R_A < R_{co}$, accretion can take place. In the latter situation, matter with angular momentum will enter the magnetosphere and flow to the neutron star surface, causing the angular velocity of rotation of the neutron star to increase. This decrease in spin period will go on until $R_A = R_{co}$, after which further accretion and spin-up will be impossible. The star will therefore settle at a spin period for which these two radii are just equal to each other. This is called the equilibrium spin period P_{eq}, which is given by:

$$P_{eq} = (2.4\text{ms})\, B_9{}^{6/7}\, R_6{}^{16/7}\, M^{-5/7} \left(\dot{M}/\dot{M}_{\text{Edd}}\right)^{-3/7} \tag{8}$$

(e.g. see [169]), where B_9 is the surface dipole magnetic field strength in units of 10^9 G, R_6 is the neutron star radius in units of 10^6 cm, M is the mass of the neutron star in solar units, and $\dot{M}_{\text{Edd}}$ is the Eddington accretion rate (maximum possible accretion rate before accreted matter is blown away by radiation pressure).

The X-ray pulsars with accretion disks seem all to be spinning near to their equilibrium spin periods. Their magnetic field strengths, inferred from X-ray cyclotron lines, are typically of order 10^{12} G [163, 95], yielding equilibrium spin periods of order of seconds, just as observed in these sources. The fact that their spin periods still show a secularly decreasing trend (spin-up) may be due either to a secular decrease of the surface magnetic field strength, or a secular increase of the accretion rate, or both [43]. A decrease of the surface magnetic field strength might be due to temporary or permanent "burying" of the field due to the accretion process (see Sect. 5 and Sect. 6).

c) Possible explanation for the Corbet relation

As shown by [194] the Corbet relation between the spin periods and orbital periods of the Be/X-ray binaries indicates that the Be stars are surrounded by a disk of matter, ejected from their equatorial regions at relatively low velocities. One may also call this a "slow wind" that carries with it the angular momentum it had at the moment the wind particles the equatorial regions of the very rapidly rotating star. When the wind density is low, the neutron star spins faster than its equilibrium spin period (see above) and accretion is

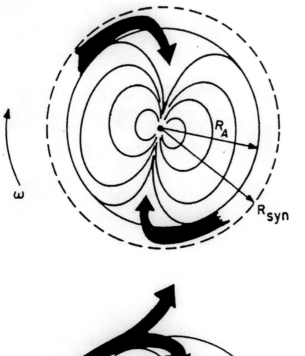

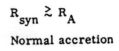

$$R_{syn} \gtrsim R_A$$

Normal accretion

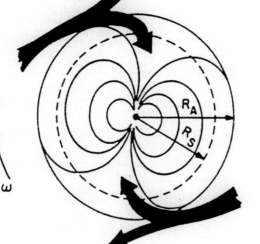

$$R_{syn} < R_A$$

Matter ejected

Fig. 7. Schematic representation of Alfven radius R_A and co-rotation radius, indicated here as R_{syn}, of a rotating magnetized neutron star. R_A depends on accretion rate $\dot{M}$, mass M and dipole magnetic field strength at neutron star surface B_9; R_{syn} depends on rotation period P and mass of the neutron star. When $R_A = R_{syn}$ the neutron star spins at its equilibrium spin period P_{eq}. If it rotates slower, accretion is possible; if it rotates faster, centrifugal forces on matter entering the magnetosphere will swing this matter out and accretion is impossible. (after Schreier 1977, *Annals of the New York Academy of Sciences*, **402**, 445).

not possible. The neutron star now spins down due to the torques that the surrounding accretion disk exerts on the magnetosphere (see [49]). When the wind density increases, the magnetosphere is compressed and the equilibrium spin period becomes shorter than the spin period of the neutron star, such that now accretion becomes possible (see Fig. 7). At this moment the neutron star turns on as an accreting X-ray "transient" source, and spin-up will occur due to the accretion of disk angular momentum. On average one expects the spin period of the neutron star in these systems to hover around a value equal to the equilibrium period corresponding to the mean density in the wind of the Be star near the orbit of the neutron star. The wider the orbit, the less dense the wind, and thus the lower the average accretion rate, and the longer the corresponding equilibrium spin period.

3 Stellar Evolution and the Formation of Compact Objects in Binary Systems

In order to understand how neutron stars, black holes and white dwarfs can be formed in binary systems, a brief overview of the basic elements of the evolution of single stars is necessary. We refer to e.g. [25] and [72] for further details.

3.1 Summary of the Evolution of Single Stars

The evolution of a star is driven by a rather curious property of a self-gravitating gas sphere in hydrostatic equilibrium, described by the virial theorem, namely that the radiative loss of energy of such a gas sphere causes it to contract and herewith, due to release of gravitational potential energy, to increase its temperature. Thus, while the star tries to cool itself by radiating away energy from its surface, it gets hotter instead of cooler (i.e. it has a 'negative heat capacity'). The more it radiates to cool itself, the more it will contract, the hotter it gets and the more it is forced to go on radiating. Clearly, this 'vicious virial cycle' is an unstable situation in the long run and explains why the star, starting out as an interstellar gas globe, must finally end its life as a compact object. In the meantime the star spends a considerable amount of time in the intermediate stages, which are called: 'the main-sequence', 'the giant branch' etc. It is important to realize that stars do *not* shine because they are burning nuclear fuel. They shine because they are hot due to their history of gravitational contraction.

A massive star ($M \geq 10\,M_\odot$) evolves through a series of cycles of nuclear burning alternating with stages of gravitational contraction of the core. In this way it exhausts one type of nuclear fuel after the other, until its core is made of iron, at which point further fusion requires, rather than releases, energy. The core mass of such a star becomes larger than the Chandrasekhar limit, the maximum mass possible for an electron-degenerate configuration ($\sim 1.4\,M_\odot$).

Table 4. End products of stellar evolution as a function of initial mass

		Final product	
Initial mass	He-core mass	Single star	Binary star
$< 2.3\,M_\odot$	$< 0.45\,M_\odot$	CO white dwarf	He white dwarf
$2.3 - 8\,M_\odot$	$0.5 - 2.1\,M_\odot$	CO white dwarf	CO white dwarf
$8 - 10(12)\,M_\odot$	$2.1 - 2.5(3.5)\,M_\odot$	O–Ne–Mg white dwarf	neutron star
$10(12) - 25\,M_\odot$	$2.5(3.5) - 8\,M_\odot$	neutron star	neutron star
$> 25\,M_\odot$	$> 8\,M_\odot$	black hole	black hole

Therefore the core implodes to form a neutron star or black hole. The gravitational energy released in this implosion (4×10^{53} erg $\simeq 0.15\,M_{\rm core}c^2$) is far more than the binding energy of the stellar envelope, causing the collapsing star to violently explode and eject the outer layers of the star, with a speed of $\sim 10^4$ km s^{-1}, in a supernova event. The final stages during and beyond carbon burning are very short lasting (~ 60 yr for a $25\,M_\odot$ star) because most of the nuclear energy generated in the interior is liberated in the form of neutrinos which freely escape without interaction with the stellar gas and thereby lowering the outward pressure and accelerating the contraction and nuclear burning.

Less massive stars ($M \leq 8\,M_\odot$) suffer from the occurrence of degeneracy in the core at a certain point of evolution. Since for a degenerate gas the pressure only depends on density and not on the temperature, there will be no stabilizing expansion and subsequent cooling after the ignition. Hence, the sudden temperature rise (due to the liberation of energy after ignition) causes a run-away nuclear energy generation producing a so-called 'flash'. In stars with $M \leq 2.3\,M_\odot$ the helium core becomes degenerate during hydrogen shell burning and, when its core mass $M_{\rm He}$ reaches $0.45\,M_\odot$, helium ignites with a flash. The helium flash is, however, not violent enough to disrupt the star. Stars with masses in the range $2.3 < M/M_\odot < 8$ ignite carbon with a flash. Such a carbon flash was believed to perhaps disrupt the entire star in a so-called carbon-deflagration supernova. However, recent observations of white dwarfs in galactic clusters that still contain stars as massive as $8^{+3}_{-2}\,M_\odot$ [126, 197] indicate that such massive stars still may terminate their life as a white dwarf. They apparently shed their envelopes in the AGB-phase before carbon ignites violently. Furthermore, stars in close binary systems, which are the prime objects in this review, will have lost their envelope as a result of mass transfer via Roche-lobe overflow. This is expected to have the very interesting effect: that in the mass range $M = 8$ - $11^{+1}_{-1}\,M_\odot$ *only* stars in binaries are expected to leave neutron stars, but not single stars, for the following reasons [110, see Fig. 8]. During the "dredge-up" phase in a single star on the Asymptotic Giant Branch (AGB) the bottom boundary of the deep convective H-rich envelope penetrates downwards into the thick Helium layer surrounding the degenerate O–Ne–Mg core. It then erodes away this helium layer, mixing the helium into the hydrogen envelope, and in this way prevents further growth by helium shell burning of the degenerate O–Ne–Mg core to the Chandrasekhar limit. The single star then blows away its very extended hydrogen envelope in an AGB "superwind" followed by a planetary nebula phase, and leaves behind an O–Ne–Mg white dwarf. In this way, all *single* stars up to $11^{+1}_{-1}\,M_\odot$ will leave white dwarfs. On the other hand, in a relatively close binary (orbital period less than a few years) the star will lose its H-rich envelope by binary mass transfer before it reaches the AGB and "dredge up", and it will leave behind the helium core with its unaffected mass, in the range $M = 2.1 - 2.5(3.5)\,M_\odot$. In these helium stars the degenerate O–Ne–Mg core is able to grow further by helium shell burning, to reach the Chandrasekhar mass and undergo an electron-capture collapse and produce a neutron star.

The possible end-products and corresponding initial masses are listed in Table 4. It should be noted that the actual values of the different mass ranges are only known approximately due to considerable uncertainty in our knowledge of the evolution of massive stars. Prime causes of this uncertainty include limited understanding of the mass loss undergone by stars in their various evolutionary stages. To make a black hole, the initial ZAMS stellar mass must exceed at least $20\,M_\odot$ [45, 46], or possibly, $25\,M_\odot$. From an analysis of black hole binaries it seems that a mass-fraction of ~ 0.35 must have been ejected in the (symmetric) stellar core collapse leading to the formation of a black hole [97]. Another fundamental problem is understanding convection, in particular in stars that consist of layers with very different chemical composition. Finally, there is the unsolved question of whether or not the velocity of convective gas cells may carry them beyond the boundary of the region of the star which is convective according to the Schwarzschild criterion. For example, inclusion of this so-called overshooting in evolutionary calculations decreases the lower mass-limit for neutron star progenitors.

Three timescales of stellar evolution

There are three fundamental timescales of stellar evolution. When the hydrostatic equilibrium of a star is disturbed (e.g. because of sudden mass loss), the star will restore this equilibrium on a so-called dynamical (or pulsational) timescale:

$$\tau_{\rm dyn} = \sqrt{R^3/GM} \simeq 50 \text{ min } (R/R_\odot)^{3/2}\,(M/M_\odot)^{-1/2} \qquad (9)$$

When the thermal equilibrium of a star is disturbed, it will restore this equilibrium on a thermal (or Kelvin–Helmholtz) timescale, which is the time it takes to emit all of its thermal energy content at its present luminosity:

$$\tau_{\rm th} = GM^2/RL \simeq 30 \text{ Myr } (M/M_\odot)^{-2} \qquad (10)$$

The third stellar timescale is the nuclear one, which is the time needed for the star to exhaust its nuclear fuel reserve (which is proportional to M), at its

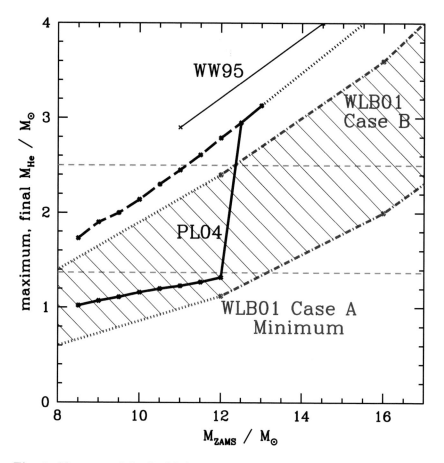

Fig. 8. The mass of the final helium core versus the initial main-sequence stellar mass, for single stars (fully drawn curve) and binary components (dashed curve), after [110]. In single stars in the mass range 8 to 11 $M_\odot$, on the Asymptotic Giant Branch "dredge-up" erodes away the helium layers surrounding the degenerate O–Ne–Mg core, which prevents the cores of single stars in this mass range growing to the Chandrasekhar limit and reaching core collapse. On the other hand, in an interacting binary the stars in this mass range lose their hydrogen-rich envelopes by mass transfer to their companion, and therefore avoid "dredge-up".They become helium stars with a mass equal to the *original* mass of the helium core, which is larger than the Chandrasekhar mass, and their degenerate O–Ne–Mg cores are able to grow to this mass limit due to helium-shell burning, and will evolve to core collapse and the formation of a neutron star. Thus, according to [110] single stars in this mass range do not produce neutron stars, but binary components do.

present fuel consumption rate (which is proportional to L), so this timescale is given by:

$$\tau_{\text{nuc}} \simeq 10 \text{ Gyr } (M/M_\odot)^{-2.5} \qquad (11)$$

In calculating the above mentioned timescales we have assumed a mass-luminosity relation: $L \propto M^{3.5}$ and a mass-radius relation for main-sequence stars: $R \propto M^{0.5}$. Both of these relations are fairly good approximations for $M \geq M_\odot$. Hence, it should also be noted that the rough numerical estimates of these timescales only apply to ZAMS stars.

3.2 The Variation of the Outer Radius During Stellar Evolution

Fig. 9 depicts the evolutionary tracks in the Hertzsprung-Russel diagram of six different stars ($50\,M_\odot$, $20\,M_\odot$, $12\,M_\odot$, $5\,M_\odot$, $2\,M_\odot$ and $1\,M_\odot$). These tracks were calculated by Tauris (see [157]) using Eggleton's evolutionary code (e.g. [111, 112]). The observable stellar parameters are: luminosity (L), radius (R) and effective surface temperature (T_{eff}). Their well-known relationship is given by: $L = 4\pi R^2 \sigma T_{\text{eff}}^4$. Fig. 10 shows the calculated stellar radius as a function of age for the $5\,M_\odot$ star. Important evolutionary stages are indicated in the figures. Between points 1 and 2 the star is in the long-lasting phase of core hydrogen burning (nuclear timescale). At point 3 hydrogen ignites in a shell around the helium core. For stars more massive than $1.2\,M_\odot$ the entire star briefly contracts between points 2 and 3, causing its central temperature to rise. When the central temperatures reaches $T \sim 10^8$ K, core helium ignites (point 4). At this moment the star has become a red giant, with a dense core and a very large radius. During helium burning it describes a loop in the HR-diagram. Stars with $M \geq 2.3\,M_\odot$ move from points 2 to 4 on a thermal timescale and describe the helium-burning loop on a (helium) nuclear timescale following point 4. Finally, during helium shell burning the outer radius expands again and at carbon ignition the star has become a red supergiant on the asymptotic giant branch (AGB).

The evolution of less massive stars ($M \leq 2.3\,M_\odot$) takes a somewhat different course. After hydrogen shell ignition the helium core becomes degenerate and the hydrogen burning shell generates the entire stellar luminosity. While its core mass grows, the star gradually climbs upwards along the red giant branch until it reaches helium ignition with a flash. For all stars less massive than about $2.3\,M_\odot$ the helium core has a mass of about $0.45\,M_\odot$ at helium flash ignition. The evolution described above depends only slightly on the initial chemical composition and effects of convective overshooting.

The core mass–radius relation for low-mass RGB stars

For a low-mass star ($\leq 2.3\,M_\odot$) on the red giant branch (RGB) the growth in core mass is directly related to its luminosity, as this luminosity is entirely generated by hydrogen shell burning. As such a star, composed of a small

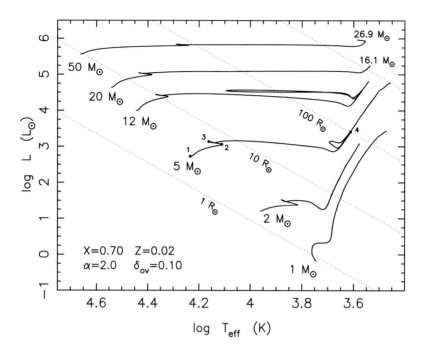

Fig. 9. Stellar evolutionary tracks in the HR-diagram (after [157]).

dense core surrounded by an extended convective envelope, is forced to move up the Hayashi track its luminosity increases strongly with only a fairly modest decrease in temperature. Hence one also finds a relationship between the giant's radius and the mass of its degenerate helium core – almost entirely independent of the mass present in the hydrogen-rich envelope [124, 196, 145]. This relationship is very important for LMXBs and wide-orbit binary pulsars since, as we shall see later on, it results in a relationship between orbital period and white dwarf mass.

3.3 The Evolution of Helium Stars

For low-mass stars, the evolution of the helium core in post main-sequence stars is practically independent of the presence of an extended hydrogen-rich envelope. However, for more massive stars ($> 2.3\,M_\odot$) the evolution of the core of an isolated star differs from that of a naked helium star (i.e. a star which has lost its hydrogen envelope via mass transfer in a close binary system).

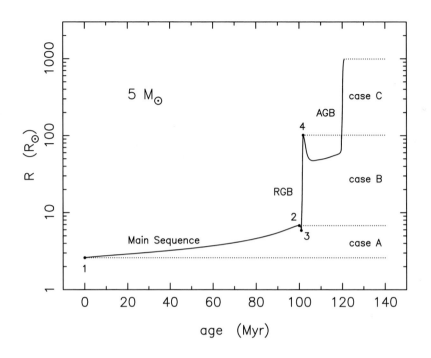

Fig. 10. Evolutionary change of the radius of the $5\,M_\odot$ star plotted in the figure above (Fig. 9). The ranges of radii for mass transfer to a companion star in a binary system according to RLO cases A, B and C are indicated – see Sect. 3.4 for an explanation.

Thus, it is very important to study the giant phases of helium star evolution. Pioneering studies in this field are those of [103, 99, 53].

Of particular interest are the low-mass helium stars ($M_{\mathrm{He}} < 3.5\,M_\odot$) since they swell up to large radii during their late evolution – see Fig. 11. This may cause an additional phase of mass transfer from the naked helium star to its companion (often referred to as so-called case BB mass transfer). Recent detailed studies of helium stars in binaries have been performed by [34]. Using helium star models ($Z = 0.03, Y = 0.97$) calculated by O. Pols (2002, private communication), we fitted the helium star ZAMS radii as a function of mass:

$$R_{\mathrm{He}} = 0.212\,(M_{\mathrm{He}}/M_\odot)^{0.654}\;R_\odot \qquad (12)$$

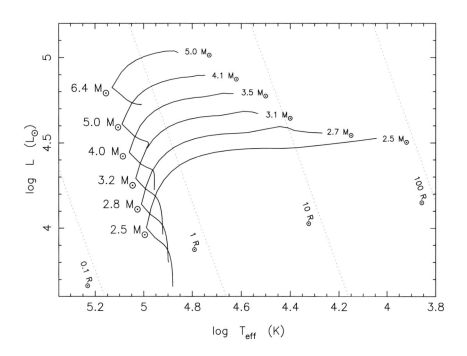

Fig. 11. Evolutionary tracks of $2.5\,M_\odot$–$6.4\,M_\odot$ helium stars ($Y = 0.98, Z = 0.02$). The final stellar mass (after wind mass loss) is written at the end of the tracks. The expansion of low-mass helium stars in close binaries often results in a second mass-transfer phase (case BB RLO). This plot was made with data provided by O. Pols (2002, private communication).

It is important to realize that helium cores in binaries have tiny envelopes of hydrogen ($< 0.01\,M_\odot$) when they detach from RLO. This has important effects on their subsequent radial evolution (e.g. [56]).

The evolution of more massive helium stars (Wolf–Rayet stars) is also quite important. There is currently not a clear agreement on the rate of intense wind mass-loss from Wolf–Rayet stars (e.g. [198, 100, 98]). A best-estimate fit to the wind mass-loss rate of Wolf–Rayet stars is, for example, given by [34]:

$$\dot{M}_{\text{He, wind}} = \begin{cases} 2.8 \times 10^{-13}\,(L/L_\odot)^{1.5}\ M_\odot\,\text{yr}^{-1}, & \log{(L/L_\odot)} \geq 4.5 \\ 4.0 \times 10^{-37}\,(L/L_\odot)^{6.8}\ M_\odot\,\text{yr}^{-1}, & \log{(L/L_\odot)} < 4.5 \end{cases} \quad (13)$$

The uncertainty in determining this rate also affects our knowledge of the threshold mass for core collapse into a black hole [132, 201, 15]. Very important in this respect is the question whether the massive ($> 3.5\,M_\odot$) helium star is "naked" or "embedded" – i.e. is the helium core of the massive star surrounded by a thick hydrogen mantle ? In the latter case this helium 'star' does not lose much mass in the form of a wind and it can go through all burning stages

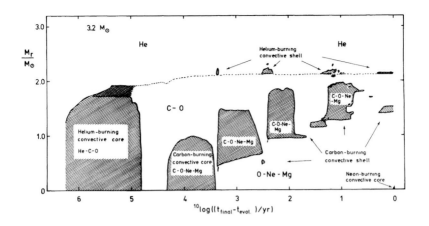

Fig. 12. The internal evolution of a $3.2\,M_\odot$ helium star. Hatched regions are convective; double hatched regions are semi-convective. The broken line indicates the region of maximum net energy generation in the helium-burning shell (approx. the boundary between the CO core and its adjacent helium layer). After [53].

and terminate as a black hole. Single star evolutionary models suggest that this happens above an initial stellar mass $\geq 19\,M_\odot$, as around this mass a sudden increase in the mass of the collapsing iron core occurs to $\geq 1.9\,M_\odot$ [200]. In order to form a black hole in a close binary, it is best to keep the helium core embedded in its hydrogen envelope as long as possible, i.e. to start from a wide "case C" binary evolution, as has been convincingly argued by [15, 198] Brown et al. (2001) and [98]. In this case common envelope evolution (Sect. 3.8) leads to a narrow system consisting of the evolved helium core and the low-mass companion. The helium core collapses to a black hole and produces a supernova in this process, shortly after the spiral-in. When the low-mass companion evolves to fill its Roche-lobe these systems are observed as Low-Mass Black-Hole X-ray Binaries, also called "soft x-ray transients" (SXTs) – see Sect. 2.

3.4 Roche-Lobe Overflow (RLO) – Cases A, B and C

The effective gravitational potential in a binary system is determined by the masses of the stars and the centrifugal force arising from the motion of the two stars around one another. One may write this potential as:

$$\Phi = -\frac{GM_1}{r_1} - \frac{GM_2}{r_2} - \frac{\Omega^2 r_3^2}{2} \tag{14}$$

where r_1 and r_2 are the distances to the center of the stars with mass M_1 and M_2, respectively; Ω is the orbital angular velocity; and r_3 is the distance to the rotational axis of the binary. It is assumed that the stars are small with respect to the distance between them and that they revolve in circular orbits, i.e. $\Omega = \sqrt{GM/a^3}$, where $M = M_1 + M_2$ and a the semi-major axis. In a binary where tidal forces have circularized the orbit, and brought the two stellar components into synchronized co-rotation, one can define fixed equipotential surfaces in a co-moving frame (see e.g. [169]). The equipotential surface passing through the first Lagrangian point, L_1 defines the 'pear-shaped' Roche-lobe – see the cross-section in Fig. 13. If the initially more massive star (the donor) evolves to fill its Roche-lobe the unbalanced pressure at L_1 will initiate mass transfer (Roche-lobe overflow, RLO) onto its companion star (the accretor). The radius of the donor's Roche-lobe, R_L is defined as that of a sphere with the same volume as the lobe. It is a function only of the orbital separation a, and the mass ratio $q \equiv M_{donor}/M_{accretor}$ of the binary components. It can be approximated as [37]:

$$\frac{R_L}{a} = \frac{0.49\, q^{2/3}}{0.6\, q^{2/3} + \ln(1 + q^{1/3})} \tag{15}$$

which is accurate to within a few per cent for all values of q. A very convenient expression valid for $q \geq 1.25$ is:

$$\frac{R_L}{a} = \frac{2}{3^{4/3}} \left(\frac{q}{1+q} \right)^{1/3}. \tag{16}$$

A star born in a close binary system with a radius smaller than that of its Roche-lobe may, either because of expansion of its envelope at a later evolutionary stage or because the binary shrinks sufficiently as a result of orbital angular momentum losses, begin RLO. The further evolution of the system will now depend on the evolutionary state and structure of the donor star at the onset of the overflow, which is determined by M_{donor} and a, as well as the nature of the accreting star. [71] define three types of RLO: cases A, B and C. In case A, the system is so close that the donor star begins to fill its Roche-lobe during core-hydrogen burning; in case B the primary star begins to fill its Roche-lobe after the end of core-hydrogen burning but before helium ignition; in case C it overflows its Roche-lobe during helium shell burning or beyond. It is clear from Fig. 10 that cases B and C occur over a wide range of radii (orbital periods); case C even up to orbital periods of ~ 10 years. The precise orbital period ranges for cases A, B and C depend on the initial donor star mass and on the mass ratio. Once the RLO has started it continues until the donor has lost its hydrogen-rich envelope (typically $\geq 70\%$ of its total mass) and subsequently no longer fills its Roche-lobe.

3.5 Orbital Changes Due to Mass Transfer and Mass Loss from the System

When a star fills its Roche lobe, matter in its outer layers can freely flow over towards the companion along the first Lagrangian point L_1. Also, some

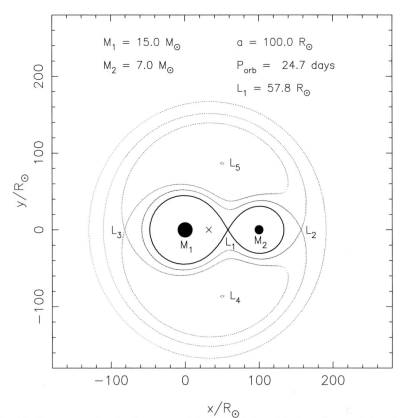

$M_1 = 15.0 \, M_\odot$ $a = 100.0 \, R_\odot$

$M_2 = 7.0 \, M_\odot$ $P_{orb} = 24.7$ days

$L_1 = 57.8 \, R_\odot$

Fig. 13. A cross-section in the equatorial plane of the critical equipotential surfaces in a binary. The thick curve crossing through L_1 is the Roche-lobe.

mass can be lost from the system, either by a stellar wind, or gas streams or otherwise, and this matter will then carry off orbital angular momentum. In order to see how these effects change the orbit of the system, we have to consider the equations for combined losses of mass and orbital angular momentum form the binary.

The orbital angular momentum of a binary system is given by:

$$J_{orb} = \frac{M_1 M_2}{M} \, \Omega \, a^2 \, \sqrt{1 - e^2} \tag{17}$$

where M_1 and M_2 are the masses of the accretor and the donor star, respectively. As mentioned earlier, tidal effects acting on a near-RLO (giant) star will circularize the orbit on a short timescale of $\sim 10^4$ yr [189]. In the following we therefore neglect any eccentricity ($e = 0$). A simple logarithmic differentiation of the above equation yields the rate of change in orbital separation:

$$\frac{\dot{a}}{a} = 2\frac{\dot{J}_{\rm orb}}{J_{\rm orb}} - 2\frac{\dot{M}_1}{M_1} - 2\frac{\dot{M}_2}{M_2} + \frac{\dot{M}_1 + \dot{M}_2}{M} \qquad (18)$$

where $\dot{J}_{\rm orb}$ is the loss rate of total orbital angular momentum from the system.

The "conservative" case

The simplest case occurs when there is no loss of mass and angular momentum from the system, that is: when all the mass lost from one star is captured by the other one. This case is called "conservative" evolution of the close binary. In that case, since $\dot{M}_1 = -\dot{M}_2$ and $\dot{J}_{\rm orb} = 0$, the Eq. (18) reduces to

$$\frac{\dot{a}}{a} = -2\dot{M}_2 \left(\frac{1}{M_2} - \frac{1}{M_1} \right) \qquad (19)$$

Since $\dot{M}_2 < 0$ this equation immediately shows that, when the mass-losing star M_2 is more massive than its companion (accretor) M_1, the orbital radius will *decrease*, and reversely, that if mass is transferred from the less massive to the more massive star, the orbital radius will *increase*. Simple integration of Eq. (19) yields:

$$\frac{a}{a_0} = \left(\frac{M_{1,0}\, M_{2,0}}{M_1\, M_2} \right)^2 \qquad (20)$$

where $M_{1,0}$ and $M_{2,0}$ are the initial masses.

Orbital changes when mass and orbital angular momentum are lost

In the case of mass loss from the binary, we are no longer dealing with a closed "conservative" system and there is, in principle, an infinite range of possibilities for the changes of the orbit depending on the amounts of mass lost and on the angular momentum lost with this mass. Since a mass element leaving the system is a third body, we are dealing here with the 3-body problem, for which no general solutions exist. For that reason one takes beforehand a physically reasonable "mode of mass loss", in which an assumption is made about the amount of orbital angular momentum that is expected to be carried off per unit mass lost. For extensive reviews of a variety of "modes" of mass and angular momentum losses from binaries, see [169, 140].

One simple case in which only angular momentum is lost, but no mass, refers to systems with very short orbital periods (generally less than 12 hours) in which orbital angular momentum is lost by the emission of gravitational waves [77, 138]:

$$\frac{\dot{J}_{\rm gwr}}{J_{\rm orb}} = -\frac{32\, G^3}{5\, c^5} \frac{M_1 M_2 M}{a^4} \quad {\rm s}^{-1} \qquad (21)$$

where c is the speed of light in vacuum. The validity of this mechanism has been beautifully demonstrated in PSR 1913+16 which is an ideal GTR-laboratory (e.g. [159]). For sufficiently narrow orbits the above equation becomes the dominant term in Eq. (18) and will cause a to decrease. Therefore, the orbits of very narrow binaries will tend to continuously shrink, forcing the components into contact. Gravitational radiation losses are a major force driving the mass transfer in very narrow binaries, such as CVs and LMXBs [41].

Of particular importance for CVs and LMXBs is also orbital angular momentum loss by "magnetic braking", a concept introduced by [188], which acts as follows. Stars with convective envelopes tend to have surface magnetic fields. These couple to the ejected ionized stellar wind material and enforce co-rotation of the wind out to several stellar radii from the surface. In semi-detached binaries tidal forces tend to keep the stellar rotation synchronous with the orbital motion, such that the spin angular momentum of the donor star carried off by the magnetically coupled wind must be compensated by the system's orbital angular momentum. An effectively coupled wind can thus enforce a decrease in the orbital separation, just like gravitational radiation. If these wind losses occur simultaneously with the RLO the mass-transfer rate is increased by the induced shrinking of the Roche lobe, just as in the case of gravitational radiation losses. Such "magnetic braking" occurs particularly in systems where the donor star is a G, K or M dwarf, as is the case in most CVs and short-period LMXBs. The observed spin-down rates of young single stars of these types provide empirical quantitative information on the spin-down torques produced by "magnetic braking" (e.g. see [121, 108]). It turns out that in most cases in CVs and LMXBs the mass-transfer rates driven by this effect are at least an order of magnitude higher than those by gravitational radiation alone. For more details, see for example [190, 157].

Orbital changes due to mass loss in X-Ray binaries

In X-ray binaries one often observes mass loss from the vicinity of the compact star in the form of double-sided relativistic jets. This is, for example, the case in SS433, Cyg X-3, Cyg X-1 and the "micro-quasar" GRS 1915+105. Here the mass is first transferred from the donor star to the vicinity of the compact star and from there it is ejected from the system, with the specific orbital angular momentum of the compact star. When such extensive mass loss from the system occurs it is usually the dominant term in the orbital angular momentum balance equation and its total effect is given by:

$$\frac{\dot{J}_{ml}}{J_{orb}} = \frac{\alpha + \beta q^2 + \delta\gamma(1+q)^2}{1+q} \frac{\dot{M}_2}{M_2} \qquad (22)$$

where α, β and δ are the fractions of mass lost from the donor in the form of a direct fast wind, the mass ejected from the vicinity of the accretor and from a circumbinary coplanar toroid (with radius, $a_r = \gamma^2 a$), respectively – see [169, 140]. The accretion efficiency of the accreting star is thus given by: $\epsilon = 1 - \alpha - \beta - \delta$, or equivalently:

$$\partial M_1 = -(1 - \alpha - \beta - \delta)\,\partial M_2 \qquad (23)$$

where $\partial M_2 < 0$. These factors will be functions of time as the binary system evolves during the mass-transfer phase.

The general solution for calculating the change in orbital separation during the X-ray phase is found by integration of the orbital angular momentum balance equation (Eq. (18)). It is often a good approximation to assume that other forms of angular momentum loss are negligible during this type of RLO, and if α, β and δ are constant in time one obtains:

$$\frac{a}{a_0} = \Gamma_{ls} \left(\frac{q}{q_0}\right)^{2(\alpha+\gamma\delta-1)} \left(\frac{q+1}{q_0+1}\right)^{\frac{-\alpha-\beta+\delta}{1-\epsilon}} \left(\frac{\epsilon q+1}{\epsilon q_0+1}\right)^{3+2\frac{\alpha\epsilon^2+\beta+\gamma\delta(1-\epsilon)^2}{\epsilon(1-\epsilon)}}$$

(24)

where the subscript '0' denotes initial values and Γ_{ls} is a factor of order unity to account for possible tidal spin–orbit couplings other than the magnetic braking. We remind the reader that $q \equiv M_2/M_1 \equiv M_{donor}/M_{accretor}$.

3.6 Effects of Explosive Mass Loss on the Orbits

a) Effects of a symmetric explosion

We consider systems with initial circular orbits and we neglect the effects of the impact of the ejected shell on the companion star, which is in general small (e.g. see [169]). We assume the explosion to be instantaneous (infinitely short duration). In this case the orbital changes can be simply expressed in terms of the ratio $m_f = \frac{M_{1f}+M_{2f}}{M_{1,0}+M_{2,0}}$ of the total mass of the system after and before the explosive mass loss [11]. We express the orbital semi-major axis a, and orbital period P in units of the initial orbital radius a_0 and initial orbital period P_0, respectively, and 0 and f indicate quantities before and after the explosion. One then obtains the following simple expressions for the orbital parameters after the explosion [11, 44]:

$$a_f = \frac{m_f}{2\,m_f - 1}$$

(25)

$$e_f = \frac{a_f - 1}{a_f} = \frac{1 - m_f}{m_f}$$

(26)

$$p_f = \frac{m_f}{(2\,m_f - 1)^{3/2}}$$

(27)

One thus observes that the system is disrupted (hyperbolic orbit) if $m_f < 0.5$. This is a consequence of the virial theorem. The runaway velocity of the center of gravity of bound systems after star 2 (with orbital velocity V_2) ejected an amount ΔM is given by

$$V_g = \frac{\Delta M\,V_2}{M_{1f} + M_{2f}}$$

(28)

Hence, the largest velocity that a system can attain after a symmetric supernova explosion is V_2.

b) Effects of asymmetric supernova explosions

As mentioned in the introduction, there is firm evidence that many newborn neutron stars receive a momentum kick at birth which gives rise to the high velocities observed (typically ~ 400 km s^{-1}), although there also appears to be a fraction ~ 10–20% that receive kicks ≤ 50 km s^{-1} in order to explain the observed neutron stars in globular clusters and the population of very wide-orbit X-ray binaries [105]. We will come back to this important point in Sect. 4, where we will argue that the low kick velocities occur preferentially for neutron stars in binaries of intermediate mass. It is still an open question whether or not black holes also receive a kick at birth. At least in some cases there are observational indications of mass ejection during their formation, as in any successful supernova explosion [61], and the recent determination of the run-away velocity (112 ± 18 km s^{-1}) of the black hole binary GRO J1655–40 [94] seems to suggest that also the formation of black holes is accompanied by a kick.

In an excellent paper Hills [59] calculated the dynamical consequences of binaries surviving an asymmetric supernova (SN). Tauris and Takens [152] generalized the problem and derived analytical formulas for the velocities of stellar components ejected from disrupted binaries and also included the effect of shell impact on the companion star. If the collapsing core of a helium star results in a supernova explosion it is a good approximation that the collapse is instantaneous compared with P_{orb}. Here we summarize a few important equations. The orbital energy of a binary is given by:

$$E_{\mathrm{orb}} = -\frac{GM_1 M_2}{2\,a} = -\frac{GM_1 M_2}{r} + \tfrac{1}{2}\mu v_{\mathrm{rel}}^2 \qquad (29)$$

where r is the separation between the stars at the moment of explosion; μ is the reduced mass of the system and $v_{\mathrm{rel}} = \sqrt{GM/r}$ is the relative velocity of the two stars in a circular pre-SN binary. The change of the semi-major axis as a result of the SN is then given by [44]:

$$\frac{a}{a_0} = \left[\frac{1 - (\Delta M/M)}{1 - 2(\Delta M/M) - (w/v_{\mathrm{rel}})^2 - 2\cos\theta\,(w/v_{\mathrm{rel}})} \right] \qquad (30)$$

where $a_0 = r$ and a are the initial and final semi-major axis, respectively; ΔM is the amount of matter lost in the SN; w is the magnitude of the kick velocity and θ is the direction of the kick relative to the orientation of the pre-SN velocity. The orientation of the kick magnitude is probably completely uncorrelated with respect to the orientation of the binary – the escaping neutrinos from deep inside the collapsing core are not aware that they are members of a binary system (see, however, [105] and Sect. 4 for the hypothesis that the kick *magnitude* may depend on the pre-SN history of the collapsing core). For each binary and a sufficiently high value of w there exists a critical angle, θ_{crit} for which a SN with $\theta < \theta_{\mathrm{crit}}$ will result in disruption of the orbit (i.e. if the denominator of Eq. (30) is less than zero).

The sudden mass loss in the SN affects the bound orbit with an eccentricity:

$$e = \sqrt{1 + \frac{2E_{\text{orb}}L_{\text{orb}}^2}{\mu G^2 M_1^2 M_2^2}} \tag{31}$$

where the orbital angular momentum can be derived from (see also Eq. (17)):

$$L_{\text{orb}} = |\boldsymbol{r} \times \boldsymbol{p}| = r\,\mu\sqrt{(v_{\text{rel}} + w\cos\theta)^2 + (w\sin\theta\,\sin\phi)^2} \tag{32}$$

Note, in the two equations above, v_{rel} is the pre-SN relative velocity of the two stars, whereas the stellar masses and μ now refer to the post-SN values.

Systems surviving the SN will receive a recoil velocity from the combined effect of instant mass loss and a kick. One can easily find this velocity, v_{sys} from conservation of momentum. Let us consider a star with mass M_{core} collapsing to form a neutron star with mass M_{NS} and hence:

$$v_{\text{sys}} = \sqrt{\Delta P_x^2 + \Delta P_y^2 + \Delta P_z^2}/(M_{\text{NS}} + M_2) \tag{33}$$

where the change in momentum is:

$$\begin{aligned}
\Delta P_x &= M_{\text{NS}}(v_{\text{core}} + w\cos\theta) - M_{\text{core}}v_{\text{core}} \\
\Delta P_y &= M_{\text{NS}}w\sin\theta\cos\phi \\
\Delta P_z &= M_{\text{NS}}w\sin\theta\sin\phi
\end{aligned} \tag{34}$$

and where M_2 is the unchanged mass of the companion star; v_{core} is the pre-SN velocity of the collapsing core, in a center of mass reference frame, and ϕ is the angle between the projection of $\boldsymbol{w}$ onto a plane $\perp$ to v_{core} (i.e. $w_y = w\sin\theta\cos\phi$). Beware, if the post-SN periastron distance, $a(1 - e)$ is smaller than the radius of the companion star then the binary will merge.

It is important to realize the difference between *gravitational* mass (as measured by an observer) and *baryonic* mass of a neutron star. The latter is $\sim 15\,\%$ larger for a typical equation-of-state. When considering dynamical effects on binaries surviving a SN this fact is often (almost always) ignored!

3.7 The Three Main Types of Close Binary Evolution and the Formation Of X-Ray Binaries

When a star is born as a member of a binary system, with a radius smaller than that of its Roche lobe it may, due to the evolutionary expansion of its envelope, after some time begin to overflow its Roche lobe. Since the lifetime of stars decreases with increasing stellar mass, the more massive star in a binary will be the first to do this. When this happens, the matter flowing out along the first Lagrangian point L_1 will fall towards the companion. The further evolution of the system will now depend on: (i) the evolutionary state of the star at the onset of the overflow, which is determined by the mass of the primary star, the orbital separation a and the mass ratio q of the system; and (ii) whether the envelope of the star at the onset of the mass transfer is in radiative or convective equilibrium. The factor (i) determines whether the binary is in one of the three cases A, B or C of binary evolution, defined in Sect. 3.4 and Fig. 10. Case A occurs only for systems with very short orbital periods (less than a few days), while cases B and C cover a very wide range of orbital periods, case C even up to periods of 10 years or more. The precise orbital period ranges for the cases A, B and C depend on the initial primary mass $M_{1,0}$ and the mass ratio q.

Response of the mass-losing star

Contact star with a radiative envelope

When the contact star loses an amount ΔM to its companion it will restore its hydrostatic equilibrium on a dynamical (sound-crossing) timescale, that is, almost instantaneously (within a few hours). In stars with radiative envelopes, the new hydrostatic equilibrium radius is smaller than its radius before the onset of the mass transfer, but as the star is now out of thermal equilibrium, its radius will begin to expand on a thermal timescale to restore this equilibrium. If the mass M_1 of the mass-losing star is larger than that of its companion (M_2), the transfer will make the orbit, and with it, the Roche-lobe radius R_L of the mass-losing star, to shrink (see Eq. (19) and Eq. (20)). If the radius of the star after readjustment to hydrostatic equilibrium is still larger than that of its Roche lobe, the mass loss will continue until so much mass has been lost that it fits within its new Roche lobe. The subsequent expansion of the star on a thermal timescale will, however, cause the mass transfer to resume, causing the Roche lobe to shrink further, and so on. As a result, the star will continue to transfer mass until itself has become the less massive component of the system. Its further expansion and mass loss will now cause, according to Eq. (19), the orbit to expand, such that finally the thermal equilibrium radius of the star may be able to fit within its Roche lobe. In systems evolving according to case B, practically only the helium core of the mass-losing star remains after the mass transfer, surrounded by a very thin hydrogen-rich outer envelope of negligible mass. The star may now, during its further nuclear evolution, either shrink or slowly expand on a nuclear timescale. The latter is the case in case A systems and in case B systems with original primary star masses $\leq 2.3\,M_{\odot}$. After the mass transfer, these systems are still in the phase of hydrogen burning: in case A systems, hydrogen still burns in the core, in the low-mass case B systems it burns in a shell surrounding a degenerate helium core of low mass ($\leq 0.45\,M_{\odot}$). The radii of these stars gradually increase when hydrogen burning advances. Therefore they continue to slowly transfer mass on a nuclear timescale. The systems in this slow mass-transfer phase are Algol-type close binaries. In the more massive case B systems the helium core is more massive than $0.45\,M_{\odot}$ and contracts further and ignites helium burning, upon which the star expels the last part of its hydrogen-rich envelope and settles on the helium-burning main sequence of helium stars. As depicted in Fig. 11 these stars have a very small radius and therefore are deep inside their Roche lobes. Fig. 14 schematically depicts the subsequent evolutionary phases of a close binary in which the more massive star at the onset of the mass transfer has a radiative envelope.

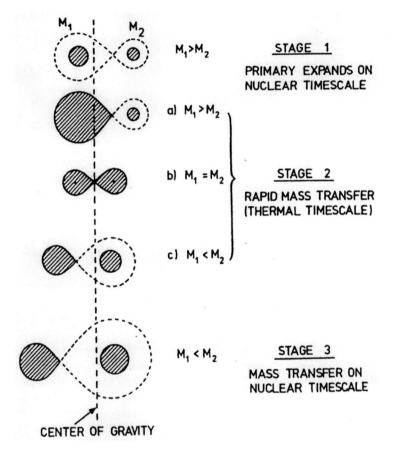

Fig. 14. Subsequent evolutionary stages and orbital dimensions of a close binary system in which the primary star has a radiative envelope at the time its radius begins to fill the Roche lobe

The order of magnitude timescale of the first phase of mass transfer (stage 2 in the figure) is simply the thermal timescale of the (initial) primary star. This leads to an order of magnitude mass-transfer timescale of (e.g. see [169]):

$$\dot{M_1} = \frac{-M_1}{\tau_{th}(M_1)} = -3 \times 10^{-8} \left(\frac{M_1}{M_\odot}\right)^3 M_\odot \, \text{yr}^{-1} \qquad (35)$$

After the reversal of the mass ratio and stabilization of the system, the mass transfer in stage 3 takes place on a nuclear timescale, leading to a mass-transfer rate of the order of (see Sect. 3.1 and [169]):

$$\dot{M}_1 = 10^{-10} \left(\frac{M_1}{M_\odot} \right)^{3.5} M_\odot \, \text{yr}^{-1} \tag{36}$$

Response of the accreting companion

Being the less massive component, it has a longer thermal timescale than its companion. Therefore the rapid transfer of mass to it causes its thermal equilibrium to be strongly disturbed, especially since the accreting material carries considerable kinetic energy which is dissipated into heat when the matter accretes on the surface. The matter also carries a fraction of the orbital angular momentum with it, which will strongly spin-up the the rotation of the accreting star. The details of the accretion process are still poorly known, but it is likely that unless the mass ratio of the system is large (≥ 0.4) the star will swell up to overflow its own Roche lobe. This occurs for radiative as well as convective envelopes of the mass-losing star. A common envelope then forms around the binary. In the next section we consider the evolution which then ensues.

Contact star with a convective envelope: common-envelope evolution

A star with a deep convective envelope has the characteristic that upon losing mass its radius expands on a dynamical timescale (see [8]), which is only of the order of hours to days. If this star is the more massive component of a close binary, its Roche lobe will shrink as a result of the mass transfer. Therefore the result is a violently unstable phase of mass transfer in which theoretically mass-transfer rates of the order of 0.01 to 0.1 solar masses per year could result. However, the companion will never in such a short time be able to accommodate these large amounts of mass dumped onto it. Therefore, as first suggested by [104, 101] in this case the expanding envelope of the primary star will completely engulf the secondary star, leading to the formation of a common envelope (abbreviated here as CE), inside which the secondary star and the core of the primary orbit each other. Due to the large frictional drag on its orbital motion inside the CE, the secondary star will rapidly spiral inwards. In this process a large amount of orbital potential energy is converted into heat by friction, causing the envelope to be expelled. As a result one expects, after the spiral in, a very close binary to remain, consisting of the secondary star and the core of the primary star. Paczynski and Ostriker proposed this type of evolution to explain the existence of the CV binaries. These have extremely short orbital periods, in general only of the order of a few hours, but they often contain massive white dwarfs, which can only be produced when stars are on the Asymptotic Giant Branch, that is: have radii of hundreds of solar radii. The original orbits of these systems, before the onset of the mass transfer, must therefore have been of the order of hundreds of days up to several years. An alternative outcome of the CE evolution, in the case where there is not enough orbital energy available to expel the envelope out of the deep gravitational potential well in which the system resides, is that the two stars will completely merge, resulting into the formation of one very rapidly rotating single star. Analytical considerations [92] as well as numerical hydrodynamical calculations of CE evolution have shown that the CE process proceeds very rapidly, on a timescale of order 10^2 to 10^3 years (e.g. [146, 149]). CE evolution is important not only for the understanding of the formation of the CV-binaries, but also of the many other types of very close binaries, ranging from the LMXBs and close double neutron stars, such as the Hulse–Taylor binary pulsar PSR 1913+16 and the close white-dwarf-pulsar system PSR 0655+64 (see Fig. 4), to the very close double white dwarfs, such as the AM CVn binaries. Apart from mass transfer from a convective envelope, formation of a CE may also ensue if the system is tidally unstable [28, 24]. And, more generally, if the mass ratio of the secondary and primary star is very low, also without invoking Common Envelopes the onset of the mass transfer will give rise to a very large reduction in orbital size of the system, resulting in a very close binary system, as was shown by [176], when they calculated the further evolution of the classical High-Mass X-ray Binary Cen X-3. They found that the outcome is a very close binary consisting of a helium star (the helium core of the massive star) and a compact object, in an orbit with a period of only a few hours. Their suggestion that 4.8 hour X-ray binary Cygnus X-3 might have resulted from such spiral in was confirmed by the discovery that the donor star in this system is a Wolf–Rayet star (helium star, [185]). The further evolution of such a system is expected to produce a double neutron star with an eccentric orbit, like the Hulse–Taylor binary pulsar [44]. Thus, Cygnus X-3 appears to be an ideal progenitor for a double neutron star (unless its compact object is a black hole, which at this moment cannot be ruled out, [40]).

A simple estimation of the reduction of the orbital separation can be found by simply equating the binding energy of the envelope of the (sub)giant donor to the required difference in orbital energy (before and after the CE-phase). Following the formalism of [195, 30], let $0 < \eta_{CE} < 1$ describe the efficiency of ejecting the envelope, i.e. of converting orbital energy into the kinetic energy that provides the outward motion of the envelope: $E_{env} \equiv \eta_{CE} \, \Delta E_{orb}$ or,

$$\frac{GM_{donor}M_{env}}{\lambda \, a_i \, r_L} \equiv \eta_{CE} \left[\frac{GM_{core}M_1}{2 \, a_f} - \frac{GM_{donor}M_1}{2 \, a_i} \right] \qquad (37)$$

yielding the ratio of final (post-CE) to initial (pre-CE) orbital separation:

$$\frac{a_f}{a_i} = \frac{M_{core} \, M_1}{M_{donor}} \frac{1}{M_1 + 2M_{env}/(\eta_{CE} \, \lambda \, r_L)} \qquad (38)$$

where $M_{core} = M_{donor} - M_{env}$; $r_L = R_L/a_i$ is the dimensionless Roche-lobe radius of the donor star so that $a_i \, r_L = R_L \approx R_{donor}$ and λ is a parameter which depends on the stellar mass-density distribution, and consequently also on the evolutionary stage of the star. The orbital separation of the surviving binaries is quite often reduced by a factor of ~ 100 as a result of the spiral-in. If there is not enough orbital energy available to eject the envelope the stellar components will merge in this process.

The binding energy of the envelope

The total binding energy of the envelope to the core is given by:

$$E_{\text{bind}} = -\int_{M_{\text{core}}}^{M_{\text{donor}}} \frac{GM(r)}{r} dm + \alpha_{\text{th}} \int_{M_{\text{core}}}^{M_{\text{donor}}} U dm \qquad (39)$$

where the first term is the gravitational binding energy and U is the internal thermodynamic energy per unit mass. The latter involves the basic thermal energy for a simple perfect gas, the energy of radiation, as well as terms due to ionization of atoms and dissociation of molecules and the Fermi energy of a degenerate electron gas [54, 55]. The value of α_{th} depends on the details of the ejection process, which is very uncertain. A value of α_{th} equal to 0 or 1 corresponds to maximum and minimum envelope binding energy, respectively. By simply equating Eqs. (37) and (39) one is able to calculate the parameter λ for different evolutionary stages of a given star.

Dewi and Tauris [33] and Tauris and Dewi [155] were the first to publish detailed calculations on the binding energy of the envelope to determine the λ-parameter (however, see also the pioneering investigations of [7]). Dewi and Tauris [33] investigated stars with masses 3–$10\, M_\odot$ and found that while $\lambda < 1$ on the RGB, $\lambda \gg 1$ on the AGB (especially for stars with $M < 6\, M_\odot$). Hence, the envelopes of these donor stars on the AGB are easily ejected; with only a relatively modest decrease in orbital separation resulting from the spiral-in. For more massive stars ($M > 10\, M_\odot$) $\lambda < 0.1 - 0.01$ (see Fig. 15) and the internal energy is not very dominant [33, 109]. This result has the important consequence that many HMXBs will not produce double neutron star systems because they coalesce during their subsequent CE-phase (leading to a relatively small merging rate of double neutron star systems, as shown by [192]).

It should be noted that the exact determination of λ depends on how the core boundary is defined (see [155] for a discussion). For example, if the core boundary (bifurcation point of envelope ejection in a CE) of the $20\, M_\odot$ in Fig. 15 is moved out by $0.1\, M_\odot$ then λ is typically increased by a factor of ~ 2.

3.8 Conservative Evolution and the Formation of HMXBs

Formation of the "standard" HMXBs

The formation of the HMXBs can be understood in terms of the "conservative" evolution of normal massive close binary systems [175, 164] in the ways

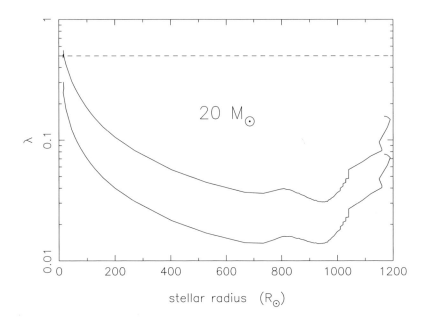

Fig. 15. The λ-parameter for a 20 $M_\odot$ star as a function of stellar radius. The upper curve includes internal thermodynamic energy ($\alpha_{\mathrm{th}} = 1$) whereas the lower curve is based on the sole gravitational binding energy ($\alpha_{\mathrm{th}} = 0$) – see Eq. (39). There is a factor $\sim$2 in difference between the λ-curves in accordance with the virial theorem. It is a common misconception to use a constant value of $\lambda = 0.5$ (marked by the dashed line). See text for details.

described in Sect. 3.7. We consider as an example the evolution of a massive case B system as depicted in Fig. 16, which started out with components of 20 and 8 $M_\odot$ and an initial orbital period of 4.70 days. The subsequent evolutionary phases, as calculated by [32] are described in the figure caption and are as follows. After 6.17×10^6 yr the primary star has terminated core-hydrogen burning and overflows its Roche lobe. In only 3×10^4 yr it transfers 14.66 $M_\odot$ to its companion and leaves behind a 5.34 $M_\odot$ helium core which becomes a helium-burning pure helium star. The orbital period is now 10.86 days and the companion has become a 22.66 $M_\odot$ star which is still practically unevolved as it has been rejuvenated by the 14.66 $M_\odot$ of unprocessed hydrogen-rich matter from its companion's envelope. As first noticed by [102] binary systems consisting of a helium-burning helium star and a massive main-sequence star exactly resemble the Wolf–Rayet binaries, of which hundreds are known in our Galaxy and the Magellanic Clouds. Wolf–Rayet stars tend to be hydrogen-deficient stars of relatively low mass for their very large luminosities: in the Wolf–Rayet binaries they are practically always some two to four times less massive than their normal O- or early B-type companions (e.g. see [183, 169]). They have a remarkable emission line spectrum indicating large mass loss by stellar winds at a rate typically of order $10^{-5} M_\odot \, \mathrm{yr}^{-1}$, and velocities of several thousands of km s^{-1}. Some 60% of Wolf–Rayet stars are found in binaries of this type. As depicted in Fig. 11 the radii of helium stars more massive than about 3.5 $M_\odot$ do not expand much until the end of their evolution, when their burnt-out core collapses to a neutron star or a black hole. For the 5.34 $M_\odot$ helium star of Fig. 16 this occurs 0.69×10^6 yr after the mass transfer. It is assumed here that it leaves a 2 $M_\odot$ neutron star (similar to the 1.85 $M_\odot$ neutron star in the HMXB Vela X-1, see [4]). Due to the explosive mass ejection in the supernova (assuming instantaneous symmetric mass ejection) the orbital period increases to 12.63 days, a small orbital eccentricity is induced, and the system is accelerated to a runaway velocity of 35 km s^{-1}. At age 10.41 Myr (about 3.5 Myr after the supernova explosion) the 22.66 $M_\odot$ O-type companion star has terminated its core-hydrogen burning phase and has become a blue supergiant star with a strong stellar wind. The accretion of matter from this wind induces the compact object to become an X-ray source, and the system becomes a HMXB resembling Vela X-1. The HMXB phase, until the supergiant overflows its Roche lobe, probably lasts only between 10^4 and 10^5 yr. After overflowing the Roche lobe it is likely that a Common Envelope will form (although, since the envelope of the supergiant is here radiative, it is also conceivable that the system will undergo heavy loss of mass and orbital angular momentum of the type calculated by [176]; see also [69]). Assuming the system to survive this phase as a binary (which in the case of this narrow system is probably doubtful, see Sect. 3.8), the outcome will be a very close binary consisting of a helium star (Wolf–Rayet star) and a compact object: neutron star or black hole. As mentioned in Sect. 2.1 we know already three systems of this type, one in our Galaxy (Cyg X-3) and two in external galaxies (see also Sect. 7). If one increases the initial orbital period of the system with an arbitrary factor, also the orbital periods of the later evolutionary phases will increase by a similar factor. In this way one can also obtain the very wide O-supergiant X-ray binaries recently discovered by INTEGRAL, one of which has an orbital period of 330 days [193, 137]. This orbital period would require an initial orbital period of the system of about 100 days.

Formation of the B-emission X-ray binaries

The formation of the B-emission X-ray binaries, a class of somewhat less massive HMXBs, and much more abundant than the "standard" HMXBs, can be explained also in terms of "conservative" evolution of binaries that started with primary star masses typically in the range 8 to 15 $M_\odot$. In this case, due to the dependence of helium core mass on initial stellar mass (e.g. see [167]):

$$M_{\mathrm{He}} = 0.073 \left(\frac{M}{M_\odot} \right)^{1.42} \tag{40}$$

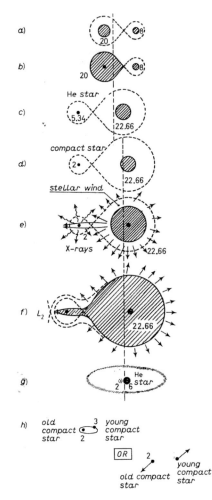

Fig. 16. Subsequent stages in the evolution of a massive close binary that started out with components of 20 and 8 $M_\odot$ in a circular orbit of 4.70 day period (first five evolutionary phases calculated by [32]).It is assumed that the supernova explosion of the primary star leaves a 2 $M_\odot$ compact star (neutron star). The initial orbital period was chosen such that the resulting HMXB resembles one of the "standard" HMXBs such as Vela X-1, which has a 1.85 $M_\odot$ neutron star [4]. (a) $t = 0$, $P = 4.70$ d, birth of the system; (b) $t = 6.17 \times 10^6$ yr, $P = 4.70$ d, onset of mass transfer; (c) $t = 6.20 \times 10^6$ yr, $P = 10.86$ d, end of mass transfer, primary star has become a 5.34 $M_\odot$ Wolf–Rayet (helium) star; (d) $t = 6.89 \times 10^6$ yr, $P = 12.63$ d, He-star (WR-star) has exploded as a supernova; (e) $t = 10.41 \times 10^6$ yr, the 22.66 $M_\odot$ companion has left the main sequence and became a blue supergiant with a very strong stellar wind, accretion of wind matter turned the compact star into a strong X-ray source. (f) $t = 10.45\,10^6$ yr, $P = 12.63$ d, onset of second stage of mass exchange, the X-ray source is extinguished and rapid orbital shrinking and large mass loss from the system begins. If the system survives this phase as a (very close) binary, this system will consist of a helium star and the compact star: (g) $t = 10.45 \times 10^6$ yr, $P = $ few hours, onset of second Wolf–Rayet phase; (h) $t = 11 \times 10^6$ yr, second helium star has exploded as a supernova, survival or disruption of the system depends on the mass of the second compact remnant and the occurrence of kicks.[Increasing the initial orbital period of the system with an arbitrary factor will increase the orbital periods of the later evolutionary phases by a similar factor].

(for chemical composition $X = 0.70$, $Z = 0.03$) in combination with Eq. (20) for conservative orbital change,the mass exchange will produce systematically longer orbital periods of the post-mass-exchange binaries, even for systems that started out with very short initial orbital periods. This explains the absence of systems with orbital periods shorter than about 15 days among the Be/X-ray binaries [167, 179, 182]. An important finding by [105] is that among the Be/X-ray binaries with well-determined orbits, there are two distinct groups: one group, which contains about half of all well-studied systems, having orbits of low eccentricity ≤ 0.25, and another group, with very eccentric orbits, up to 0.9. Also the three known radio pulsars with an early B-type companion have very eccentric orbits. As will be shown below, if there are no velocity kicks imparted to the neutron stars formed in the Be/X-ray binaries, eccentricities larger than 0.2 cannot be produced in these systems. Therefore, the high orbital eccentricities observed in the second group require that a velocity kick of several hundreds of km s^{-1} is imparted to their neutron stars in their birth events [181], whereas the systems in the first-mentioned group apparently hardly received any velocity kick in their birth events, in any case less than 50 km s^{-1} ([105]; see Sect. 4). Fig. 17 depicts the evolutionary model for the formation of a typical B-emission X-ray binary, as calculated by [52, 53]. The system started out as a normal early B-type spectroscopic binary with a primary star of mass $13\,M_\odot$, a mass ratio 0.5 and an initial orbital period of 2.58 days. After 12 million years the primary star has left the main sequence and begins to overflow its Roche lobe. 2×10^4 years later it has transferred its entire H-rich envelope to its companion, which increased its mass to $17\,M_\odot$, while only the remaining helium core of the primary star has a mass of $2.5\,M_\odot$. The orbital period after the conservative mass transfer is 20.29 days. Due to the orbital motion, the transferred mass had a lot of angular momentum, and flowed inwards to the secondary through an accretion disk. This will have increased the rotation rate of the accreting star to the break-up limit, causing this star to have become an extremely rapid rotator (systems that are observed in the phase of mass transfer, such as β Lyrae, show indeed the accreting star to be surrounded by a thick accretion disk, and to be an extremely rapid rotator, e.g. see references in [169]). Since during its later evolution the radius of a helium star of $2.5\,M_\odot$ expands very much, this star, at age 1.4×10^7 years overflows its Roche lobe again and a phase of so-called "case BB mass transfer" ensues, in which it transfers $0.3\,M_\odot$ towards its companion, which increases the orbital period to 25.1 days. 4000 years later the core of the helium star collapses to a neutron star and $0.8\,M_\odot$ are expected to be explosively ejected, which in the case of symmetric mass ejection, increases the orbital period to 30.63 days, induces an orbital eccentricity of 0.04 and a runaway velocity of the center of mass of the system of 7.3 km s^{-1}. The final system, consisting of a very rapidly rotating early B-type main-sequence star and a neutron star, will first closely resemble the binary systems consisting of a young radio pulsar and a B-emission star, such as PSR 1259-63 ([62]; although that system has a much larger orbital eccentricity and orbital period). Later in life, when the rotation rate of the neutron star has much slowed down, such that accretion will become possible, it will resemble the B-emission X-ray binaries with nearly circular orbits. Immediately after the supernova explosion, when the neutron star is still a very rapidly spinning energetic young pulsar, like the Crab pulsar, the system will resemble the peculiar eccentric orbit B-emission X-ray binary $0236+610$/LSI+61 303, which shows recurrent radio, X-ray and gamma-ray outbursts with a period equal to that of its orbital period of 26.45 days. This system also shows relativistic jets. We thus see that for the B-emission systems we observe in our Galaxy binaries in all evolutionary phases between the supernova explosion and the B-emission X-ray binary phase.

Formation of double neutron stars: descendants of B-emission X-ray binaries

Calculations of Common-Envelope evolution, using the equations given in Sect. 3.7, predict that due to their small orbital separations, the narrow-orbit "standard" HMXBs are unlikely to survive the second phase of mass transfer and spiral-in as the orbits are expected to shrink by a factor 100 or more [195, 149]; see, however, also the remarks made in Sect. 3.8: if these systems would, because of the presence of a radiative envelope in the donor star, not enter a CE phase, they perhaps might still survive spiral-in as a very narrow binary, in the sense of [176]). On the other hand, the wide B-emission X-ray binaries such as X Per (orbital period $P = 250$ days) are expected to survive spiral-in as systems depicted in the right-hand part of Fig. 18: short-period binaries consisting of a helium star and a compact star. When the helium star terminates its evolution with a supernova explosion, the system is either disrupted or a double neutron star with a very narrow and eccentric orbit is formed, closely resembling the 8 double neutron stars presently known in the galactic disk (see Table 3 and Fig. 4). Thus it is most likely that these binary pulsars are the descendants of the B-emission X-ray Binaries [177, 168, 169], because of the steep shape of the initial mass function, these lower-mass HMXBs are much more numerous than the "standard" massive ones. Some authors have expressed doubts as to whether or not neutron stars can survive the spiral-in process through the envelope of a massive companion star, without turning into a black hole due to highly super-Eddington-limited accretion resulting from heavy neutrino-cooling [21, 14]. However, there are various arguments indicating that indeed the neutron stars in the Be/X-ray binaries can survive this spiral-in. First of all, we have the close binary pulsars with circular orbits in which the companion of the pulsar is a massive white dwarf, such as PSR 0655+64 ($P = 1.04$ days). The progenitor of its about $1\,M_\odot$ white dwarf must have been a star of $\geq 5\,M_\odot$, and the neutron star clearly survived its spiral-in through the at least $4\,M_\odot$ envelope of this star. A further strong argument against the neutron star in Be/X-ray binaries turning into black holes during spiral-in is that in that case the formation rate of close binary pulsars consisting of a black hole and a young neutron star would be

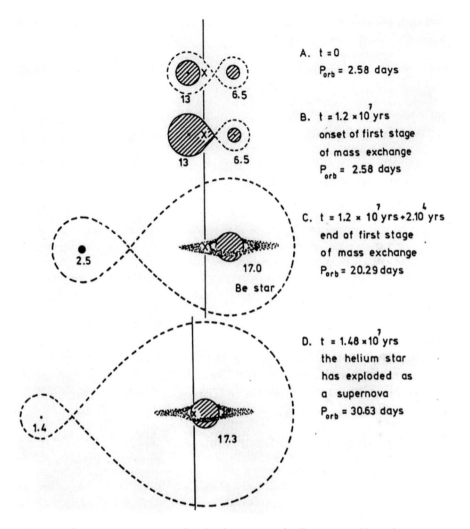

A. $t = 0$
    $P_{orb} = 2.58$ days

B. $t = 1.2 \times 10^7$ yrs
    onset of first stage
    of mass exchange
    $P_{orb} = 2.58$ days

C. $t = 1.2 \times 10^7$ yrs $+ 2.10^4$ yrs
    end of first stage
    of mass exchange
    $P_{orb} = 20.29$ days

D. $t = 1.48 \times 10^7$ yrs
    the helium star
    has exploded as
    a supernova
    $P_{orb} = 30.63$ days

Fig. 17. Conservative scenario for the formation of a B-emission X-ray binary out of a close pair of early B stars with masses of 13.0 and 6.5 $M_\odot$. The numbers in the figure indicate mass, in solar units. After the end of the mass transfer the Be star presumably has a circumstellar disk or shell associated with its rapid rotation (induced by the previous accretion of matter with high angular momentum; from [52, 53]).

of similar order as the formation rate of the Be/X-ray binaries (as a sizeable fraction of these systems will surely survive spiral-in as binaries). The estimate is that there are several thousands of Be/X-ray-like systems in our Galaxy [91, 168, 169]. These systems live shorter than 10^7 yrs, so their formation rate is of order 10^{-4} to 10^{-3} per year. This implies that the formation rate of close binaries consisting of a black hole and a young pulsar would be at least of order 10^{-4} per year, which is about 0.01 of the pulsar birth rate. Presently some 1600 normal "garden variety" pulsars are known, so we would expect some 16 of these to be radio pulsars orbiting black holes. However, no such system is known. This is a strong argument against the scenario in which the neutron stars of HMXBs would not survive the spiral-in as neutron stars. Fig. 18 shows the various ways in which – using the Common Envelope approach – the persistent ("standard") and B-emission X-ray binaries are expected to terminate their evolution. If the Be star is less massive than about $8\,M_\odot$ its burned-out core is not massive enough to leave a neutron star and becomes a CO white dwarf. In that case a very close binary radio pulsar with a circular orbit will remain, consisting of a (recycled) neutron star and a massive white dwarf. PSR $0655 + 64$ (orbital period $P = 1.04$ days) is thought to be such a system.

4 Double Neutron Stars and the Birth Kick Velocities of Neutron Stars: Evidence for Two Different Neutron-Star Formation Mechanisms: Iron Core Collapse Versus Electron-Capture in Degenerate O–Ne–Mg Cores

4.1 The Birth Kick Velocities of Neutron Stars

As mentioned above, [105] made the very important discovery that there exists a separate class of B-emission X-ray binaries with wide orbits of low eccentricity (≤ 0.25). The systems in this class tend to have low X-ray luminosities ($< 10^{34}$ erg s^{-1}).

A well-known example is X-Per, in which the neutron star has an almost circular orbit with a period of 250 days. About half of all Be/X-ray binaries with known orbits appear to belong to this class and the relatively low X-ray luminosities of these sources imply that these systems are on average considerably nearer to us than the high-eccentricity Be/X-ray binaries (which during outbursts can reach a luminosity of 10^{38} erg s^{-1}). Therefore, as [105] pointed out, the systems in the low-eccentricity class probably form the bulk of the Be/X-ray binary population, since the known numbers of sources in both classes are about the same. These authors pointed out that the neutron stars in the low-eccentricity systems cannot have received a kick velocity at their birth exceeding 50 km s^{-1}. Until the discovery of this class of X-ray binaries it was generally thought that all neutron stars receive a high kick velocity at their birth, of order at least a few hundred km s^{-1} (see e.g.: [84]; [57, 60]).

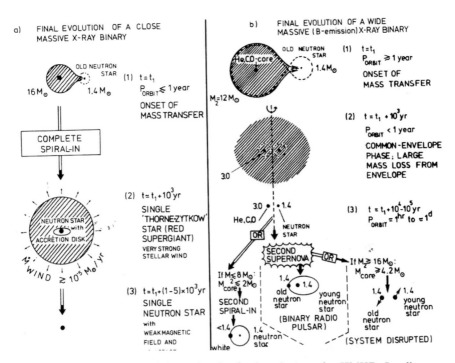

Fig. 18. The various possibilities for the final evolution of a HMXB. In all cases where the compact star is a neutron star, the onset of Roche-lobe overflow is expected to lead to the formation of a common envelope and the occurrence of spiral-in. (a) In systems with orbital periods less than about 100 days (to one year) there is probably not enough energy available in the orbit to eject the common envelope and the neutron star probably spirals down into the core of the companion, and after the loss of the envelope a single neutron star or black hole will stay behind. (b) In systems with orbital periods longer than 100 days to 1 year the common envelope is ejected during spiral in and a very close binary can be left, consisting of the neutron star and the core of the companion, composed by helium and/or heavier elements. Companions more massive than about $8\,M_\odot$ leave cores that explode as a supernova, leaving an eccentric-orbit binary pulsar or two runaway pulsars. Companions less massive than $8\,M_\odot$ leave close binaries with a circular orbit and a massive white-dwarf companion, like PSR 0655+64. [In systems where the companion at the onset of the mass transfer still has a radiative envelope, possibly the formation of a common envelope can be avoided; then the evolution may follow a somewhat different path, but also leads to the formation of very close systems in the end, see [176, 69].]

Often a Maxwellian distribution is used to represent the observed distribution of pulsar velocities, and the characteristic velocity of these Maxwellians is typically around 300–400 km s^{-1} [57].

A recent very detailed study by [60] of the accurately determined proper motions of 233 radio pulsars shows that there is no room for a separate population of low-velocity single pulsars. Particularly, these authors found that the velocity distribution of young pulsars (age < 3 million years) is very well represented by a single Maxwellian with a characteristic velocity of about 400 km s^{-1}, and there is no evidence for a bimodal velocity distribution as had been argued by [23].

On the other hand, [105] showed, by means of population synthesis calculations that include the evolution of binaries and the presence of birth kicks imparted to neutron stars, that with the assumption of only one Maxwellian with a high characteristic velocity (several hundred km s^{-1}) one can reproduce the high-eccentricity population of the Be/X-ray binaries, but one totally fails to reproduce the presence of a large population of systems with low eccentricities. They convincingly showed that the only way in which both the observed high-e and the low-e populations of the Be/X-ray binaries can be reproduced is: by assuming that there are two distinct populations of neutron stars: one population that receives hardly any kick velocity at birth ($v_k < 50$ km s^{-1}) and another which receives the "canonical" high velocity kick of order several hundreds of km s^{-1} at birth.

4.2 Double Neutron Stars and the Low Kick-Velocity Neutron Star Population

At present, 9 double neutron stars are known, 8 of them in the galactic disk and one in a globular cluster. The eight systems in the galactic disk are listed in Table 3. As the Table shows, the double neutron stars tend to have very narrow orbits. As pointed out in the last section they are the later evolutionary products of Be/X-ray systems with orbital periods > 100 days, which evolved through Common-Envelope evolution into a very close binary consisting of a helium star (the core of the Be star), together with the neutron star, with an orbital period of only a few hours. Due to the large frictional and tidal effects during spiral in the orbit of the system is expected to have become perfectly circular. The helium star generates its luminosity by helium burning, which produces C and O, and subsequently by carbon burning, producing Ne and Mg.

If the helium star has a mass in the range 1.6 to about 3.5 $M_\odot$ (corresponding to a main-sequence progenitor in the range 8 to 11 ($\pm$1) $M_\odot$, the precise limits of this mass range depending on metallicity and on the assumed model for convective energy transport [144, 93, 110]) it will during carbon burning develop a degenerate O–Ne–Mg core, surrounded by episodic C-and He-burning shells (e.g. [99, 53]). When such a degenerate core develops, the envelope of the helium star begins to expand, causing in a binary the onset of mass transfer by Roche-lobe overflow [53, 35]. Roche-lobe overflow leads to the formation of an accretion disk around the neutron star and accretion of matter with angular momentum from this disk will cause the spin frequency of the neutron star to increase. Therefore one expects that during the later evolution of these helium stars of relatively low mass the first-born neutron star in the system will be "spun up" to a short spin period. This neutron star had already a long history of accretion: first when it was in a wide binary with an early-type (presumably Be) companion; subsequently during the spiral-in phase into the envelope of its companion and now as companion of a Roche-lobe overflowing helium star. Since all binary pulsars which had a history of mass accretion (so-called "recycled" pulsars [119]) tend to have much weaker magnetic fields than normal single pulsars, it is thought that accretion in some way causes a weakening of the surface dipole magnetic field of neutron stars [147] and several theories have been put forward to explain this accretion-induced field decay [6] (see [9] for a review; [202, 27]). With a field weakened to about 10^{10} Gauss (as observed in the recycled components of the double neutron stars (see Table 3), and an Eddington-limited accretion rate of helium (rate 4×10^{-8} $M_\odot$ yr^{-1}) a neutron star can be spun-up to a shortest possible spin period of a few tens of milliseconds [139].

When the helium star finally explodes as a supernova, the second neutron star in the system is born. This is a newborn neutron star without a history of accretion and is therefore expected to resemble the "normal" strong-magnetic field single radio pulsars [142], which have typical surface dipole magnetic fields strengths of $10^{12} - 10^{13}$ Gauss. This was nicely confirmed by the discovery of the double pulsar system PSRJ0737-3039AB, which consists of a recycled pulsar (star A) with a very rapid spin ($P = 23$ ms) and a weak magnetic field (7×10^9 G), together with a normal strong-magnetic-field ($1, 2 \times 10^{12}$ G) pulsar (star B) with a "normal" pulse period of 2.8 sec [16, 85]. The explosive mass loss in the second supernova has made the orbit eccentric and since the two neutron stars are basically point masses, tidal effects in double neutron star systems will be negligible and there will be no tidal circularization of the orbit. (On timescales of tens of millions of years the orbits may be circularized by a few per cent due to the emission of gravitational waves in the shortest-period system of PSRJ0737-3039, but in all the other double neutron stars this is a negligible effect, except in the final stages of spiraling together, see e.g. [138]). In case of spherically symmetric mass ejection in the second supernova there is a simple relation between the orbital eccentricity and the amount of mass ΔM_{sn} ejected in the supernova (see Eq. (26)):

$$e = \Delta M_{sn}/(M_{ns1} + M_{ns2}) \qquad (41)$$

where M_{ns1} and M_{ns2} are the masses of the first- and the second-born neutron stars. The "conventional" kick velocities of neutron stars of about 400 km s^{-1} [60] are quite similar to the orbital velocities of the neutron stars in close double neutron stars such as the Hulse–Taylor binary pulsar PSRB1913+16 ($P_{orb} = 7.75$ hours). Therefore, a kick velocity of this order produces a major disturbance of the orbit and – unless it is imparted in a very specific direction – will in general impart a large eccentricity to the orbit, of order 0.5 or more. The Hulse–Taylor binary pulsar has a large eccentricity $e = 0.617$ and the same is true for the system PSRJ1811-1736 ($e = 0.828$), which indeed might be due to such large kick velocities. However, as Table 3 shows, very surprisingly all of the other 6 double neutron stars in the galactic disk have very small orbital eccentricities, in the range 0.088 to 0.27. Such eccentricities are the ones which one expects from the pure sudden mass loss effects in the supernova explosion, given by Eq. (41), but not in case a randomly directed kick velocity of order 400 km s^{-1} is imparted to the second-born neutron star at birth. (In particular, the small orbital eccentricities of the two relatively wide double neutron stars PSRJ1518+4909 and PSRJ1829+2456 are impossible to reconcile with high kick velocities). Furthermore, [36, 172] have pointed out that the relation between spin period of the recycled neutron star and orbital eccentricity observed in double neutron star systems [42] can only be understood if the second-born neutron stars in these systems received a negligible velocity kick in their birth events. (Interestingly, the Hulse–Taylor binary pulsar PSRB1913+16 and PSRJ1811-1736 also fit this relation, which also suggests that their high orbital eccentricities were purely due to the effects of the sudden mass loss in the second supernova. And indeed, since their first-born neutron stars are quite strongly recycled, they must have had a quite extended episode of disk accretion. This implies an extended episode of stable Roche-lobe overflow from the helium star progenitor of the second-born neutron star. And this in turn suggests that these helium stars had a degenerate O–Ne–Mg core, as only the development of such cores causes the envelopes of helium stars to expand).

It thus appears that the second-born neutron stars in these 6 low-eccentricity systems belong to the same "kick-less" class as the neutron stars in the low-eccentricity class of Be/X-ray binaries [171, 172, 173]. The same holds for the young strong-magnetic-field pulsar in the eccentric radio-pulsar binary PSRJ1145-6545 which has a massive white dwarf as a companion [68, 3, 2]).

The orbital eccentricity of 0.172 of this binary shows that the neutron star was the last-born object in the system ([113, 156]; formation of a white dwarf cannot introduce an orbital eccentricity). The low value of its eccentricity would be hard to understand if the neutron star received the canonical 400 km s^{-1} kick at its birth.

4.3 The Low Masses of the Low-Kick Second-Born Neutron Stars in the Double Neutron Star Systems and in PSRJ1145-6545

In the eccentric white-dwarf/neutron-star system of PSRJ1145-6545 the mass of the neutron star is known from the measurement of relativistic effects to be $1.28(2)\,M_\odot$ [2] (the number within parentheses indicates the 95% confidence uncertainty of the last digit; the total mass of the system is $2.30\,M_\odot$ and the mass of the white dwarf is at least one solar mass). Also in two of the low-eccentricity double neutron stars the masses of both stars are accurately known from measured relativistic effects (see [143]):

(i) in PSRJ0737-3039 the second-born neutron star has $M_B = 1.250(3)\,M_\odot$ and the first-born one has $M_A = 1.330(3)\,M_\odot$ [85].

(ii) in PSRJ1756-2251 the second-born neutron star has a mass of 1.18(3) $M_\odot$ and the first-born one a mass of 1.40(3) $M_\odot$ [42].

In most of the other double neutron stars the masses of the stars are not yet accurately known, but in two of the other low-eccentricity systems the second-born neutron stars must be less massive than $1.30\,M_\odot$ for the following reasons. In all double neutron star systems the relativistic parameter that can be measured most easily is the General Relativistic rate of periastron advance, which directly yields the sum of the masses of the two neutron stars (e.g. see [143]). In the low-eccentricity systems of PSRJ1518+4904, PSRJ1829+2456 and PSRJ1906+0746 the resulting sum of the masses turns out to be 2.62, 2.53 and 2.61 $M_\odot$, respectively. The individual masses of the neutron stars in these systems are still rather poorly determined, but in the first two of these three systems the already crudely determined other relativistic parameters indicate that the second-born neutron star has the lowest mass of the two (see references in [171]). As in these systems the sum of the masses is around 2.60 $M_\odot$, the second-born neutron stars in these two systems cannot be more massive than $1.30\,M_\odot$.

Thus we find that in at least four of these six systems the second-born neutron star has a low mass, in the range 1.18 to $1.30\,M_\odot$ and belongs to the low-kick category. And the same holds for the second-born neutron star in the low-eccentricity white-dwarf-neutron-star binary PSRJ 1145-6545, which has a mass of only $1.28\,M_\odot$. Also in the system of PSRJ1909+0746 the masses of the neutron stars cannot differ much from 1.30 $M_\odot$. We thus see that in at least five cases a low (or no) kick velocity is correlated with a low neutron star mass of on average around $1.25(\pm0.06)\,M_\odot$.

A neutron star of $1.25\,M_\odot$ corresponds to a pre-collapse mass of about 1.44 $M_\odot$, as during the collapse the gravitational binding energy of the neutron star of about 0.20 $M_\odot$ (slightly depending on the assumed equation of state of neutronized matter) is lost in the form of neutrinos. So apparently the cores, which collapsed to these second-born neutron stars, had a mass very close to the Chandrasekhar mass.

4.4 Formation Mechanisms of Neutron Stars and Possible Resulting Kicks

It is long known [93, 144] that there are two basically different ways in which neutron stars are expected to form, i.e.:

(i) In stars which originated in the main-sequence mass range between 8 and about 11 (±1) $M_\odot$, which in binaries produce helium stars in the mass range 1.6 to about 3.5 $M_\odot$ [53, 35], the O–Ne–Mg core which forms during carbon burning becomes degenerate and when its mass approaches the Chandrasekhar mass, electron captures on Mg and Ne cause the core to collapse to a neutron star. Since these stars did not reach Oxygen and Silicon burning, the baryonic mass of the neutron star, which forms in this way, is expected to be purely determined by the mass of the collapsing degenerate core, which is the Chandrasekhar mass. The gravitational mass of this neutron star is then the Chandrasekhar mass minus the gravitational binding energy of the neutron star, which is about $0.20 \, M_\odot$. Thus a neutron star with a mass of about $1.24 \, M_\odot$ is expected to result.

(ii) In stars initially more massive than $11 \, (\pm 1) M_\odot$, the O–Ne–Mg core does not become degenerate and these cores proceed through Oxygen and Silicon burning to form an iron core. When the mass of this iron core exceeds a critical value it collapses to form a neutron star. The precise way in which here neutrino transport during core bounce and shock formation results in a supernova explosion is not yet fully understood. It appears that first the shock stalls and then several hundreds of milliseconds later, is revitalized. Some fall back of matter from the layers surrounding the proto neutron star is expected to occur (see [46]) such that the neutron star that forms may be substantially more massive than the mass of the collapsing Fe-core. In fact there are two expected mass regimes for the resulting neutron stars: for stars with initial main-sequence masses in the range $11(\pm 1) \, M_\odot$ to $19 \, M_\odot$ the collapsing cores are expected to be about $1.3 \, M_\odot$, whereas for stars more massive than $19 \, M_\odot$ the collapsing iron core is expected to have a mass $> 1.7 \, M_\odot$ [162], leading to the formation of neutron stars with (gravitational) masses $> 1.6 \, M_\odot$. Taking some fall-back of matter into account, the neutron stars formed from these types of iron cores may be expected to have gravitational masses $> 1.3 \, M_\odot$ and $> 1.7 \, M_\odot$, respectively.

The fact that the pre-collapse masses of the low-mass, low-kick neutron stars were very close to the Chandrasekhar limit suggests that these neutron stars are the result of the electron-capture collapse of the degenerate O–Ne–Mg cores of helium stars that originated in the mass range 1.6 to $3.5 \, M_\odot$ (initial main-sequence mass in the range 8 to $11(\pm 1) \, M_\odot$). Can one understand why such neutron stars would not receive a birth kick whereas those formed by the collapse of an iron core would? While in the past neutron-star kicks in general were ascribed to asymmetric neutrino emission (e.g. [17]), in recent years the ideas have shifted towards hydrodynamic instabilities during the explosion. For example, [133, 134] found large-scale hydrodynamic instabilities to develop in the layers surrounding the proto neutron star during the explosion of a $15 \, M_\odot$ star with a collapsing iron core, which imparted velocities up to 1000 km s^{-1} to the neutron star. On the other hand, for collapsing O–Ne–Mg cores, [73] did not find large neutron-star velocities. This is ascribed to the facts that (a) here the ejecta in the immediate vicinity of the proto neutron star is very small, and (b) the explosion of the O–Ne–Mg core by neutrino heating occurs very fast (much faster than for iron cores, where the development of the explosion takes hundreds of milliseconds), not allowing hydrodynamic instabilities to develop. It thus appears that a difference in purely hydrodynamic effects during these very different types of explosions may explain the differences in the kick velocities of the resulting neutron stars.

4.5 Why There Are No Low-Velocity Single Pulsars: O–Ne–Mg Core Collapse is Restricted to Binary Systems!

As mentioned in Sect. 3.1, [110] recently argued that *single* stars in the mass range 8 to 11 ($\pm$1) $M_\odot$ do not produce neutron stars, for the reasons explained in Fig. 8. These stars produce helium cores in the mass range 1.6 to about 3.5 $M_\odot$, but when they ascend the Asymptotic Giant Branch (AGB), their convective envelopes during the "dredge-up" phase penetrate the helium layers surrounding their degenerate O–Ne–Mg cores, and erode these helium layers away. Therefore the degenerate cores of these stars can no longer grow by helium shell burning. These stars lose their envelopes due to the heavy wind mass loss during the AGB phase, and are expected to leave behind their degenerate O–Ne–Mg cores as white dwarfs. Only single stars more massive than about $11(\pm1)\,M_\odot$ will leave neutron stars, formed in this case by iron core collapse. As argued above, these neutron stars will be of the high-kick class, so all single neutron stars are expected to be high-velocity objects, as is indeed observed [60]. On the other hand, as argued by [110], an 8 to about $11(\pm1)\,M_\odot$ star in an interacting binary system cannot reach the AGB, as it will have already lost its hydrogen envelope to its companion (or out of the system) by Roche-lobe overflow. Therefore, in binaries such stars will leave helium stars with masses in the range 1.6 to 3.5 $M_\odot$, which will evolve to e-capture core collapse, ending with a formation of a low-velocity neutron star. One therefore expects these low-velocity neutron stars to *only* be born in binary systems. This provides a very elegant way to explain why only in binaries one finds evidence for "kick-less" neutron stars.

4.6 Summarizing the Evidence for Two Types of Core Collapse Derived From B-Emission X-Ray Binaries and Double Neutron Stars

The combination of observations indicating that: *(i) among the Be/X-ray binaries and the double neutron stars there is a substantial group with low orbital eccentricities, suggesting that their last-born neutron stars received hardly any velocity kick at birth, (ii) the low-kick second-born neutron stars in the double neutron star systems have a low mass, about 1.25 $M_\odot$, and (iii) the absence of low-velocity neutron stars in the young radio pulsar population* can be consistently explained if the low-mass low-kick neutron stars originate from the electron-capture collapse of the degenerate O–Ne–Mg cores of stars that started out with main-sequence masses in the range 8–11 $M_\odot$, while the high-kick-velocity neutron stars originated from the iron-core collapses of stars that started out with masses in excess of about 11 $M_\odot$. Such an explanation is fully consistent with the model proposed by [110] according to which neutron star formation by electron-capture collapse can *only* occur in interacting binaries and *not* in single stars.

4.7 The Origins of Wide Low-Mass X-ray Binaries and of Neutron Stars in Globular Clusters: Accretion-Induced Collapse of O–Ne–Mg White Dwarfs?

The existence of Low-Mass X-ray Binaries and large numbers of millisecond pulsars in globular star clusters would seem hard to reconcile with the generally high kick velocities, of order 400 km s^{-1}, observed in the young radio pulsar population [60], since the escape velocity from a globular cluster is of order of at most a few tens of km s^{-1}. The discovery of the existence of a low-kick population of neutron stars, allegedly formed by the collapse of the degenerate O–Ne–Mg core which grows to the Chandrasekhar limit, can very well explain the existence of globular cluster neutron stars, if one assumes that these originated from O–Ne–Mg white dwarfs, which captured a companion star by the usual capture mechanisms operating in globular clusters (e.g. see [191] for a review). In such a binary, the mass of the white dwarf may grow by mass transfer from its companion and thus reach electron capture and undergo an Accretion-Induced-Collapse (AIC) to form a neutron star. Since little or no kick velocity is imparted to the neutron star in this case, the binary will only receive a small kick due to the loss of binding energy of the neutron star (equivalent to some $0.15\,M_\odot$), which will not exceed a few tens of km s^{-1}. This system will then later on evolve into a LMXB and still later into a millisecond radio pulsar. Because of their large masses ($\geq 1.2\,M_\odot$) the O–Ne–Mg white dwarfs are much favoured for capturing a companion, since after the first few billion years of the life of the cluster, they will be the most massive cluster stars, and such massive stars will sink to the dense core of the cluster where, because of the high star density capturing a companion will frequently occur. The model in which the globular cluster neutron stars are formed by the AIC of O–Ne–Mg white dwarfs at the same time explains the otherwise very puzzling fact that a few of the globular cluster radio pulsars have relatively strong magnetic fields (10^{10} to 10^{12} G), suggesting that they are relatively young.

Similarly, the existence of the very wide LMXBs which were the progenitors of the wide binary radio pulsar systems (see Fig. 4) is very puzzling. For example, the progenitor LMXBs of the circular orbit binary radio pulsars PSR 0820+02 ($P_{\rm orb} = 1232.47$ d), PSR 1800-27 ($P_{\rm orb} = 406.78$ d) and PSR 1310+18 ($P_{\rm orb} = 255.8$ d) must have had orbital periods of about 400 d, 60 d and 40 d, respectively (see next section), and a donor star of about $1\,M_\odot$. If the neutron star in these systems had received a kick of several 100 km s^{-1} at its birth, these wide systems would certainly have been disrupted. We do know, however, excellent progenitors for such systems in the form of the recurrent novae, such as RS Oph, RR Tel, etc. These systems consist of a low-mass red giant and a massive white dwarf in an orbit of at least several months period. If the massive white dwarfs in these systems are O–Ne–Mg ones, and some of the transferred mass that gives rise to the outbursts is retained every outburst cycle, these white dwarfs can grow to the Chandrasekhar limit and collapse to a kick-less neutron star in a wide orbit around the low-mass giant companion. These systems will then, when mass transfer from the giant resumes, become wide LMXBs like the system of GX 1+4, which consists of a red giant and an X-ray pulsar. The tidal forces in systems with a Roche-lobe filling red giant are strong and will rapidly circularize the orbits after the explosion. The final outcome of these systems will then be a wide circular radio pulsar binary in which the companion is a low-mass helium white dwarf (M $\leq 0.45\,M_\odot$), like the ones depicted in the right-hand part of Fig. 4.

5 Formation of LMXBs and Millisecond Pulsar Binaries

5.1 Scenarios for LMXBs and IMXBs

Fig. 19 depicts the formation of an LMXB and millisecond pulsar system. There are now more than 40 binary millisecond pulsars (BMSPs) known in the galactic disk. They can be roughly divided into three observational classes [151]: class A contains the wide-orbit ($P_{\rm orb} > 20$ days) BMSPs with low-mass helium white dwarf companions ($M_{\rm WD} < 0.45\,M_\odot$), whereas the close-orbit BMSPs ($P_{\rm orb} \leq 15$ days) consist of systems with either low-mass helium white dwarf companions (class B) or systems with relatively heavy CO/O-Ne-Mg white dwarf companions (class C). The latter class evolved through a phase with significant loss of angular momentum (either common envelope evolution or extreme mass transfer on a sub-thermal timescale) and descends from IMXBs with donors: $2 < M_2/M_\odot < 8$, see [154].

The single MSPs are believed to originate from tight class B systems where the companion has been destroyed or evaporated – either from X-ray irradiation when the neutron star was accreting, or in the form of a pulsar radiation/wind of relativistic particles (e.g. [180, 128, 106, 38, 136, 158]). Observational evidence for this scenario is found in eclipsing MSPs with ultra light companions – e.g. PSR 1957+20 ($P_{\rm orb} = 0.38$ day; $M_2 \simeq 0.02\,M_\odot$) and the planetary pulsar: PSR 1257+12 [199].

For LMXBs it has been shown by [117, 118] that an orbital bifurcation period separates the formation of converging systems (which evolve with decreasing $P_{\rm orb}$ until the mass-losing component becomes degenerate and an ultra-compact binary is formed) from the diverging systems (which finally evolve with increasing $P_{\rm orb}$ until the mass-losing star has lost its envelope and a wide detached binary is formed). This important bifurcation period is about 2–3 days depending on the strength of the magnetic braking torque.

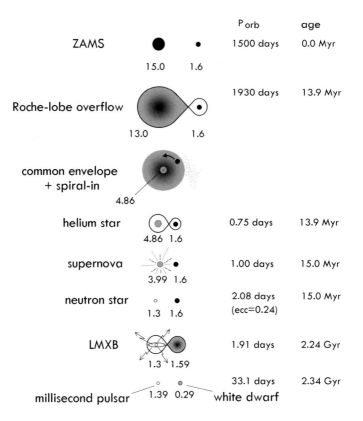

Fig. 19. Cartoon depicting the evolution of a binary system eventually leading to an LMXB and finally the formation of a binary millisecond pulsar. Parameters governing the specific orbital angular momentum of ejected matter, the common envelope and spiral-in phase, the asymmetric supernova explosion and the stellar evolution of the naked helium star all have a large impact on the exact evolution. Parameters are given for a scenario leading to the formation of the observed binary millisecond pulsar PSR 1855+09. The stellar masses given are in solar units.

5.2 Formation of Wide-Orbit Binary Millisecond Pulsars

In LMXBs with initial $P_{orb} > 2$ days the mass transfer is driven by internal thermonuclear evolution of the donor star since it evolves into a (sub)giant before loss of orbital angular momentum dominates. These systems have been studied by many authors (see [196, 145, 131, 63, 122] and recently [39, 153]) and [108]. For a donor star on the red giant branch (RGB) the growth in core mass is directly related to the luminosity, as this luminosity is entirely generated by hydrogen shell burning. As such a star, with a small compact core surrounded by an extended convective envelope, is forced to move up

Table 5. Stellar parameters for giant stars with $R = 50.0 R_\odot$.

$M_{\mathrm{initial}}/M_\odot$	1.0*	1.6*	1.0**	1.6**
$\log L/L_\odot$	2.644	2.723	2.566	2.624
$\log T_{\mathrm{eff}}$	3.573	3.593	3.554	3.569
$M_{\mathrm{core}}/M_\odot$	0.342	0.354	0.336	0.345
$M_{\mathrm{env}}/M_\odot$	0.615	1.217	0.215	0.514

* Single star (X=0.70, Z=0.02 and $\alpha = 2.0$, $\delta_{\mathrm{ov}} = 0.10$).
** Binary star (at onset of RLO: $P_{\mathrm{orb}} \simeq 60$ days and $M_{\mathrm{NS}} = 1.3 M_\odot$).
After [153].

the Hayashi track its luminosity increases strongly with only a fairly modest decrease in temperature. Hence one also finds a relationship between the
giant's radius and the mass of its degenerate helium core – almost entirely
independent of the mass present in the hydrogen-rich envelope (see Table 5).

It has also been argued that the core-mass determines the rate of mass-
transfer [196]. In the scenario under consideration, the extended envelope
of the giant is expected to fill its Roche-lobe until termination of the mass
transfer. Since the Roche-lobe radius, R_L only depends on the masses and
separation between the two stars it is clear that the core mass, from the
moment the star begins RLO, is uniquely correlated with P_{orb} of the system.
Thus also the final orbital period (~ 2 to 10^3 days) is expected to be a function
of the mass of the resulting white dwarf companion [131]. Tauris and Savonije
[153] calculated the expected ($P_{\mathrm{orb}}, M_{\mathrm{WD}}$) correlation in detail and found an
overall best fit:

$$M_{\mathrm{WD}} = \left(\frac{P_{\mathrm{orb}}}{b}\right)^{1/a} + c \tag{42}$$

where, depending on the chemical composition of the donor,

$$(a, b, c) = \begin{cases} 4.50\ 1.2 \times 10^5\ 0.120 & \text{Pop.I} \\ 4.75\ 1.1 \times 10^5\ 0.115 & \text{Pop.I+II} \\ 5.00\ 1.0 \times 10^5\ 0.110 & \text{Pop.II} \end{cases} \tag{43}$$

Here M_{WD} is in solar mass units and P_{orb} is measured in days. The fit is valid
for BMSPs with helium white dwarfs companions and $0.18 \leq M_{\mathrm{WD}}/M_\odot \leq$
0.45. The formula depends slightly on the adopted value of the convective
mixing-length parameter. It should be noted that the correlation is *independent* of β (the fraction of the transferred material lost from the system – see
Sect. 3.5), the mode of the mass loss and the strength of the magnetic braking
torque, since the relation between giant radius and core mass of the donor star
remains unaffected by the exterior stellar conditions governing the process of
mass transfer. However, for the *individual* binary P_{orb} and M_{WD} do depend
on these parameters. In Fig. 20 we have plotted a theoretical ($P_{\mathrm{orb}}, M_{\mathrm{WD}}$)
correlation and also plotted evolutionary tracks calculated for four LMXBs.

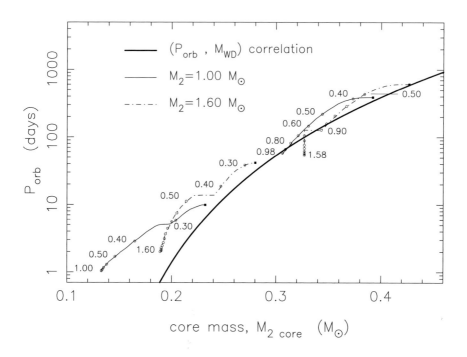

Fig. 20. Evolutionary tracks of four LMXBs showing P_{orb} as a function of M_{core} of the donor star. The initial donor masses were 1.0 and 1.6 $M_\odot$ (each calculated at two different initial P_{orb}) and the initial neutron star mass was 1.3 $M_\odot$. The total mass of the donors during the evolution is written along the tracks. At the termination of the mass-transfer process the donor only has a tiny ($\leq 0.01\,M_\odot$) H-envelope and the end-points of the evolutionary tracks are located near the curve representing the $(P_{\mathrm{orb}}, M_{\mathrm{WD}})$ correlation for BMSPs. After [153].

Although clearly the class of BMSPs with helium white dwarf companions is present, the estimated masses of the BMSP white dwarfs are quite uncertain, since they depend on the unknown orbital inclination angle and the pulsar mass, and no clear observed $(P_{\mathrm{orb}}, M_{\mathrm{WD}})$ correlation has yet been established from the current observations. In particular there may be a discrepancy for the BMSPs with $P_{\mathrm{orb}} \geq 100$ days [151].

5.3 Formation of Close-Orbit Binary Millisecond Pulsars

In LMXBs with initial $P_{\mathrm{orb}} < 2$ days the mass transfer is driven by loss of angular momentum due to magnetic braking and gravitational wave radiation. The evolution of such systems is very similar to the evolution of CVs (see e.g. [141, 190, 39]).

6 From X-ray Binaries to Binary Radio Pulsars: the Concept of "Recycling"

6.1 Introduction

In the foregoing sections we have seen that the double neutron stars and the neutron stars with massive white dwarf companions in circular orbits are the remnants of HMXBs and IMXBs, respectively, and that the binary pulsars with low-mass helium white dwarf companions (and other low-mass companions, such as ablated hydrogen dwarfs) are the remnants of LMXBs. The neutron stars in all these types of systems have gone through an extensive accretion phase, which particularly in the LMXBs was very long lasting: in the close LMXBs, where mass transfer is driven by angular momentum loss, the typical timescales are of order 10^8 to 10^9 years, in the wide LMXBs, where nuclear evolution of a low-mass giant companion drives the transfer the timescale is typically of order 10^8 yrs (except in very wide systems, such as the progenitor of PSR 0820+02, where it is close to 10^7 yrs). Since in the LMXBs this was disk accretion, the neutron stars in these systems will have acquired a large amount of angular momentum. The observed spin-up behavior of the X-ray pulsars that are accreting from a disk (Sect. 2.4 and Fig. 6) show indeed that one expects such neutron stars to evolve towards shorter and shorter spin periods. If at the same time the magnetic field has weakened to the values observed in the double neutron stars(about 10^{10}G) or in the binary pulsars with white dwarf companions (10^8 to 10^{10} G) (see Fig. 3), then as Eq. (8) shows, the neutron stars can be spun up to periods of the order of 20 ms in the case of double neutron stars, and 1 ms in the case of LMXBs. There is overwhelming evidence that the magnetic fields in these binary neutron stars have decayed to the values quoted above. Possible reasons for the field decay are briefly discussed in Sect. 6.4.

6.2 Evolution of a Neutron Star Born in a Binary System: Recycling

The ingredients discussed in the last section allow us to describe the evolution in the diagram of surface dipole magnetic field strength B_s versus spin period P, for a neutron star that is born in a close binary system. This is depicted in Fig. 21. This diagram is in fact a different representation of the $\dot{P}$ versus P diagram of Fig. 3, as $B_s = \text{const}\,(\dot{P}\,P)^{1/2}$. The neutron star is born in the upper left-hand part of the diagram as a strongly magnetized rapidly spinning pulsar. It will within a few million years spin down along a horizontal track through the diagram and cross the "deathline". To the right of this line no radio pulsars can exist as the polar cap electric field has become too weak to create pairs, and no more pulsar wind can be produced, causing the radio pulsar emission to stop [20, 10]. The region to the right of the "deathline" is the "graveyard". During its stay in the graveyard its further spin down in the weak wind of the companion star (or other processes) may make its magnetic dipole moment decay, causing it to move downwards in the B_s versus P diagram. When finally the companion begins to overflow its Roche lobe and begins to transfer matter through a disk, the rotation rate of the neutron star will rapidly increase, causing it to move to shorter and shorter spin periods, that is: towards the left in the B_s versus P diagram. Thus, as shown in Fig. 21, it will cross the deathline from the right and move again into the region of the "living" radio pulsars in the diagram. However, in this phase it is still an X-ray source, and will not generate radio pulses. It can at maximum be spun up to the spin period corresponding to the "spin-up line", indicated in the figure, which corresponds to the shortest spin period which an accreting neutron star with a given surface dipole field strength B_s can attain, that is, for the maximum possible accretion rate, which is the Eddington rate. Thus, according to Eq. (8) the equation for the spin-up line is:

$$P_{\min} = (2.4\text{ms})\,(B_9)^{6/7}\,R_6{}^{16/7}\,M^{-5/7} \qquad (44)$$

where B_9, R_6 are the surface dipole magnetic field strength and radius of the star in units of 10^9 G and 10^6 cm, respectively, and M is its mass in solar masses. Assuming the radius and mass of all neutron stars to be the same, this gives the "spin-up line"("spin-up limit") depicted in Fig. 21 and Fig. 3. The importance of the spin-up line was first realized by [119, 120] (see also [142]) and by [1]. Only after the accretion phase is terminated and the donor star itself has become a compact star (neutron star, white dwarf or black hole), or if the accretion is otherwise interrupted, the rapidly spinning neutron star will become observable as a radio pulsar. Because of its return from the "graveyard" such a pulsar is called "recycled" (this term was coined by Radhakrishnan). The recycling model has been very successful in explaining the peculiar position of the bulk of the binary and millisecond radio pulsars in the $\dot{P}$ versus P diagram (Fig. 3). The model predicts that the recycled pulsars can only be found between the spin-up line and the deathline, and this is indeed exactly the region in the $\dot{P}$ versus P diagram where they are found. It also predicts that the companions should in general be dead stars(neutron stars, white dwarfs and possibly also black holes) which, with the exception of a few systems with "evaporating" little companions, is indeed the case. It furthermore predicts that the orbits of the recycled neutron stars with a neutron star companion should be eccentric, because of the second supernova explosion, but that, on the other hand, the orbits of recycled pulsars with a white dwarf companion must be circular, because of the tidal and frictional circularization of the orbit during the spiral-in phase and the long-lasting mass transfer phase. All of these predictions fit excellently with the observations, confirming "recycling" to be the cause of the the the formation of three of the four classes of binary pulsars mentioned in Sect. 2.3. The fourth class, in which the neutron star was born after the white dwarf, causing the orbit to be eccentric, is represented by systems such as PSR J1145-6545 and PSR 2303+46. Also the large number of millisecond pulsars found in globular clusters nicely confirms

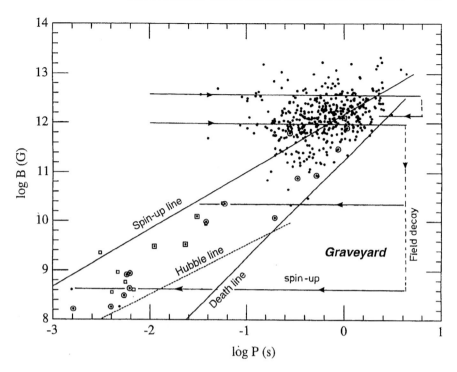

Fig. 21. Surface dipole magnetic field strength B_s v. pulse period diagram of pulsars, with possible evolutionary tracks of pulsars indicated. Pulsars are born in the left upper part of the diagram and – if no field decay occurs – move towards the right along horizontal tracks (fully drawn). In the graveyard the magnetic field of a single pulsar probably does not decay (see text), but that of a binary neutron star does decay, presumably due to external circumstances (accretion, spin-down). Furthermore, after the start of mass transfer from its companion, the accretion of matter with angular momentum causes these neutron stars in binaries to be "spun-up" towards the left in the diagram along the indicated lines until they reach the spin-up line (spin-up limit, see also Fig. 3). After the companion has itself become a compact star (white dwarf or neutron star) or has disappeared, the spun-up neutron star becomes observable as a radio pulsar. It will then slowly spin down, that is, move to the right again in the diagram. Dots indicate normal non-recycled pulsars. Circles indicate some well-known radio pulsars in binaries, squares with dots are some globular cluster binary pulsars and open squares are some single pulsars in globular clusters.

the recycling model. As a matter of fact, because of the exceptional richness in Low-Mass X-ray Binaries of globular clusters, in combination with the "recycling" model, A.G. Lyne was inspired to his search for millisecond pulsars in globular clusters, which led to the discovery of the first cluster millisecond pulsar: the 3 ms pulsar in the cluster M28 [83]. Since then over 40 globular cluster radio pulsars have been found, most of them millisecond pulsars (e.g. see [191]).

6.3 Why Did the Magnetic Fields of the Recycled Neutron Stars Decay?

It seems now well established that the magnetic fields of non-accreting neutron stars do not spontaneously decay on a timescale of order of 10^7 years, as used to be thought in the past. Even on very long timescales, of order 10^9 years they appear not to undergo much spontaneous decay. The first system in which this was discovered was the recycled pulsar PSR 0655+64 which has a $1\,M_{odot}$ white dwarf companion with a cooling age of order billions of years. As the white dwarf was the second-formed object in this system, this neutron star is even older but still has a 10^{10} G magnetic field [75]; similarly, the observed number of millisecond pulsars in our Galaxy indicates that their magnetic fields of between 10^8 and 10^9 G do not substantially decay on a timescale shorter than a billion year [178]. Also careful studies of the population characteristics of radio pulsars in general now no longer give support to the idea that the magnetic fields of isolated (non-accreting) neutron stars would undergo substantial decay (e.g. [9, 58]). Since recycled pulsars all appear to have much weaker magnetic fields than the "garden variety" radio pulsars, and all have undergone substantial accretion, it seems likely that their field decay is either (i) directly related to the accretion process, or (ii) is otherwise related to the fact that they are members of a binary system. The fact that the two observed last-born neutron stars in double neutron star binaries (PSRJ 0737-3039B and PSRJ 1906+0746) as well as the neutron star in the eccentric white-dwarf neutron-star system PSRJ 1145-6545, all three have "canonical" strong magnetic fields of order 10^{12} G, like normal single "garden variety" pulsars, indicates that the fact that the neutron star is born in a binary does not affect the value of its magnetic field strength at birth. The only remaining possibility is therefore that the accretion process has somehow caused the field of recycled pulsars to decay. A number of models has been proposed for this, each with its advantages and disadvantages. Space does not permit an extensive discussion here. I refer to the reviews by [9, 127]. The discovery of the "Magnetars", i.e. neutron stars with surface dipole magnetic field strengths up to 10^{15} G (see Fig. 3; [161]), excludes the possibility that the field would be confined only to the neutron star crust, as the crust alone is not strong enough to confine a field of this magnitude. The field therefore must be anchored in the superfluid and superconducting interior of the neutron star. Bhattacharya and Srinivasan have proposed a model of magnetic field evolution of neutron stars based on the fact that the magnetic field in a superconductor is concentrated in fluxtubes, which are expected to be pinned to the quantized vortices in the superfluid of the neutron star core. This model works roughly as follows: when the neutron star's rotation slows down, magnetic flux tubes are carried outwards from the core into the crust by outward moving quantized superfluid core vortices. As the crust is conductive, the fields of these tubes can then decay. As in this way magnetic field is really destroyed in the spin-down process, it cannot come back when later in life the neutron star is spun up back to a very rapid rotation rate. In this model the 10^8 to 10^{10} G fields observed in the recycled pulsars are really an extension of the core magnetic field of the neutron star. It appears that for this model to work well, the neutron star has to be spun down to a rotation period of hundreds of seconds. For a single neutron star this will not easily happen, but for a neutron star in a binary, even a weak wind from a solar-type companion can, on a timescale of hundreds of millions of years, spin the rotation of a neutron star companion to periods of this order. For details about this model and further references and alternative models, we refer the reader to reviews by [9, 127].

7 Relativistic Binaries with One or Two Black Hole Components

Since the most massive stars are expected to leave black holes as remnants, one expects that the most massive X-ray binaries may leave very close compact-star binaries in which one or both of the stars is a black hole.

7.1 Final Evolution of Well-Known HMXB which will Produce at Least One Black Hole

The final evolution of the systems of Cygnus X-1 and LMC X-1: production of a close double black hole or a black hole with a young neutron star

Cygnus X-1 consists of O9.7Iab blue supergiant with a mass of about 35 $M_\odot$ and a black hole with a mass of about 15 $M_\odot$ in a 5.6 day orbit (e.g. [50, 51]). The blue supergiant is very close to filling its Roche lobe. When it overflows this lobe one expects, since the envelope is still in radiative equilibrium, and since the mass ratio of the system is not extreme, that the system will not go into common-envelope evolution, but rather that Roche-lobe overflow on the thermal timescale of the blue supergiant will ensue. This timescale is about 10^4 yrs, so a mass-transfer rate of order $2 \times 10^{-3} M_\odot$/yr will ensue, which is 4 orders of magnitude larger than the Eddington limit of the black hole. Therefore, one expects over 99.9 per cent of this mass to be ejected from the vicinity of the black hole, carrying off the specific orbital angular momentum of that star. In this case the change of the orbit during the mass transfer is given by Eq. (24), in which the constants α, δ and γ are equal to zero, and $\beta = 1$. From this equation, taking the limit for β approaching unity one finds that the change in orbital radius is given by (see [169]):

$$a/a_0 = (q_0/q)^2 \left((1 + q_0)/(1 + q)\right) exp2 \left(q - q_0\right) \tag{45}$$

where $q = M_1/M_{2,0}$ and $q_0 = M_{1,0}/M_{2,0}$, in which M_1 and $M_{1,0}$ are the actual and initial masses of the donor star, and $M_{2,0}$ is the mass of the compact object, which stays constant. This mass loss will go on until only the $15\,M_\odot$ helium core of the supergiant is left, so the final value of the mass ratio q in this equation is $q = 1$, while the initial value is $q_0 = 35/15$. With these values of q and q_0, Eq. (45) yields that the orbital radius is reduced by a factor 0.63, yielding an orbital radius after the mass transfer of 31.1 solar radii, corresponding to an orbital period $P_f = 3.64$ d. The resulting system consists of a $15\,M_\odot$ Helium star (Wolf–Rayet star) and a $15\,M_\odot$ black hole. With a wind mass loss rate of about $10^{-5}M_\odot$/yr and a lifetime of about 5×10^5 yr, typical for such a WR star, this star at the end of its life will have a mass of about $10\,M_\odot$ when its core collapses. Due to this wind mass loss the orbital radius increases to 37.3 solar radii, corresponding to an orbital period of 5.24 d.

Assuming the helium star to collapse to a black hole, with no mass ejection in the process, the system will leave behind a double black hole binary with components of 10 and $15\,M_\odot$ and the same orbital period. If the helium star would leave behind a $1.4\,M_\odot$ neutron star, some $8.6\,M_\odot$ will be explosively ejected and if no kick to the neutron star is imparted, the final semi-major axis, orbital period and eccentricity will be 78.9 solar radii, 19.7 d and 0.502, respectively. Binaries of the latter type, in which first formed a black hole and later a neutron star we will call BHNS binaries (see [192]). Using the equations for orbital decay by gravitational radiation losses (e.g. [138]) one finds that in both case the orbital decay timescale by losses of gravitational radiation is very much longer than the Hubble time, so binaries descending from systems like Cyg X-1 are not expected to coalesce within the age of the Universe. The main reason for this wide final orbit is the large mass of the black hole in the Cyg X-1 system.

If the black hole had a mass of only $6\,M_\odot$, like in the system of LMC X-1 [150], the orbit will shrink much more than in the Cyg X-1 case, due to the exponential factor in Eq. (45). Starting with the orbital period of 4.22 d of LMC X-1, and assuming here the O7-9 III donor star to have a mass of $30\,M_\odot$ with a helium core of $12\,M_\odot$ one finds that the resulting binary consisting of a $12\,M_\odot$ helium star and the $6\,M_\odot$ black hole has an orbital radius of only 1.13 solar radii. At the end of the life of the helium star, that has lost $4\,M_\odot$ in the form of stellar wind, the orbital radius will have increased to $1.45\,R_\odot$, corresponding to an orbital period of 1.29 h. If the $8\,M_\odot$ helium star collapses to a black hole with the same mass, a double black hole will be formed with this same orbital period. If the $8\,M_\odot$ helium star would explode in a symmetric supernova and leave a $1.4\,M_\odot$ neutron star, the final orbital radius, period and eccentricity would be $14.14\,R_\odot$, 49.9 h and 0.89, respectively. In case that the close double black hole system is left, the orbital decay time by GW emission is about 4×10^7 years, in the case with a neutron star remnant it is about 3×10^9 years. In both cases, therefore, the remnants of the black hole HMXBs like LMC X-1 are expected to contribute to the production of bursts of GWs that can be detected by LIGO.

One may wonder how the evolution of these two systems would proceed if one would simply use the Common-Envelope formalism of Eq. (38). If we assume $\eta_{CE}\,\lambda$ to be equal to 0.3, the final orbital radius of Cyg X-1 after spiral-in would be only $1.12\,R_\odot$, and the Roche-lobe radius of the $15\,M_\odot$ helium star only $0.45\,R_\odot$, which is only half the radius of this helium star. One would therefore in this case expect the spiral-in to continue and the black hole to end up in the center of the helium star, which would be completely destroyed in the process. The same is the case - even more so - for the system of LMC X-1. However, in view of the radiative envelopes of the donors in these two systems, in combination with their not extreme mass ratios before Roche lobe overflow, we expect that the evolution with mass transfer and heavy mass loss, as described by Eq. (45), to more closely represent the real final evolution of these systems, implying that they will produce relativistic binaries composed of two compact objects.

Production of a black hole with a recycled neutron star

Following [192] we will call these systems NSBH binaries. Relatively wide HMXBs consisting of a massive donor star and an X-ray pulsar will certainly go into common-envelope evolution, as by the time the donor overflows its Roche lobe, its envelope will be convective. Also the extreme mass ratios of these systems make it impossible to avoid the CE phase. We now know at least two such systems, namely 4U 1223-62/Wray977, which consists of a blue hypergiant (B1.5Ia) in an eccentric orbit with $P = 41.5$ d (e.g. see [67]), and the shrouded INTEGRAL HMXB with an orbital period of 330 days and an O-supergiant donor discovered by [137]. We first consider the system of 4U1223-62/Wray977. The hypergiant in this system is likely to have a mass of about $35\,M_\odot$, which again will have a helium core of mass of $15\,M_\odot$. Assuming the neutron star to have a mass of $1.8\,M_\odot$ like that of Vela X-1, and $\eta_{CE}\,\lambda$ to be equal to 1.0, one finds a final orbital period after spiral-in of about 2 hours, in which the helium star can stay just inside its Roche lobe. [Had one assumed a value of 0.3 instead of 1.0 for this parameter, the final Roche-lobe radius had been half the radius of the helium star, and the neutron star would have spiraled completely into the center of the donor, leaving no binary system]. If finally, after wind mass loss, etc. the helium star collapses to a $10\,M_\odot$ black hole, one will have a very close binary with a period of about 3.5 hours, consisting of this black hole and the *recycled* neutron star. If there was no explosive mass ejection during the formation of the black hole, this system will have a circular orbit. If there was mass ejection, the orbit will be eccentric. The orbital decay time of this system by GW emission will be of order of a billion years, so systems of this type will contribute to the LIGO rate of bursts of GWs. The second system, with its orbital period of 330 days and an O-supergiant donor, will surely survive the spiral-in of the neutron star, also if the CE-efficiency parameter $\eta_{CE}\,\lambda$ has the more realistic value 0.2 or 0.3. In the latter case the final system left behind will have a period similar to the one left behind by 4U1223-62 for the case $\eta_{CE}\,\lambda = 1.0$. Therefore, if the $15\,M_\odot$ helium star leaves behind a black hole, the final system here will again be a very close binary consisting of a black hole and a recycled pulsar, which will coalesce on a timescale of order a billion years.

The Wolf–Rayet X-ray binaries: an intermediate phase between the black-hole HMXBs and double compact objects

Presently three close X-ray binaries consisting of a helium star (Wolf–Rayet star) and a compact object are known: Cygnus X-3 ($P_{orb} = 4.8$ h; [185]), the extragalactic sources IC10 X-1 ($P_{orb} = 34.8$ h; [115]; ATel 955) and NGC 300 X-1 ($P_{orb} = 32.8$ h; [19]). For a variety of reasons, such as their X-ray characteristics, the latter two systems most probably harbour a black hole. One may wonder what initial configuration of a black hole HMXB may produce systems with these orbital periods. As indicated above: with a black-hole mass of $15\,M_\odot$ the systems do not spiral in much, whereas for a black hole mass of $6\,M_\odot$ the spiral-in is very deep. Therefore, the orbital periods of the latter two WR X-ray binaries suggest that their black holes have masses intermediate between 6 and $15\,M_\odot$. Assuming a donor mass of $35\,M_\odot$ and the 5.6 d orbital period of Cyg X-1, but a $10\,M_\odot$ black hole, and using again Eq. (45) formalism, one finds that helium star plus black hole system resulting after spiral-in has an orbital period 14.4 h, which due to the stellar wind mass loss from the helium star results in an orbital period of 22.7 hours at the end of its evolution. So, if one had started out with a Cyg X-1 like HMXB with an orbital period about 1.5 to 2 times that of Cyg X-1, one would have obtained a Wolf–Rayet X-ray binary with an orbital period like those of IC10 X-1 and NGC 300 X-1. It thus appears most likely to us that the black holes in these systems have a mass around 10 solar masses or somewhat higher. Applying the same reasoning to the Cyg X-3 system, with its period of only 4.8 hours, one finds that here, if the compact star is a black hole, it most probably has a relatively small mass, around 6 to $8\,M_\odot$. The compact star here might, however, also be a neutron star (e.g. [82]).

8 The Galactic Formation Rate of Relativistic Binaries

It is very important to constrain the local merging rate of NS/BH-binaries in order to predict the number of events detected by LIGO. This rate can be determined either from binary population synthesis calculations, or from observations of galactic NS-NS systems (binary pulsars). Both methods involve considerable uncertainties (e.g. [65] and references therein). The current theoretical estimates from population synthesis for the galactic merger rate of NS-NS systems cover a wide range: $10^{-6} - 10^{-3}$ yr^{-1}.

Therefore, it is better to use the *observed* galactic double neutron star population to constrain this rate. The discovery of the very tight binary pulsar system PSR J0737–3039 and the very young non-recycled binary pulsar J 1906+746 indicate that the galactic NS-NS merger rate is almost an order of magnitude larger than previously estimated [16, 66]. The last-mentioned authors, using a detailed model of the galactic pulsar population and correcting for selection effects, obtain as the most likely value of the galactic merger rate of double neutron stars: 180 Myr^{-1}. With 68% confidence they find the rate to be between 80 and 370 Myr^{-1}, and with 95% confidence: between 40 and 660 Myr^{-1}. This implies in a steady state of star formation, a galactic formation rate of close double neutron stars of about one per 5500 years. This is close to what one expects from their alleged progenitor population, the wide B-emission X-ray binaries. The galactic population of these systems is estimated to be several thousands (see Sect. 3.8). Assuming a galactic population of these systems of 2000, and a typical lifetime of the 12 $M_\odot$ Be-type companion of about 10^7 yr, one would obtain, if all Be/X-ray binaries produce a close double neutron star, a galactic formation rate of these systems of about one per 5000 years! However, the number of Be/X-ray binaries in the Galaxy may be quite a bit higher. Van den Heuvel [168] estimates it to be of order 10^4. And also, probably only a fraction of these systems survive spiral-in as a close binary (e.g. see [192]). Taking these facts into account, [168] estimated the galactic formation rate of close double neutron stars to be $(2-4) \times 10^{-4}$ yr^{-1}, which still is very close to the observational estimate of [66].

In order to extrapolate the galactic coalescence rate out to the volume of the Universe accessible to LIGO, one can either use a method based on star formation rates or a scaling based on the B-band luminosities of galaxies. Using the latter method [65] found a scaling factor of $(1.0-1.5) \times 10^{-2}$ Mpc^{-3}, or equivalently, ~ 400 for LIGO I (out to 20 Mpc for NS-NS mergers). Since LIGO II is expected to look out to a distance of 300 Mpc (for NS-NS mergers), the volume covered by LIGO II is larger by a factor of $(300/20)^3$ and thus the scaling factor in this case, relative to the coalescence rates in the Milky Way, is about 1.3×10^6. Therefore, the expected rate of detections from NS-NS inspiral events is roughly between 2–500 yr^{-1} for LIGO II. However, LIGO I may not detect any NS/BH inspiral event.

Voss and Tauris [192] also investigated the formation rates of BHNS, NSBH and BHBH systems by means of population synthesis calculations. They find a relatively high galactic merger rate of the BHBH systems, compared to NSNS systems and mixed NS/BH systems, and estimate a BHBH merging detection rate of 840 yr^{-1} for LIGO II.

They found the galactic BHBH merger rate to be about 7 times larger than the NSNS merger rate, while they found the BHNS merger rate and the

NSBH (systems in which the NS formed before the BH) merger rate to be 3 and about 20 times lower than the NSNS merger rate, respectively. It should be kept in mind, however, that these calculated rates depend very strongly on the many input assumptions of population synthesis. Particularly important is the input assumption in these calculations that *all* neutron stars received a large kick velocity at birth which, as we have seen in Sect. 4 is not a good assumption, *particularly for the formation of the double neutron stars.* New population synthesis calculations with the revised ideas about neutron star kicks as presented in Sect. 4 should therefore be carried out. Finally, one should also be aware that compact star mergers in globular clusters probably contribute significantly to the total merger rates [114].

9 Acknowledgements

This research was supported in part by the National Science Foundation under Grant No. PHY05-51164 to the Kavli Institute for Theoretical Physics, UC Santa Barbara, California. I am grateful to the Kavli Institute for Theoretical Physics, the Leids Kerkhoven-Bosscha Fonds and the Netherlands Research School for Astronomy NOVA for providing financial support for my travel and stay in Santa Barbara for 3 months in 2006 and 4 months in 2007. I thank also the many colleagues which helped me with input for this review, particularly Thomas Tauris, Chris Fryer and Fernando Camilo.

References

1. Alpar, M.A., Cheng, A.F., Ruderman, M.A. and Shaham, J. (1982). *Nature* **300**, 728
2. Bailes (2005), in *Binary Radio Pulsars*, Eds. F.A.Rasio and I.H.Stairs (Astronomical Society of the Pacific Conf. Series Vol 328) pp 33–36
3. Bailes, M., Ord, S.M., Knight, H., and Hotan, A.W. (2003). *ApJ* **595**, L49
4. Barziv et al. (2001). *A&A* **377**, 925
5. Bildsten, L., Chakrabarty, D., Chiu, J., et al. (1997). *ApJS* **113**, 367
6. Bisnovatyi-Kogan, G.S. and Komberg, B.V. (1974). *Astron. Zh.* **51**, 373
7. Bisscheroux, B. (1999). *Master's thesis*, University of Amsterdam (1999).
8. Bhattacharya, D. and van den Heuvel, E.P.J. (1991). *Phys. Reports* **203**, 1
9. Bhattacharya, D. and Srinivasan, G. (1995), in *X-ray Binaries*, eds. W.H.G. Lewin, J. van Paradijs and E.P.J. van den Heuvel. (Cambridge University Press)
10. Bjornsson, C.-I. (1996). *ApJ* **471**, 321
11. Blaauw, A. (1961). *Bull. Astr. Inst. Neth.* **15**, 265
12. Blundell, K.M. (2007) *Private Communication*
13. Boriakoff, V., Buccheri, R. and Fauci, F.(1983). *Nature* **304**, 417
14. Brown, G.E. (1995). *ApJ* **440**, 270
15. Brown, G.E., Lee, C.H. and Bethe, H.A. (1999). *New Astronomy* **4**, 313

16. Burgay, M., D'Amico, N., Possenti, A., et al. (2003). *Nature* **426**, 531
17. Burrows, A. and Hayes, J. (1996). *Phys. Rev. Letters* **76**, 352
18. Camilo, F. (2007). *Private Communication*
19. Carpano, S., Pollock, A.M.T., Prestwich, A., et al. (2007). *A&A* **466**, L17
20. Chen, K.Y., and Ruderman, M.A. (1993). *ApJ* **402**, 264
21. Chevalier, R.A. (1993). *ApJ* **411**, L33
22. Corbet, R.H.D. (1984). *A&A* **141**, 91
23. Cordes, J., and Chernoff, D.F. (1998). textitApJ **505**, 315
24. Counselman (1973). *ApJ* **180**, 307
25. Cox, J.P. and Giuli, R.T. (1968). *Stellar Structure, vols. I and II*, (Gordon and Breach, New York)
26. Cumming, A., Zweibel, E. and Bildsten, L. (2001). *ApJ* **557**, 958
27. Cumming, A. (2005), in *Binary Radio Pulsars*, Eds. F.A.Rasio and I.H.Stairs (Astron. Soc. Pacific Conf. Series Vol 328) pp 311–316
28. Darwin, G.H. (1879). *Proc. Roy. Soc. London* **29**, 168
29. Davidson, K. and Ostriker, J.P. (1973). *ApJ* **179**, 585
30. de Kool, M. (1990). *ApJ* **358**, 189
31. de Loore, C., De Greve, J.R. and De Cuyper, J.P. (1975a). textitApSpSci **36**, 219
32. de Loore, C.W.H., De Greve, J.P., van den Heuvel, E.P.J. and De Cuyper, J.P. (1975b). textitMem.Soc.Astron. Ital. **45**, 893
33. Dewi, J.D.M. and Tauris, T.M. (2000). *A&A* **360**, 1043
34. Dewi, J.D.M., Pols, O.R., Savonije, G.J. and van den Heuvel, E.P.J. (2002). *MNRAS* **331**, 1027
35. Dewi, J.D.M. and Pols, O.R. (2003). *MNRAS* **344**, 629
36. Dewi, J.D.M., Podsiadlowski, Ph., and Pols, O.S. (2005). *MNRAS* **363**, L71
37. Eggleton, P.P. (1983). *ApJ* **268**, 368
38. Ergma, E. and Fedorova, A.V. (1991). *A&A* **242**, 125
39. Ergma, E., Sarna, M.J. and Antipova, J. (1998). *MNRAS* **300**, 352
40. Ergma, E. and Yungelson, L.R. (1998). *A&A* **333**, 151
41. Faulkner, J. (1971). *ApJ* **170**, L99
42. Faulkner, A.J., Kraemer,M., Lyne, A.G. et al. (2005). textitApJ **618**, L119
43. Finger, M. (1998), in *The Many Faces of Neutron Stars*, Eds. R. Buccheri, J.van Paradijs and A.Alpar. (Kluwer, Dordrecht) pp 369–384
44. Flannery, B.P. and van den Heuvel, E.P.J. (1975). *A&A* **39**, 61
45. Fryer, C.L. (1999). *ApJ* **522**, 413
46. Fryer, C. L. (2006). *New Astron. Rev.* **50**, 492
47. Fryxell, B.A. and Taam, R.E. (1988). *ApJ* **335**, 862
48. Ghosh, P. (2007). *Rotation and Accretion Powered Pulsars* (World Scientific, London), 772 pp
49. Ghosh, P. and Lamb, F.K. (1979). *ApJ* **234**, 296
50. Gies, D. and Bolton, C.T. (1982). *ApJ* **260**, 240
51. Gies, D. and Bolton, C.T. (1986). *ApJ* **304**, 371
52. Habets, G.M.H.J. (1985). *Advanced Evolution of Helium Stars and Massive Close Binaries* (Ph.D.Thesis Univ. of Amsterdam)
53. Habets, G.M.H.J. (1986). *A&A* **167**, 61
54. Han, Z., Podsiadlowski, P. and Eggleton, P.P. (1994). *MNRAS* **270**, 121
55. Han, Z., Podsiadlowski, P. and Eggleton, P.P. (1995). *MNRAS* **272**, 800
56. Han, Z., Podsiadlowski, P., Maxted, P.F.L., et al. (2002). *MNRAS* **336**, 449

57. Hansen, B.M.S. and Phinney, E.S. (1997). *MNRAS* **291**, 569
58. Hartman, J.W. (1997). *A&A* **322**, 127
59. Hills, J. (1983). *ApJ* **267**, 322
60. Hobbs, G., Lorimer, D.R., Lyne, A.G. and Kramer, M. (2005). *MNRAS* **360**, 974
61. Iwamoto, K., Mazzali, P.A., Nomoto, K., et al. (1998). *Nature* **395**, 672
62. Johnston, S., et al. (2001). *MNRAS* **326**, 643
63. Joss, P.C., Rappaport, S.A. and Lewis W. (1987). *ApJ* **319**, 180
64. Kalogera, V. and Baym, G. (1996). *ApJ* **470**, L61
65. Kalogera, V., Narayan, R., Spergel, D.N. and Taylor, J.H. (2001). *ApJ* **556**, 340
66. Kalogera, V., Kim, C., Lorimer, D.R., et al. (2004). *ApJ* **601**, L179
67. Kaper, L., van der Meer, A., van Kerkwijk, M.H. and van den Heuvel, E.P.J. (2006). *A&A* **457**, 595
68. Kaspi, V.M., Lyne, A.G. and Manchester, R.N. (2000). *ApJ* **543**, 321
69. King, A.R. and Begelman, M.C. (1999). *ApJ* **519**, L169
70. King, A.R. and Ritter, H. (1999). *MNRAS* **309**, 253
71. Kippenhahn, R. and Weigert, A. (1967). *Z. Astrophys.* **65**, 251
72. Kippenhahn, R. and Weigert, A. (1990). *Stellar Structure and Evolution*, (Springer, Heidelberg)
73. Kitaura, F.S., Janka, H.-Th., and Muller, E. (2006). *A&A* **450**, 345
74. Kolb, U., Davies, M.B., King, A.R., Ritter, H. (2000). *MNRAS* **317**, 438
75. Kulkarni, S.R. (1986). *ApJ* **306**, L85
76. Lamb, F.K., Pethick, C.J. and Pines, D. (1973). *ApJ* **184**, 271
77. Landau, L.D. and Lifshitz, E. (1958). *The Classical Theory of Fields*, Pergamon Press, Oxford)
78. Lewin, W.H.G., van Paradijs, J.A. and van den Heuvel, E.P.J. (1995). *X-ray Binaries* (Cambridge University Press) 662pp
79. Lewin, W.H.G. and van der Klis, M. (2006). *Compact Stellar X-ray Sources* (Cambridge University Press) 690pp
80. Liu, Q.Z. (2001). *Private Communication*
81. Liu, Q.Z., van Paradijs, J. and van den Heuvel, E.P.J. (2007). *A & A* **409**, 807
82. Lommen, D., Yungelson, L., van den Heuvel, E., Nelemans, G. and Portegies Zwart, S. (2005). *A&A* **443**, 231
83. Lyne, A.G., Brinklow, A., Middleditch, J., Kulkarni, S.R. and Backer, D.C. (1987). *Nature* **328**, 399
84. Lyne, A.G. and Lorimer, D.R. (1994). *Nature* **369**, 127
85. Lyne, A.G., Burgay, M., Kramer, M., et al. (2004). *Science* **303**, 1153
86. Lyne, A.G. and Graham-Smith, F. (1990). *Pulsar Astronomy*, (Cambridge University Press) 274pp
87. Manchester, R.N. and Taylor, J.H. (1977). *Pulsars*, (Freeman, San Francisco)
88. Maraschi, L., Treves, A. and van den Heuvel, E.P.J. (1976). *Nature* **259**, 292
89. McClintock, J.E. and Remillard, R.A. (1986). *ApJ* **308**, 110
90. McClintock, J.E. and Remillard, R.A. (2006), in *Compact Stellar X-ray Sources*, Eds. W.H.G.Lewin and M. van der Klis. (Cambridge University Press) p 157–213
91. Meurs, E. and van den Heuvel, E.P.J. (1988). *A&A* **226**, 88
92. Meyer, F. and Meyer-Hofmeister, E. (1978). *A&A* **78**, 167
93. Miyaji, S., Nomoto, K., Yokoi, K., and Sugimoto, D. (1980) *PASJ* **32**, 303

94. Mirabel, I.F., Mignami, R., Rodrigues, I., et al. (2002). *A&A*, in press
95. Nagase, F. (1989). *PASJ* **41**, 1
96. Nauenberg, M. and Chapline, G. (1973). *ApJ* **179**, 277
97. Nelemans, G., Tauris, T.M. and van den Heuvel, E.P.J. (1999). *A&A* **352**, L87
98. Nelemans, G. and van den Heuvel, E.P.J. (2001). *A&A* **376**, 950
99. Nomoto, K. (1984). *ApJ* **277**, 791
100. Nugis, T. and Larmers, H.J.G.L.M. (2000). *A&A* **360**, 227
101. Ostriker J.P. (1976), in *Structure and Evolution of Close Binary Systems*, eds. P.P. Eggleton et al. (Reidel, Dordrecht) p. 206
102. Paczynski, B. (1967). *Acta Astron.* **17**, 355
103. Paczynski B. (1971). *Acta Astron.* **21**, 1
104. Paczynski B. (1976), in *Structure and Evolution of Close Binary Systems*, eds. P.P. Eggleton et al. (Reidel, Dordrecht) p. 75
105. Pfahl, E. Podsiadlowski, P., Rappaport, S.A. and Spruit, H. (2002). *ApJ* **574**, 364
106. Podsiadlowski, P. (1991). *Nature* **350**, 136
107. Podsiadlowski, P. and Rappaport, S.A. (2000). *ApJ* **529**, 946
108. Podsiadlowski, P., Rappaport, S.A. and Pfahl, E. (2002). *ApJ* **565**, 1107
109. Podsiadlowski, P., Rappaport, S.A. and Han, Z. (2002). *MNRAS*, submitted (astro-ph/0207153)
110. Podsiadlowski, Ph., Langer, N., Poelarends, A.J.T., Rappaport, S., Heger, A., and Pfahl, E. (2004). *ApJ* **612**, 1044
111. Pols, O.R., Tout, C.A., Eggleton, P.P. and Han, Z. (1995). *MNRAS* **274**, 964
112. Pols, O.R., Schröder, K.P., Hurley, J.R., et al. (1998). *MNRAS* **298**, 525
113. Portegies Zwart, S.F. and Yungelson, L.R. (1999). *MNRAS* **309**, 26p
114. Portegies Zwart, S.F. and McMillan, S.L.W. (2000). *ApJ* **528**, L17
115. Prestwich, A. et al. (2007). *ATel*, Nr.955
116. Psaltis,D. (2006), in *Compact Stellar X-ray Sources*, Eds. W.H.G.Lewin and M.van der Klis. (Cambridge Univ. Press), pp 1–38
117. Pylyser, E. and Savonije, G.J. (1988). *A&A* **191**, 57
118. Pylyser, E. and Savonije, G.J. (1989). *A&A* **208**, 52
119. Radhakrishnan, V. and Srinivasan, G. (1982). *Current Science* **51**, 1096
120. Radhakrishnan, V. and Srinivasan, G. (1984), in *Proc. Second Asian-Pacific IAU Regional Meeting, Bandung Indonesia 24-29 Aug.1981* Eds. B.Hidayat and M.W.Feast. (Tira Pustaka, Jakarta) p. 423
121. Rappaport, S.A., Verbunt, F. and Joss, P.C. (1983). *ApJ* **275**, 713
122. Rappaport, S.A., Podsiadlowski, P., Joss, P.C., et al. (1995). *MNRAS* **273**, 731
123. Rappaport, S.A. and van deen Heuvel, E.P.J. (1982), in *B-e stars, Proceedings of a Symposium in Munchen*, (Reidel, Dordrecht)
124. Refsdal, S. and Weigert, A. (1971). *A&A* **13**, 367
125. Reimers, D. (1975), in *Problems in Stellar Atmospheres and Envelopes*, eds. B. Bascheck, W.H.Kegel, G. Traving. (Springer, New York) p. 229
126. Reimers, D. and Koester, D. (1988). *ESO Messenger* **54**, 47
127. Ruderman, M.A. (1998), in *The Many Faces of Neutron Stars*, eds. R. Buccheri, J. van Paradijs and A. Alpar. (Kluwer, Dordrecht) p. 77
128. Ruderman, M.A., Shaham, J. and Tavani, M. (1989). *ApJ* **336**, 507
129. Savonije, G.J. (1978). *A&A* **62**, 317
130. Savonije, G.J. (1983), in *Accretion Driven Stellar X-ray Sources*, eds. W.H.G. Lewin and E.P.J. van den Heuvel. (Cambridge Uni. Press) p. 343

131. Savonije, G.J. (1987). *Nature* **325**, 416
132. Schaller, G., Schaerer, D., Meynet, G. and Maeder, A. (1992). *A&AS* **96**, 269
133. Scheck, L., Plewa, T., Janka, H.-Th., Mueller, E. (2004). *Phys.Rev.Lett.* **92**, id.011103
134. Scheck, L., Kifonidis, H., Janka, H.-Th., Mueller, E. (2006). *A&A* **457**, 963
135. Schreier, E., Levinson, R., Gursky, H., et al. (1972). *ApJ* **172**, L79
136. Shaham, J. (1992), in *X-ray Binaries and Recycled Pulsars*, eds. E.P.J. van den Heuvel and S.A. Rappaport. (Kluwer, Dordrecht) p. 375
137. Sidoli, L., Paizis, A. and Mereghetti, S. (2006). *A&A* **450**, L9
138. Shapiro, S.L. and Teukolsky, S.A. (1983). *Black Holes, White Dwarfs and Neutron Stars*, (Wiley-Interscience, New York), 645pp
139. Smarr, L.L. and Blandford, R.D. (1976). *ApJ* **207**, 574
140. Soberman, G.E., Phinney, E.S. and van den Heuvel, E.P.J. (1997). *A&A* **327**, 620
141. Spruit, H.C. and Ritter, H. (1983). *A&A* **124**, 267
142. Srinivasan, G. and van den Heuvel, E.P.J. (1982). *A&A* **108**, 143
143. Stairs, I.H. (2004). *Science* **304**, 547
144. Sugimoto, D. and Nomoto, K. (1980). *Space Sc. Rev.* **25**, 155
145. Taam, R.E. (1983). *ApJ* **270**, 694
146. Taam, R.E. and Bodenheimer, P. (1991). *ApJ* **373**, 246
147. Taam, R.E. and van den Heuvel, E.P.J. (1986). *ApJ* **305**, 235
148. Taam, R.E. and Fryxell, B.A. (1988). *ApJ* **327**, L73
149. Taam, R.E. and Sandquist, E.L. (2000). *ARA&A* **38**, 113
150. Tanaka, Y. and Lewin, W.H.G. (1995), in *X-Ray Binaries*, Eds. W.H.G.Lewin, J.van Paradijs and E.P.J.van den Heuvel. (Cambridge Univ. Press), p. 126
151. Tauris, T.M. (1996). *A&A* **315**, 453
152. Tauris, T.M. and Takens, R. (1998). *A&A* **330**, 1047
153. Tauris, T.M. and Savonije, G.J. (1999). *A&A* **350**, 928
154. Tauris, T.M., van den Heuvel, E.P.J. and Savonije, G.J. (2000). *ApJ* **530**, L93
155. Tauris, T.M. and Dewi, J.D.M. (2001). *A&A* **369**, 170
156. Tauris, T.M. and Sennels, T. (2000). *A&A* **355**, 236
157. Tauris, T.M. and van den Heuvel, E.P.J. (2006), in *Compact Stellar X-ray Sources*, Eds. W.H.G. Lewin and M. van der Klis. (Cambridge Univ. Press), p. 623
158. Tavani, M. (1992), in *X-ray Binaries and Recycled Pulsars*, eds. E.P.J. van den Heuvel and S.A. Rappaport. (Kluwer, Dordrecht) p. 387
159. Taylor, J.H. and Weisberg, J.M. (1989). *ApJ* **345**, 434
160. Terquem, C., Papaloizou, J.C.B., Nelson, R.P. and Lin, D.N.C. (1998). *ApJ* **502**, 588
161. Thompson, C. and Duncan, R. (1995). *MNRAS* **275**, 255
162. Timmes, F.X., Woosley, S.E. and Weaver, T.A. (1996). *ApJ* **457**, 834
163. Truemper, J, Pietsch, W., Reppin, C., Voges, W., Staubert, R. and Kenditzorra, E. (1978). *ApJ* **219**, L105
164. Tutukov, A.V. and Yungelson, L.R. (1973). *Nauchnye Informatsii* **27**, 70
165. van den Heuvel, E.P.J. (1981a). *Vistas in Astronomy* **25**, 95
166. van den Heuvel, E.P.J. (1981b), in *Fundamental Problems in the Theory of Stellar Evolution*, Eds. D.Sugimoto, D.Q.Lamb and D.N.Schramm. (Reidel, Dordrecht) p 155
167. van den Heuvel, E.P.J. (1983), in *Accretion-driven Stellar X-ray Sources*, Eds. W.H.G. Lewin and E.P.J.van den Heuvel. (Cambridge Univ. Press), p 303

168. van den Heuvel, E.P.J. (1992), in *X-ray Binaries and Recycled Pulsars*, Eds. E.P.J.van den Heuvel and S.A.Rappaport. (Kluwer Acad. Publishers, Dordrecht), p 233

169. van den Heuvel, E.P.J. (1994a), in *Interacting Binaries*, Saas-Fee course 22, Eds.H. Nussbaumer and A. Orr. (Springer, Heidelberg) p. 263

170. van den Heuvel, E.P.J. (1994b), in *The Evolution of X-ray Binaries, AIP Conf. Proceedings, Vol. 308*, Eds. S.Holt and C.S.Day. (American Institute of Physics Press, New York), p 18

171. van den Heuvel, E.P.J. (2004), in *Proc. 5th INTEGRAL Workshop, ESA SP-552*, Eds. V. Schoenfelder, G. Lichti and C.Winkler. (ESA Publ. Div. ESTEC, Noorwijk) p 185

172. van den Heuvel, E.P.J. (2005), in *The Electromagnetic Spectrum of Neutron Stars*, Eds. A. Baykal et al. (Springer, the Netherlands) p 191

173. van den Heuvel, E.P.J. (2006). *Advances in Space Res.* **38**, 2667

174. van den Heuvel, E.P.J. (2007). *Astro-ph/* 0704.1215v2

175. van den Heuvel, E.P.J. and Heise, J. (1972). *Nature – Physical Science* **239**, 67

176. van den Heuvel, E.P.J. and de Loore, C. (1973). *A&A* **25**, 387

177. van den Heuvel, E.P.J. and Taam, R.E. (1984). *Nature* **309**, 235

178. van den Heuvel, E.P.J., van Paradijs, J.A. and Taam, R.E. (1986). *Nature* **322**, 153

179. van den Heuvel, E.P.J. and Rappaport, S.A. (1987), in *Physics of Be-stars*, Proc. IAU Colloq. 92 (Cambridge Uni. Press) p. 291

180. van den Heuvel, E.P.J. and van Paradijs, J. (1988). *Nature* **334**, 227

181. van den Heuvel, E.P.J. and van Paradijs, J. (1997). *ApJ* **483**, 399

182. van den Heuvel, E.P.J., Portegies Zwart, S.F., Bhattacharya, D. and Kaper, L. (2000). *A&A* **363**, 563

183. van der Hucht, K.A. (2006). *A&A* **458**, 453

184. van der Klis, M. (2006), in *Compact Stellar X-ray Sources*, Eds. W.H.G.Lewin and M. van der Klis. (Cambridge Univ. Press), p 39

185. van Kerkwijk, M.H., Charles, P.H., Geballe, T.R., King, D.L., Miley, G.K., Molnar, L.A. van den Heuvel, E.P.J., van der Klis, M. van Paradijs, J. (1992). *Nature* **355**, 703

186. van Kerkwijk, M.H. and Kulkarni, S.R. (1999). *ApJ* **516**, L25

187. van Paradijs, J. and McClintock, J.E. (1995), in *X-ray Binaries*, Eds. W.H.G.Lewin, J. van Paradijs and E.P.J.van den Heuvel. (Cambridge Univ. Press), 58

188. Verbunt, F. and Zwaan, C. (1981). *A&A* **100**, L7

189. Verbunt, F. and Phinney, E.S. (1995). *A&A* **296**, 709

190. Verbunt, F. and van den Heuvel, E.P.J. (1995), in *X-ray Binaries*, eds. W.H.G. Lewin, J. van Paradijs and E.P.J. van den Heuvel (Cambridge Uni. Press)

191. Verbunt, F. and Lewin, W.H.G. (2006), in *Compact Stellar X-ray Sources*, Eds. W.H.G. Lewin and M. van der Klis. (Cambridge Univ. Press), 341

192. Voss, R. and Tauris, T.M. (2003). *MNRAS*, **342**, 1169

193. Walter, R., Heras, Z., Bassani, L. et al. (2006). *A&A* **453**, 133

194. Waters, L.B.F.M. and van Kerkwijk, M.H. (1989). *A&A* **223**, 196

195. Webbink, R.F. (1984). *ApJ* **277**, 355

196. Webbink, R.F., Rappaport, S.A. and Savonije, G.J. (1983). *ApJ* **270**, 678

197. Weidemann, V. (1990). *ARA&A* **28**, 103

198. Wellstein, S. and Langer, N. (1999). *A&A* **350**, 148
199. Wolszczan, A. (1994). *Science* **264**, 538
200. Woosley, S.E. and Weaver, T.A. (1995). *ApJS* **101**, 181
201. Woosley, S.E., Langer, N. and Weaver, T.A. (1995). *ApJ* **448**, 315
202. Zhang, C.M. (1998). *A&A* **330**, 195

Dynamical Formation and Evolution of Neutron Star and Black Hole Binaries in Globular Clusters

Monica Colpi and Bernadetta Devecchi

Department of Physics G. Occhialini, University of Milano Bicocca, Piazza della Scienza 3, 20126 Milano, Italy `colpi@mib.infn.it`

In the crowded stellar environment of a globular cluster, binaries hosting a collapsed object find their path of formation and evolution. When a neutron star is present, a Low-Mass X-ray Binary or a Millisecond Pulsar is observed. Detected in large numbers, these binaries are considered to be the most crystalline outcome of close gravitational encounters. This chapter describes the role of dynamical interactions in shaping the population of compact binaries, exploring the evolution of neutron stars and stellar-mass black holes inside our galactic globular clusters.

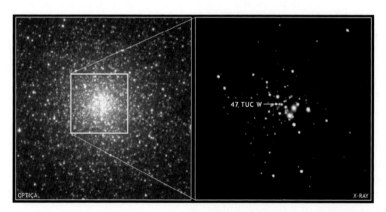

Fig. 1. Optical (left panel) and X-ray (right panel) images of the galactic globular cluster 47 Tucanae. 47 Tuc W is a binary Millisecond Pulsar spinning at 2.35 ms. Courtesy of http://chandra.harvard.edu/photo/2005/47tuc/index.html

Fig. 1 depicts the beauty of the globular cluster 47 Tucanae, as it appears in optical and X-rays. Compact stars, in the dense stellar core, reveal their nature through their X-ray emission, providing invaluable information on their true nature and on their strong dynamical coupling with cluster stars.

1 Introduction

Since the earliest discovery of a number of bright Low Mass X-ray Binaries (LMXBs) in galactic globular clusters GCs [33, 34, 12] and the detection of the first Millisecond Pulsar (MSP) (in M82 [55]), GCs have been considered to be preferred sites for the formation of exotic binaries.

LMXBs are binaries in which a neutron star, accreting from a companion star, is spun up to very short periods [1]. When accretion comes to a halt, the rapidly rotating, weakly magnetized neutron star turns on as a radio pulsar evolving into a binary MSP (bMSP). It is known that LMXBs and bMSPs form in the galactic field from primordial binaries and in GCs, but what surprises most is that GCs host more X-ray sources and MSPs than one would expect from their mass. Indeed LMXBs are at least a hundred times more numerous per unit mass than in the galactic field, and this excess has been interpreted to be a consequence of the high GC stellar densities which lead to the creation of binaries with a neutron star as a result of close stellar encounters [11, 54]. In the GC cores neutron stars can be tidally captured during encounters with single stars or be exchanged with lighter stars in pre-existing binaries, enhancing the probability of their recycling [100].

At present, the neutron star census in the galactic GCs comprise thirteen *bright* ($L > 10^{36}$ erg s^{-1}) LMXBs [107] in which the neutron star is actively accreting from a low-mass companion star via Roche Lobe overflow. Six of these are *transient* soft X-ray sources, currently in outburst with recurrence times from months to years. Orbital periods have been identified for a number of LMXBs, and three are found below 1 hr, indicating the presence of a white dwarf as companion star [107]. In addition to the bright sources, GCs host a large number of dim, *quiescent* LMXBs (qLMXBs: $L \sim 10^{31-33}$ erg s^{-1}). More than 100 are expected to exist in the galactic GC system [47], some representing the quiescent counterparts of the soft X-ray transients. Accretion is not active or significantly reduced in these sources: their X-ray luminosity is thought to derive from the re-radiation of heat generated by pycno-nuclear reactions occurring during major outbursts [7, 13, 9], and/or from magneto-spheric emission of a turn-on pulsar [9].

Up to now 140 MSPs have been observed in radio (and some have X-ray counterparts as the one depicted in Fig. 1) with more than 10, 20 and 30 sources in the clusters M28 [102], 47 Tucanae [8], and Terzan 5 [95], respectively. These MSPs, single (having lost the companion star in dynamical interactions) or in binaries (bMSPs) offer the unprecedented possibility of carrying out accurate statistical studies inside individual GCs, to reveal the details of their evolutionary link with the underlying population of neutron stars in (q)LMXBs.

Pulsars in relativistic binaries with a neutron star companion, that are a subject of this book, are also expected to form dynamically in GCs [106] and to outnumber those of primordial origin [86, 58]. At present, the only known GC "double neutron star" (DNS) is PSR B2127+11C living in the outskirts of M15 [112]. PSR B2127+11C is a 30.5 ms pulsar in a relativistic 8 hr orbit around a neutron star for which there is a direct measurement of orbital period decay [53].

If neutron stars dynamically couple to cluster stars, the same should hold true for the stellar-mass black holes, present in the GCs as relic of the most massive stars. So far, no stellar-mass black hole candidates have been identified in the galactic GCs, neither as bright LMXBs nor as qLMXB, and this paucity has been interpreted again as an outcome of dynamical encounters that led to their ejection.

Given the ubiquitous presence of MSPs and X-ray binaries in GCs, the joint study of the formation and evolution of neutron stars and stellar–mass black holes has become of utmost importance, as it is a theoretically well posed and very challenging problem. The target time for direct N-body codes to address the dynamical and internal (stellar) evolution of a million of stars is around 2020 [51]. Meantime, various approaches have been proposed and developed that provide a clear and deep view on this problem (see [28] for original references on the Monte Carlo code).

In Sect. 2, we set the context, exploring the role of two-body relaxation on the cluster as a whole; in particular unstable mass segregation play a key role during the approach to equipartition of kinetic energies. In Sect. 3, we focus attention on close encounters that are central for understanding the formation of compact binaries in GCs, surveying over 3 and 4-body encounters among stars and binaries. In Sect. 4, we summarize how neutron stars are produced and retained in GCs, outlining the most important dynamical channels and evolutionary events that transform a mass-transferring binary into a recycled MSP. In Sect. 5, we explore the dynamics of stellar-mass black holes in GCs considering the possibility of creating intermediate-mass black holes in GCs. We also discuss the formation of some exotica that could be added to the list of relativistic binaries, suitable for testing gravity theories. In the same section we take the liberty of discussing a number of open issues regarding relativistic binaries hosting either neutron stars or black holes in GCs and their ejected products, after reviewing main-stream findings.

2 Dynamical Context

2.1 Equilibria

GCs can be described as self-gravitating equilibria of point-mass stars [5], to zeroth order. Equilibria are constructed assuming that: (i) the gravitational potential ϕ is a slowly varying function of the position vector; (ii) spherical symmetry applies so that ϕ is just a function of radius r; (iii) stars are described, in the point-mass limit, by an isotropic distribution function f that depends upon velocity and position just through the constants of motion, i.e. the energy per unit mass $E = (1/2)v^2 + \phi(r)$, and the modulus of the angular momentum vector.

Under these simplifying assumptions, a variety of models have been used to describe the equilibria in the continuum limit, and in particular: (1) the King model [63] with a surface density profile (projected on the sky) that fits satisfactorily the photometric data of GCs; (2) the Plummer model [87] which is favorite in theoretical studies of GC gravothermal collapse. Both models have a distribution function f depending on E only and on a number of free parameters. King's models are an extension of the Singular Isothermal Sphere (SIS) described by a Maxwellian distribution function with the line-of-sight velocity dispersion σ. SIS has a density profile $n = \sigma^2/(2\pi m G r^2)$, a constant mean square velocity $\langle v^2 \rangle = 3\sigma^2$, and is characterized by an infinite mass and infinite central density. Compared to SIS, King's models are "lowered isothermal models" determined by three parameters: σ (as in SISs), the central density n_c, and the concentration parameter c (or tidal radius R_t). The distribution function is a Maxwellian only at energies $E < E_{es}$ (where E_{es} is the escape energy), and zero elsewhere, so that total mass and radius are finite, contrary to the case of SIS. The Plummer model has a distribution function $f \propto (-E)^{7/2}$ and is characterized by a scale radius a and total mass M and these two parameters alone define uniquely the properties of the equilibrium, such as the one-dimensional velocity dispersion, central density, and core and virial radii [46].

In order to account for the "granularity" of matter, it is necessary to modify this background model, in order to include:
• stars with different masses according to the initial mass function (IMF);
• the presence of primordial binaries;
• stellar evolution (mass loss and formation of compact objects like white dwarfs, neutron stars and stellar-mass black holes);
• stellar encounters (single–single, single–binary, binary–binary, and higher order interactions);
• physical collisions and star mergers.

All these effects can be treated as *perturbations* over the background equilibrium model. Stellar evolution is extremely important in the early phases of the cluster life due to supernova explosions and mass loss from stellar winds that alter the gravitational potential and eventually unbound the entire system [66]. On the contrary, stellar encounters play a role also at late times, i.e. when all the massive stars have died, and the lighter evolve less violently.

GCs are expected to transit along different near-equilibrium states, due to stellar encounters that alter the energy of single stars and in turn of the system as a whole as energetic stars are ejected from the cluster. GCs display the tendency to collapse, i.e. to develop large central densities, and in this phase exotic binaries have the highest probability to form. *Distant* two-body encounters change the velocities of individual stars slightly but their collective and repeated action gradually alter the underlying equilibrium. *Close* encounters, in contrast, lead to sudden (albeit rarer) changes in the star's motion reversing core collapse and causing (a) formation and/or destruction of binary stars; (b) formation of systems such as LMXBs and MSPs where a neutron star is involved; (c) formation of double neutron stars or stellar–mass black holes binaries; (d) ejection of stars and binaries into the galactic field.

2.2 Relaxation

GCs are considered to be *collisional* systems as *distant* two-body encounters have time to play a role during their lifetime. The degree of *collisional relaxation* is measured by the timescale t_r. Over this time, the energy of stars is altered by the (two-body) long-distant interactions with other stars.

For a population of stars of equal mass m, density n and one-dimensional velocity dispersion σ, the relaxation time is

$$t_r = 0.34 \frac{\sigma^3}{G^2 m^2 n \ln \Lambda} \tag{1}$$

where $\ln \Lambda \sim 10$ is the Coulomb logarithm [5]. Depending on the local values of σ and n the relaxation time varies of several orders of magnitude in the different regions of the cluster. For this reason it is necessary to evaluate t_r at different characteristic radii. For GCs, t_r is computed at the half-mass radius r_h. Since $n(r_h) \equiv 3M/(8\pi m r_h^3)$ (where M_{GC} is the total cluster mass) and $\sigma^2 \simeq 0.4GM/(3r_h)$ (from the virial theorem):

$$t_{rh} \simeq \frac{7 \times 10^8 \text{ yr}}{\ln \Lambda} \left(\frac{M_{GC}}{10^5 M_\odot} \right)^{1/2} \left(\frac{M_\odot}{m} \right) \left(\frac{r_h}{1 \text{ pc}} \right)^{3/2} \tag{2}$$

A second possibility is to evaluate t_r at the core radius $r_c = (9\sigma^2/4\pi Gmn_c)^{1/2}$, where n_c is the core stellar density

$$t_{rc} \simeq 3 \times 10^6 \text{yr} \left(\frac{\sigma}{7 \text{ km s}^{-1}} \right)^3 \left(\frac{M_\odot}{m} \right)^2 \left(\frac{10^5 \text{ pc}^{-3}}{n_c} \right) ; \tag{3}$$

t_{rc} is measuring the degree of relaxation in the denser regions of a GC that are experiencing collapse during the GC lifetime. While t_{rh} remains rather stable during the aging of the GC, t_{rc} can vary by orders of magnitude [39].

2.2.1 Escape in Single-Mass Clusters

The effect of stellar interactions is known to be many-fold: encounters lead to global variations in the near equilibrium structure of the GC, and also to changes in the stellar content. As mentioned earlier, *distant* encounters introduce small, repeated changes in the specific energy E of the stars, while *close* encounters alter suddenly and significantly the energy of the star in a single event. Both these two interactions can cause the *escape* of stars from the GC via two different mechanisms:

(i) *Evaporation*: a series of weak distant encounters can increase gradually the energy of a star, until a single weak encounter leads the star to escape just with a slightly positive energy ($E \sim 0$).

(ii) *Ejection*: a close encounter of a star with another single star or with a binary can produce a velocity change comparable to or in excess of the initial velocity thereby releasing the star into an escape orbit. The star that escapes suddenly carries away an energy $E < 0$ from the cluster.

In single-mass star clusters, core collapse appears to be a two-phase process. The first phase is driven by mass loss due to evaporation: encounters (occurring preferentially in the core where the density is the highest) cause stars to diffuse to higher and higher kinetic energies (i.e. to lower total energies E). The diffusion in energy space causes the stars to gradually populate the cluster halo, and in response to the expansion of the outer region, the core contracts and heats up, as required by energy conservation and virial equilibrium. An energy gradient is established that determines the energy flux from the core to the halo. Acting alone, evaporation would cause the collapse of the core on a timescale $t_c \sim 100 t_{rh}$. However, after a time $\simeq 3 t_{rh}$ a second process intervenes. The second phase starts when the core that continues to lose energy becomes denser and dynamically hotter to the point that its local relaxation time t_{rc} becomes much shorter than t_r at the larger radii where the heat flux drove the initial contraction. The hot central region of the cluster then decouples dynamically from the outside and becomes "thermally isolated". At this stage, collapse in not anymore driven by the heat flux across the core–halo structure, but it is self-regulated by local relaxation inside the core. As this nearly isothermal region contracts further and rapidly, the density ($n \propto r^{-2}$) becomes singular at the origin (as in a SIS [5],[69]). In a real cluster, this state of nominal "infinite" density is never reached since the star's dynamics is largely determined by the physics of binary stars. In the high stellar environment of a contracting core, close-body interactions with binary stars act as a source of energy which is capable of halting and reversing the collapse. It is also at this time when the core densities are the highest, that *relativistic binaries*, as double neutron star systems are expected to form [35].

Core collapse in single-mass clusters can be seen as a manifestation of the inability of self-gravitating systems to develop a Maxwellian distribution. This distribution is never attained because a Maxwellian allows arbitrarily high speeds (i.e. $E > 0$) for a small fraction of stars. These stars can not remain in the system, but rather escape driving core contraction.

2.2.2 Mass Segregation in Multi-Component Clusters

As stars form over a variety of masses, only multi-component systems describe real clusters, and in sharp contrast with the collapse in single-mass clusters, it is *mass segregation* that drives their gravothermal collapse.

Mass segregation is a relaxation process describing the tendency of stars to reach *equipartition of kinetic energy* [46, 5]. For the simplest case of stars with masses m_1 and m_2 alone, equipartition ideally leads to the condition $m_1\langle v_1^2\rangle = m_2\langle v_2^2\rangle$ [62]. This is a consequence of the fact that the heavier stars have the tendency to transfer their excess energy to the lighter stars, through relaxation.

In more detail, the tendency to equipartition of kinetic energies is the result of the competing action of two-body relaxation and *dynamical friction*, i.e. relaxation among stars with different masses. In the weak scattering approximation (i.e. large impact parameters) and for an isotropic (spherical) background of N stars with distribution function f, the velocity change, over time t, experienced by a single more massive star of mass m_1 due the collective action of independent distant two-body encounters with stars of mass m_2 (with $m_1 >> m_2$) can be expressed in the form [46]

$$\frac{1}{t}\sum \delta\mathbf{v}_1 = -4\pi G^2 m_2(m_1+m_2)\ln\lambda\, N \frac{\mathbf{v}_1}{|\mathbf{v}_1|^3}\int_{v_2<v_1} f(v_2)d^3\mathbf{v}_2. \quad (4)$$

The star m_1 is subject to a *deceleration*, called *dynamical friction*, caused only by the stars moving slower than m_1 [10, 5]. This deceleration increases with the mass of the star m_1, and can be viewed as the response (back-reaction) of m_1 to the gravitational pull exerted by the stars m_2 that create a density wake behind the trail of m_1 due to their gravitational interaction with m_1.

Since the frictional drag depends on m_1, this implies that the heavier stars tend to slow down. Thus, if there is an excess in the kinetic energy of the massive component, heavier stars transfer, in statistical terms, kinetic energy to the lighter stars. To contrast this drag that can be viewed as a cooling process, two-body relaxation acts to increase the mean square velocity $\langle v_1^2\rangle$ of m_1, at a rate

$$\frac{1}{t}\sum(|\delta\mathbf{v}_1|)^2 = +8\pi G^2 m_2^2\ln\lambda\, N\left(\frac{1}{v_1}\int_{v_2<v_1}f(v_2)d^3\mathbf{v}_2+\int_{v_2>v_1}\frac{1}{v_2}f(v_2)d^3\mathbf{v}_2\right)$$
$$(5)$$

that unlike dynamical friction does not depend on the mass m_1.

The sum of the changes in the specific kinetic energy of m_1 caused by all encounters

$$\sum \delta E_1 = \mathbf{v}_1\cdot\sum\delta\mathbf{v}_1+\sum(\delta\mathbf{v}_1)^2/2 \quad (6)$$

can thus be written in terms of two competing effects

$$\frac{1}{t}\sum \delta E_1 = 4\pi G^2 m_2 \ln \lambda N \left(-\frac{m_1}{v_1} \int_{v_2 < v_1} f(v_2)d^3\mathbf{v}_2 \right.$$
$$\left. + m_2 \int_{v_2 > v_1} \frac{1}{v_2} f(v_2)d^3\mathbf{v}_2 \right), \tag{7}$$

the first, proportional to $m_2 m_1$ and to the number of stars moving slower than v_1, that decreases the kinetic energy, and the second, proportional to m_2^2 and to the number of stars moving faster than v_1, that increases the kinetic energy of m_1. For Maxwellian distributions, the mean rate of change of the kinetic energy $m_1 E_1$ of stars of mass m_1 is

$$\left\langle \frac{d}{dt}(m_1 E_1) \right\rangle = \frac{4\sqrt{3\pi}G^2 m_1 m_2 n_2 \ln(N\lambda)}{(\langle E_1 \rangle - \langle E_2 \rangle)^{3/2}} (m_2 \langle E_2 \rangle - m_1 \langle E_1 \rangle) \tag{8}$$

where n_2 is the number density of stars of mass m_2 (equation for stars m_2 can be written switching label 1 with 2). Thus, encounters tend to drive the species toward equipartition since the sign of this rate of change (eq. [8]) depends on the comparison between $m_1 \langle E_1 \rangle$ and $m_2 \langle E_2 \rangle$. Tendency to equipartition occurs on an e-folding time

$$t_{eq} = \frac{(m_1 \langle E_1 \rangle + m_2 \langle E_2 \rangle)^{3/2}}{4(3\pi)^{1/2} G^2 m_1 m_2 \log(N\lambda)(n_1 + n_2)}. \tag{9}$$

For many galactic GCs t_{eq} is shorter than the age of the cluster, suggesting that many GCs have experienced mass segregation [64].

Each star in a GC moves in the mean field potential $\phi(r)$ generated by all the other stars, thus any loss or gain of kinetic energy occurs in a region of space where self-gravity is important. Accordingly, stars losing energy settle into more bound fastly moving orbits, and thus segregate in the core. By contrast stars gaining energy settle into less bound peripheral orbits. Thus, in the attempt to reach equipartition of kinetic energies, mass segregation leads to a *spatial segregation*. Dynamical friction acting on the heavier stars drives more rapid core contraction, and the GC can evolve into states that are progressively farther "away" from equipartition [101].

2.2.3 Spitzer Instability

The "failure" to reach equipartition can be explained by considering, for simplicity, a two-mass component system, comprising N_1 stars with mass m_1, and N_* stars of mass m_*; further, let $m_1 > m_*$ and the total masses $M_1 < M_*$, according to an IMF poorer in massive stars. Consider that stars m_* are distributed inside a homogeneous core of mass density ρ_* and size R_*. In the underlying harmonic potential, the characteristic velocity of stars m_* is $\langle v_* \rangle = (4\pi G\rho_*/3)^{1/2}R_*$. Assume that equipartition is attained, so that $\langle v_1^2 \rangle = (m_*/m_1)\langle v_*^2 \rangle$. In the self-gravitating background generated by the M_* stars, the heavier stars segregate toward the center, and in order to attain equilibrium they settle into a region of size $R_1 \simeq \sqrt{m_*/m_1}R_*$. Inside R_1 there is a mass, in heavy stars, equal to M_1, and a mass $M_*(R_1) \simeq (m_*/m_1)^{3/2} M_*$ in light stars. If $M_1 > M_*(R_1)$, i.e. if

$$M_1 > M_*(m_*/m_1)^{3/2} \tag{10}$$

the heavier stars feel increasingly the effect of their own self-gravity. Slowed down by the interaction with the lighter stars, they contract and the increase in their velocity dispersion drives them away from equipartition. The heavier stars then evolve into a self-gravitating system that decouples dynamically from the rest of the stars of mass m_*. The inner core, emptied of all light stars, evolves as if in isolation, leading to the so-called mass-segregation instability.

Spitzer [101] was the first to highlight the occurrence of this instability in two-component self-gravitating systems. Using analytic methods and a number of simplifying assumptions, he determined a simple criterion for the onset of this instability. Keeping the notation introduced above, a runaway mass-segregation instability occurs if

$$S \equiv (M_1/M_*)(m_1/m_*)^{3/2} > 0.16. \tag{11}$$

Watters et al. [110], using numerical simulations, obtained a more accurate empirical condition

$$W \equiv (M_1/M_*)(m_1/m_*)^{2.4} > 0.32. \tag{12}$$

Multi-mass clusters accelerate core collapse via mass segregation, and this leads to the development of densities as high as 10^5–10^6 stars pc^{-3} so that the core becomes the preferred site where close encounters among stars occur.

3 Close Encounters

Due to the high stellar densities that develop in GCs following mass segregation and core collapse, *close gravitational encounters* start playing a key role in *shaping the stellar populations* and in affecting the large-scale dynamics of the GC itself. Close single-binary and binary–binary encounters in the cluster core can (i) lead to the formation of new exotic binaries, and (ii) affect the GC energy balance since kinetic energy can be extracted from the internal degrees of freedom of the binary and deposited in the GC.

3.1 3-Body Encounters

3.1.1 End-States

In GCs, binary stars can either be primordial, i.e. born with the cluster, or have dynamical origin, i.e. the result of a tidal interaction between two stars or the outcome of a 3-body encounter involving unbound states of three stars at a time. In this last case, the encounter ends with the formation of a binary star and a single star:

$$1+2+3 \rightarrow (1,2) + 3.$$

Once a binary is present, the binary itself can act as seed for further interactions with single stars, and is a catalyst for the formation of exotic binaries as LMXBs, bMSPs and double neutron star or black hole binaries.

In the context of 3-body encounters, we can distinguish between *prompt* interactions where the mean squared distance between the stars has only one minimum [52, 99], and *resonant* interactions where there is more than one minimum. The end-state of the binary–single star encounter can then be classified in [99]:

1. **Fly-by:** the end-state is identical to the initial one

$$(1,2)+3 \rightarrow (1,2)+3;$$

2. **Exchange:** the field star exchanges with one of the two binary stars becoming a new member of the binary

$$(1,2)+3 \rightarrow (1,3)+2$$
$$(1,2)+3 \rightarrow (3,2)+1;$$

3. **Ionization:** the three stars are released on unbound orbits

$$(1,2)+3 \rightarrow 1+2+3;$$

4. **Merger:** the two stars in the binary collide following the encounter with a third star. A large increase of the eccentricity can bring the stars so close that their surfaces touch and eventually coalesce, i.e.

$$(1,2)+3 \rightarrow (\text{merger})+3;$$

5. **Triplet:** three stars form an unstable bound system

$$(1,2)+3 \rightarrow (1,2,3).$$

Figs. 2, 3 and 4 depict a series of encounters between a neutron star (of $1.4 \, M_\odot$) and an equal-mass binary (of total mass of $1.4 \, M_\odot$) to illustrate the different outcomes: fly-bys and exchanges (Figs. 2 and 3 respectively) that can either be prompt (left panel) or resonant (right panel), and prompt ionization (Fig. 4). For the focus of this chapter, the interactions that play a key role are fly-bys and exchanges.

3.1.2 Energetics

Energy transfer is central to our understanding of binary formation and evolution under the influence of dynamical encounters. The energy $E_{\text{tot,b}}$ of a binary with components of mass m_1 and m_2 is given by the sum of the kinetic energy of the center of mass $E_{\text{kin,b}}$ and of the binary "internal" energy E_{b} :

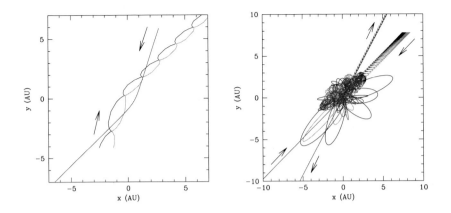

Fig. 2. Fly by: an equal-mass binary star (of total mass 1.4 $M_\odot$) interacting with a 1.4 $M_\odot$ neutron star in a prompt (left panel) and resonant (right panel) encounter. Arrows indicate the incoming and outgoing states. The neutron star is in red. Initially the incoming star and the binary orbit in the xy plane. SEE COLOUR PLATES

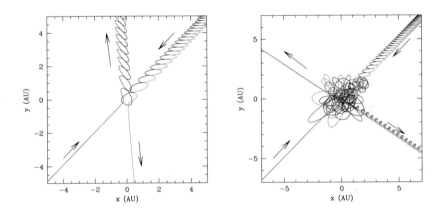

Fig. 3. Exchange: an equal-mass binary star (of total mass 1.4 $M_\odot$) interacting with a 1.4 $M_\odot$ neutron star in a prompt (left panel) and resonant (right panel) encounter. Arrows indicate the incoming and outgoing states. The neutron star is in red. Initially, the incoming star and the binary orbit in the xy plane. SEE COLOUR PLATES

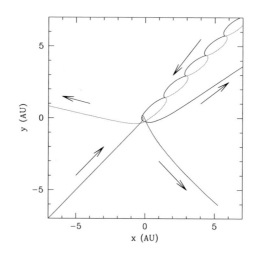

Fig. 4. Ionization: an equal-mass binary star (of total mass 1.4 $M_\odot$) interacting with a 1.4 $M_\odot$ neutron star in a prompt encounter. Arrows indicate the incoming and outgoing states. The neutron star is in red. Initially, the incoming star and the binary orbit in the xy plane. SEE COLOUR PLATES

$$E_{\text{tot,b}} = (m_1 + m_2)\frac{V^2_{\text{cm,b}}}{2} - \frac{1}{2}\frac{Gm_1m_2}{a} \tag{13}$$

where a is the binary semi-major axis. If we add a third body of mass m_3, initially at a large distance from the binary, then the total energy in the center of mass of the triple system is

$$E_{\text{tot,3}} = E_{\text{b}} + \frac{1}{2}\frac{(m_1 + m_2)m_3}{m_T}V^2 \tag{14}$$

where $m_T = m_1 + m_2 + m_3$, and V is the speed of m_3 relative to the barycenter of the binary. The last term in Eq. (14) represents the kinetic energy of the intruder star and of the binary center of mass.

During single-binary interactions, energy from (to) the binary can be transferred to (from) the single star. This leads to a change in the binary semi-major axis and eccentricity, in V, and also to the possibility of changing the binary membership.

The effect of an interaction can be estimated computing the fractional change in the internal energy of the binary, i.e. $\Delta = (E_{\text{b}}^{\text{f}} - E_{\text{b}}^{\text{i}})/E_{\text{b}}^{\text{i}}$ where the superscript f and i refer to the final and initial states, respectively. If $\Delta > 0$, the internal energy of the binary after the interaction has decreased, i.e. the binary has become more bound. Thus, according to energy conservation, the single star exits the encounter with a final speed greater than the initial speed.

This is due to the fact that a fraction of the binding energy ($\equiv |E_b|$) of the binary has been transferred from the internal degrees of freedom of the binary to the third star. Also the barycenter of the binary can acquire kinetic energy due to linear momentum conservation. On the contrary, if $\Delta < 0$ the energy of the binary increases (i.e. the binary is less bound), and the star exits the encounter with a lower speed.

The fractional energy transfer Δ in a single star-binary encounter is found to be sensitive to the dimensionless parameter p/a [49], where p represents the pericenter of the relative hyperbolic orbit, i.e. the minimum distance between the incoming star and the center of mass of the binary. The dependence on p/a can be exponential: $\Delta \propto (p/a)^{3/2} \exp[-B(p/a)^3]$, (with B a parameter related to the stellar masses and asymptotic velocity [41]). This indicates that little energy and angular momentum are exchanged when p is larger than the semi-major axis of the binary, i.e. $p > a$. In this case a gentle fly-by occurs. By contrast, when $p \leq a$, the energy exchanged among the three stars is non-negligible and in this case we need to distinguish between two regimes, i.e. when the binary is *hard* or when it is *soft*.

As a rule of thumb, a binary is *hard* (*soft*) when its binding energy $|E_b|$ is larger (smaller) than the kinetic energy of the incoming star, i.e. the dividing line separating hard and soft binaries is when

$$\frac{G(m_1 m_2)}{2a} \sim \frac{(m_1 + m_2)m_3 V^2}{2m_T}. \tag{15}$$

There is also a minimum asymptotic velocity for the star to ionize the binary, obtained for $E_{tot,3} = 0$, so that for $V > V_{ion}$ the binary can be dissociated:

$$V_{ion} = \left(\frac{Gm_T}{a} \frac{m_1 m_2}{(m_1 + m_2)m_3} \right)^{1/2}. \tag{16}$$

There is clear evidence, supported by extensive numerical experiments, that encounters tend to establish equipartition of "kinetic" energies. Accordingly, it is possible to show, in statistical terms, that [46]

soft binaries become softer - hard binaries become harder

In more detail, we can distinguish different outcomes:

• **Soft binaries -**

(i) *Ionizations* occur when the single star is approaching the binary with $V > V_{ion}$. In this case, the perturbation can be so large that one of the binary components (the closest that has been approached) increases its velocity above escape, and the binary ionizes.

(ii) *Fly-bys* occur when the single star is approaching the binary with $V < V_{ion}$, but V is still large compared to the orbital velocity of the two stars in the binary, that one (or the two) star(s) is(are) boosted to a higher velocity. In this case, the internal energy of the binary changes by a positive amount and the binary softens.

(iii) *Exchanges* occur when the incoming star comes sufficiently close to one binary component and with the right velocity that m_3 takes its place.

• **Hard binaries** -

(i) *Fly-bys*: If the binary is *hard* and kinetic energy is transferred from the binary to the single star that recedes with a speed higher than the speed of approach. In response to this energy transfer the binary further hardens, and, as guiding rule, the fractional binding energy increase is of the order of

$$\Delta \sim \xi \frac{m_3}{(m_1 + m_2)}, \tag{17}$$

with ξ varying between 0.2 and 1 [49]. For a binary with $m_1 \gg m_2$, the fractional energy exchange depends on m_2 only, and reads: $|\Delta E_{\rm b}| \simeq \xi(Gm_2 m_3/2a)$.

In fly-bys off a hard binary, the star m_3 exits the interaction with a velocity, at infinity, given by

$$V^2_{\infty,\rm out} = V^2_{\infty,\rm in} + \xi \frac{Gm_1 m_2}{(m_1 + m_2)a,} \tag{18}$$

which in the limit of $V_{\infty,\rm out} \gg V_{\infty,\rm in}$ (where $V_{\infty,\rm in}$ is the incoming relative velocity of m_3 prior to the encounter). For $m_1 \gg m_2$, $V_{\infty,\rm out}$ is proportional to the circular velocity of a test particle orbiting around m_2:

$$V^2_{\infty,\rm out} \sim \xi \frac{Gm_2}{a}. \tag{19}$$

In the same limit, the binary center of mass recoils with a velocity

$$V^{\rm bin}_{\infty,\rm rec} \sim \left(\frac{m_3}{m_1}\right) V_{\infty,\rm out} \tag{20}$$

due to linear momentum conservation.

(ii) *Exchange:* The star m_3 may also exchange with one of the binary components, if it comes in vicinity of one of the two stars. First, the star m_3 temporarily transfers its kinetic energy to the binary. A triplet form which is unstable so that the final end-state is again a single star plus a binary. In general, the probability of exchange of m_3 with one binary member is 2/3 in case of equal mass stars. When the stars have unequal masses the most probable end-state is the one in which the binary is composed by the two heavier stars [42].

3.1.3 Cross Sections and Rates

Cross sections are critical for determining the rate of occurrence of 3-body encounters in GCs. Gravity has an infinite radius of influence, thus cross sections refer to specific outcomes and are computed directly from scattering experiments.

The physical parameters characterizing the relative orbits between the three stars of given mass require knowledge of nine independent parameters that can be generated according to specific distribution functions [52, 99]. Cross sections are thus inferred selecting the desired end-states.

It is possible to estimate the cross section relative to encounters for which the energy transfer is non negligible (i.e. encounters with $p \sim a$) approximating the binary–single star encounter as a two-body interaction between the binary and the single star.

Conservation of energy and angular momentum lead to a relation between to the impact parameter b, the pericenter p and the pre-encounter relative velocity V_∞:

$$ p = \frac{Gm_T}{V_\infty^2} \left(\sqrt{1 + b^2 \left(\frac{V_\infty^2}{Gm_T} \right)^2} - 1 \right). \tag{21} $$

For small impact parameters, i.e. for $b \ll Gm_T/V_\infty^2$, Taylor expansion gives $p \sim b^2 V_\infty^2/(2Gm_T)$. Setting $p \sim a$ and solving for b, the total cross section, defined as πb^2, reads

$$ \Sigma \equiv \pi b^2 \sim \pi a^2 \left(1 + \frac{2Gm_T}{aV_\infty^2} \right). \tag{22} $$

Eq. (22) shows that the geometrical cross section of the binary, πa^2, is enhanced by *gravitational focusing* by an amount $2Gm_T/aV_\infty^2$. The condition for this increase to be important is that the incoming star has an asymptotic velocity $V_\infty^2 < 2Gm_T/a$. The reaction rate R, i.e. the number of "close" encounters per unit time with the binary then reads

$$ R \sim nV_\infty \Sigma \tag{23} $$

where n is the number density of single stars and nV_∞ the flux of stars impinging on the binary; for stars in a GC having velocity $V_\infty \sim \sigma$,

$$ R \sim 2\pi an \frac{Gm_T}{\sigma}. \tag{24} $$

Thus, if $\tilde{g}(X)$ is the fraction of scattering experiments associated with process X (e.g. fly-by, or exchange, or ionization), one can define the cross section relative to the specific outcome X as $\Sigma_X = \Sigma \tilde{g}(X)$ [99].

3.1.4 Binary Hardening

Hard binaries evolve inside GCs: their evolution is determined by the number of close encounters experienced during their lifetime. Hardening causes a progressive decrease of the semi-major axis and random changes in the eccentricity [43]. In the case of ordinary stars, hardening can lead to unstable mass transfer ending with stellar coalescence. If the binary is composed of *two compact objects*, the binary can enter the domain where gravitational wave emission becomes important, guiding the inspiral of the two objects down to coalescence.

If we consider the case of a binary of two compact objects in interaction with a light star of mass m_3, we can model the fractional energy exchange Δ as in Eq. (17), and compute the rate of change of the binding energy due to "repeated" close encounters with stars of density $\rho = m_3 n$:

$$|\dot{E}_\mathrm{b}| \sim \xi \frac{m_3}{m_1 + m_2} |E_\mathrm{b}| R \sim 2\pi\xi \frac{G\rho}{\sigma} G m_1 m_2. \tag{25}$$

In the hypothesis that the stellar density ρ and velocity dispersion σ are independent of time, hardening of the binary occurs at a constant rate. The semi-major axis a decreases with time according to

$$\dot{a} = -H \frac{G\rho}{\sigma} a^2, \tag{26}$$

where with H ($\sim 2\pi\xi$) we bracket uncertainties in the estimate of the averaged fractional energy transfer related to the characteristics of the binary (mass ratio and eccentricity) and to the range of initial impact parameters (H varying between 1.3 to 16: [94, 76]). The rate of change of a decreases as a^2, so that the hardening time

$$t_\mathrm{hard} \equiv |a/\dot{a}| \sim \sigma/(HG\rho a) \tag{27}$$

increases as the binary shrinks mainly because of the decrease of the binary cross section.

Gravitational wave back reaction accelerates the orbital decay that occurs on a timescale [84]

$$t_\mathrm{GW} = \frac{5}{256} \frac{c^5}{G^3} \frac{a^4(1 - e^2)^{7/2}}{m_1 m_2 (m_1 + m_2)}. \tag{28}$$

In the stellar background of a GC, a binary exposed to close encounters can reach a characteristic semi-major axis a_GW below which gravitational wave emission dominates: this length scale can be calculated imposing $t_\mathrm{GW} = t_\mathrm{hard}$ and reads

$$a_\mathrm{GW} = \left(\frac{256}{5} \frac{G^2}{c^5} \frac{\sigma}{H\rho} \frac{m_1 m_2 (m_1 + m_2)}{(1 - e^2)^{7/2}} \right). \tag{29}$$

In various circumstances (e.g. in the case of double neutron stars or double black holes formed dynamically) $t_\mathrm{GW}(a_\mathrm{GW})$ is less than the GC lifetime so that exotic binaries can coalesce inside GCs.

3.1.5 Binary–Binary Encounters and the Formation of Stable Hierarchical Triplets

GCs are believed to be rich in binaries, in particular in their youth and during core collapse. Thus, binary–binary encounters can be very frequent over the GC lifetime.

Binary–binary interactions possess more degrees of freedom than single-binary interactions so that their end-states are richer than in the 3-body case. An interesting outcome refers to the formation of a stable hierarchical triplet

$$(1,2)+(3,4) \rightarrow ((1,2),3)+4.$$

A hierarchical triplet is a configuration arranged in a pair of "nested" binaries: more precisely it consists of an "inner" binary of components m_1, m_2 with semi-major axis $a_{\rm in}$, orbited by a third star m_3 so that the semi-major axis of the outer binary satisfies the condition $a_{\rm out} >> a_{\rm in}$. Fig. 5 shows such a configuration.

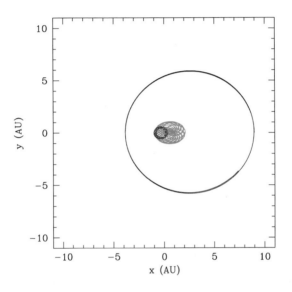

Fig. 5. A hierarchical triple system consisting of an inner binary of a $10\,M_\odot$ black hole and a $1.4\,M_\odot$ neutron star, orbited by a $1\,M_\odot$ star.

These triple systems are seen to form in N-body simulations of star clusters [104]. As the presence of the inner close binary can slow down the computational time considerably, it is also of practical importance to be able to determine if the triplet is stable or not. Indeed in a stable triplet, the inner binary can be considered as "dormant" and treated as a single object until the next close encounter. Different criteria have been developed, in order to test their stability, most of which are derived from numerical simulations [23, 109]. Recently Mardling and Aarseth [72] have developed a new theoretical criterion that allows to determine the long-term stability of these systems. They found that stability requires the following inequality:

$$\frac{R_p^{\text{out}}}{a_{\text{in}}} > C \left[(1 + q_{\text{out}}) \frac{1 + e_{\text{out}}}{\sqrt{1 - e_{\text{out}}}} \right]^{2/5} \tag{30}$$

where q_{out}, e_{out} and R_p^{out} are the outer mass ratio eccentricity and pericenter of the "outer" binary, respectively. Note that Eq. (30) does not depend on the inner eccentricity and mass ratio. Given this finding, it is clear that an interaction with a fourth star can affect the stability turning a triplet into an unstable configuration.

It is important to notice that the stability condition does not imply that all the binaries parameters remain unchanged. Indeed, first (and higher) order perturbations, induced by the presence of the outer body of the triplet, can affect the inner quantities even if the triplet remains secularly stable and lives as if in isolation. Of particular relevance is the so-called "Kozai cycle" [65], namely the fact that, for certain configurations, the inner eccentricity can oscillate over a timescale

$$T_{\text{Kozai}} = \frac{T_{\text{out}}^2}{T_{\text{in}}} \left(\frac{1 + q_{\text{out}}}{q_{\text{out}}} \right) (1 - e_{\text{out}}^2)^{3/2} g(e_{\text{in}}, w_{\text{in}}, i) \tag{31}$$

where T_{out} and T_{in} are the outer and inner period respectively, w_{in} is the argument of periapsis of the inner binary, i is the inclination and g is nearly 1 unless $e_{\text{in}} \leq 0.1$ and $i > 40^o$. The Kozai cycle derives from a first order perturbation analysis, and higher order study can highlight the presence of further oscillations, for example in the inner semi-major axis. For inclinations of the inner orbit relative to the outer one greater than 40°, it has been shown that e_{in} can reach values very close to unity. The inner binary can thus merge as the result of a direct collision between the two components if the inner pericenter becomes lower than the physical dimension of the two, or as a result of the emission of gravitational waves. This can be relevant when studying black hole evolution in GCs.

3.1.6 Reversing Core Collapse

The formation of a binary and its subsequent hardening are exothermic reactions that play a key role during core contraction and collapse.

There are mainly two mechanisms that "heat" the core and prevent its collapse [101]. The first is "kinetic heating" resulting in a net deposition of kinetic energy in the GC core by those stars that have experienced a 3-body sling-shot. Kinetic heating occurs if the energy transferred to the single star is only slightly higher than the mean kinetic energy: the warmed up star remains in the core, and thus deposits its excess energy locally. If heating of the core is faster than heat conduction across the core–halo boundary, the core expands and the collapse is reversed. On the contrary, if the star is scattered to such a high velocity that it escapes from the core, heating of the core is "gravitational". The ejection of very bound stars decreases the energy of the core. The core becomes less bound having lost stars with large negative energies and thus halts its collapse [5, 46].

For some configurations an instability develops and the subsequent evolution of the cluster is characterized by the so called "gravothermal oscillations": after a first maximum expansion, a re-collapse occurs and the process repeats itself as the core continues to contract and re-expand [44].

3.2 Linking Theory to Observations

The large data-set of X-ray binaries in GCs have been used recently to test the hypothesis on their collisional origin. This section introduces first the concept of encounter frequency necessary for the analysis, and later summarizes shortly the key observational findings.

3.2.1 Encounter Frequency

If we know the number density of binaries n_b in a GC, we can compute the encounter rate per unit volume, i.e. $n_b R$, and the rate of binary–single star interactions per GC, Γ, obtained integrating $n_b R$ over the cluster volume. This rate can be estimated for any selected outcome X and type of binary stars.

To describe the degree of "dynamical" activity in "transforming" binaries (of a specific type) in a GC, one can introduce a number of simplifying assumptions: that (i) the number density of each participant (e.g. the number density of binaries and of single stars) scales with the mass density ρ, (ii) their relative velocity V_∞ scales with the (1D) velocity dispersion σ, (iii) the cross section is dominated by gravitational focusing so that $\Sigma \propto \sigma^{-2}$, and (iv) the rate is dominated by encounters inside the dense core of the GC.

Under these simplifying assumptions, the binary encounter rate $\Gamma(X)$ relative to channel X reads [27, 90]

$$\Gamma(X) = \int d^3\mathbf{r} \; nn_b \, \Sigma(X)V \propto \rho_c r_c^3 \sigma^{-1}, \tag{32}$$

scaling with the core radius r_c and core density ρ_c. If one eliminates the velocity dispersion using the virial theorem, $\sigma \propto r_c\sqrt{\rho_c}$, one has

$$\Gamma(X) \propto \rho_c^{1.5} r_c^2 \equiv \Gamma'. \tag{33}$$

where Γ' is referred to as "collision number".

Thus, the expected number of binaries of a given type formed through process X and characterized by a life-time $\tau(X)$ would be $N(X) = \Gamma(X)\tau(X) \propto \Gamma'\tau(X)$. From the above scaling and definitions, Γ' gives a measure of the degree of dynamical activity present in the GC that can be inferred, observationally, correlating the number of sources that are believed to form dynamically to Γ', defined but a normalization. In addition, since the number of sources is expected to increase with the mass of the GC, a better measure of the degree of dynamical interaction of the GC is the correlation between $N(X)/M_{GC}$ versus $\Gamma(X)/M_{GC}$ where M_{GC} is the GC mass.

3.2.2 Observations

Thanks to the the discovery of short-period X-ray binaries and of a large
number of faint X-ray binary systems in the galactic GCs, it has been possible
to quantify the role of dynamical interactions in the formation of LMXBs (see
Pooley et al. [90] for details and references therein).

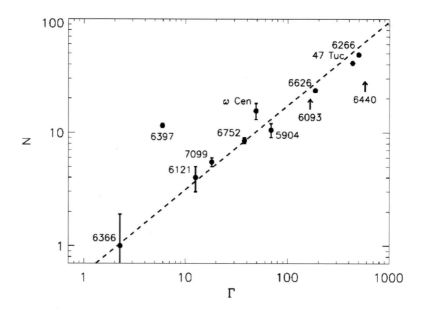

Fig. 6. Number of observed GC X-ray sources with $L_X > 4 \times 10^{30}$ erg s^{-1} versus
the normalized encounter rate Γ from [90]. The normalization has been chosen such
that $\Gamma/100$ is approximately the number of LMXBs in a cluster or, for the cases
$\Gamma < 100$, the per cent probability of the GC to host an LMXB. For NGC 6440 and
NGC 6093, the arrows refer to *Chandra* observations above the limiting luminosity.

Chandra has observed to date more than 20 GCs to a luminosity limit of
a few $\times 10^{30}$ erg s^{-1} detecting more than 800 sources within the half-light
radii of the GCs: these are Cataclysmic Variables (CVs), MSPs and qLMXBs,
besides a number (~ 150) of background sources that contaminate the sample
[90]. Classifying these low-luminosity sources is a difficult task as it requires
the search for an optical or radio counterpart. qLMXBs appear softer than
CVs in X-rays and have luminosities a few time 10^{31} erg s^{-1}. With these
selective criteria, Pooley et al. [90] found that the number of qLMXBs follows almost a linear relation with the encounter frequency (per unit mass)

Figures in Colour

Page 209

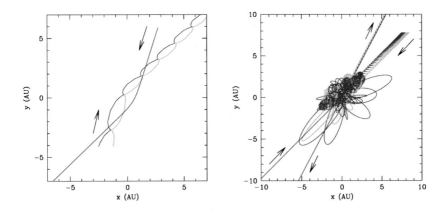

Fig. 2. Fly by: an equal-mass binary star (of total mass 1.4 $M_\odot$) interacting with a 1.4 $M_\odot$ neutron star in a prompt (left panel) and resonant (right panel) encounter. Arrows indicate the incoming and outgoing states. The neutron star is in red. Initially the incoming star and the binary orbit in the xy plane.

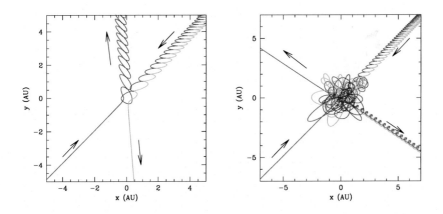

Fig. 3. Exchange: an equal-mass binary star (of total mass 1.4 $M_\odot$) interacting with a 1.4 $M_\odot$ neutron star in a prompt (left panel) and resonant (right panel) encounter. Arrows indicate the incoming and outgoing states. The neutron star is in red. Initially, the incoming star and the binary orbit in the xy plane.

Page 210

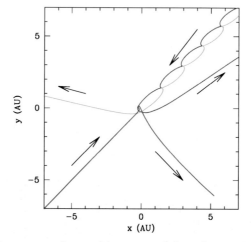

Fig. 4. Ionization: an equal-mass binary star (of total mass 1.4 $M_\odot$) interacting with a 1.4 $M_\odot$ neutron star in a prompt encounter. Arrows indicate the incoming and outgoing states. The neutron star is in red. Initially, the incoming star and the binary orbit in the xy plane.

Page 334

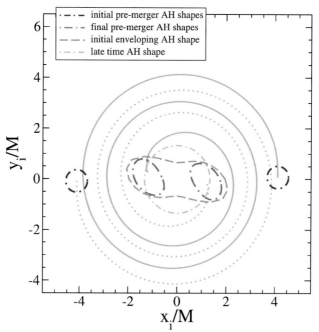

Fig. 1. A depiction of the trajectories of the black holes from a merger simulation (the "d=16" Cook–Pfeiffer case, from [55]). The green lines are the centers of the apparent horizons of each black hole. The trajectories end once a common horizon is found. Also shown are the coordinate shapes of the apparent horizons at several key moments

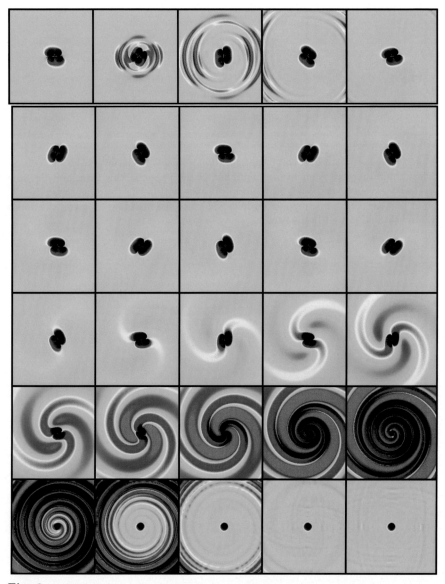

Fig. 2. A depiction of the gravitational waves emitted during the merger of two equal mass black holes (specifically "d=19" Cook–Pfeiffer initial data [55]). Shown is a color-map of the real component of the Newman–Penrose scalar Ψ_4 multiplied by r along a slice through the orbital plane, which far from the black holes is proportional to the second time derivative of the plus polarization (green is 0, toward violet (red) positive (negative)). The time sequence is from top to bottom, and left to right within each row. Each image is $25M$ apart, and a common apparent horizon is first detected at $t = 529M$ (i.e., the "merger"), which is a little after the frame in row 5, column 3. In the first several frames the spurious radiation associated with the initial data, and how quickly it leaves the domain, is clearly evident. The width/height of each box is around $100M$

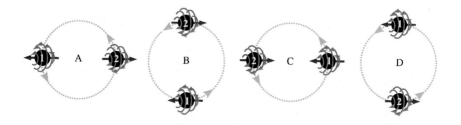

Fig. 5. A depiction of the orbital configuration resulting in the largest kick velocities. The orbital angular momentum points out of the paper in this case, and the spin vectors for each black hole is in the orbital plane within the paper as shown by the solid blue vectors. The grey curved lines illustrate the dragging of spacetime about the black hole caused by its spin.

defined in Sect. 3.2.1, providing compelling evidence for their genuine dynamical origin [90]. Fig. 6 shows the number of X-ray sources per unit mass versus the encounter rate Γ inferred in the previous section. Fig. 6 shows the clear correlation between these two quantities.

4 Neutron Stars in Globular Clusters

Neutron stars are extraordinary tools for studying stellar evolution and for exploiting dynamical processes inside GCs. In the core where the density is highest, the neutron stars may exchange with close binaries or collide with red-giants to experience a phase of mass transfer that recycles the neutron star to millisecond spin periods. The neutron star powered either by accretion or electro-magnetic braking thus becomes a probe of the "collisional" state of the GC.

The largest challenge for the study on the formation and evolution of neutron stars is at present mainly computational. A GC of 10^6 stars is expected to contain only a few hundred neutron stars after 11 Gyr. Current models do not have enough statistics to infer bulk properties of this population but despite this difficulty a basic picture is emerging for the formation, retention and evolution of neutron stars in GCs.

4.1 Formation

Young radio pulsars are known to receive at birth natal kicks resulting from recoil in symmetric supernova explosions in a binary, or/and from "asymmetries" seeded in the nascent neutron star and in the surrounding ejecta that can imprint natal velocities as large as $\sim 1000\,\mathrm{km\,s^{-1}}$. The kick velocity distribution of isolated pulsars in the galactic field follows a distribution that can be fitted with a Maxwellian with mean three-dimensional velocity of $400\,\mathrm{km\,s^{-1}}$ [56]. Since these kicks largely exceed the escape speed from a GC, there has been the need to consider all possible channels of neutron star formation to circumvent the problem of their leakage at the time of birth, given the large of number of (q)LMXBs and (b)MSPs currently observed in the galactic GCs [59].

Besides the canonical channel of neutron star formation in core collapse supernovae (CCSe) which occurs after a massive iron core is formed, "electron capture" supernovae (ECSe) have been considered in recent studies with increasing detail. An ECS occurs when a degenerate ONeMg reaches a critical mass $M_{\mathrm{ECS}} = 1.38\,M_\odot$, above which electron capture on $^{24}\mathrm{Mg}$ and $^{20}\mathrm{Ne}$ starts before subsequent nuclear burnings proceed to produce an iron core ([58, 103] and references therein). The explosion energy of these events is significantly lower than that inferred from CCSe [19], and in this case the natal kick can be considerably weaker (i.e. ~ 10 times smaller). It is the initial mass of the star's progenitor that discriminates between the CCS and ECS mode of formation, the preferred window for ECS being 8-10 $M_\odot$ in the case of single stars. Both ECSe and CCSe are sensitive to metallicity, and ECS is favored in GCs than in the field, by 30%.

Also stars in binaries can explode following an ECS: the progenitor can be (a) an evolving massive star (see [88] for details), (b) an ONeMg white dwarf driven above M_{ECS} by accretion (Accretion Induce Collapse), or (c) a double "merging" white dwarf system with total mass in excess of M_{ECS} (Merging Induced Collapse). All these channels produce a neutron star with a low natal kick.

Studies by Ivanova et al. 2007 [58] have shown that the ECS channel is sufficient to explain the bulk of the neutron stars born-and-retained in the GCs of the Milky Way. After 11 Gyr several hundred neutron stars are in the cluster, and a thousand in a GC as massive as 47 Tucanae. Those resulting from the AIC or MIC channels are abundant in relation to the number of primordial binaries (set arbitrarily in the simulation experiment) and to those that form dynamically over the cluster lifetime (see [58] for details).

MIC leads to a single neutron star, while AIC in a tight binary likely ends with the neutron star still bound to the binary companion as mass loss is not important. Neutron star formation from CCSe and ECSe terminates with a violent explosion that unbinds the binary, and only a small fraction (of a few per cent) of the neutron stars remain in the progenitor binary.[1] Thus, the role of close encounters in GCs becomes central for the formation of binaries with a neutron star.

Binary–single neutron star encounters represent a pathway to (q)LMXB formation given the large cross section offered by the incoming binary. The heavy neutron star exchanges off a star of the binary. For a (q)LMXB to exist requires furthermore the occurrence of an episode of mass transfer that spins the neutron star up to millisecond periods. When a single neutron star becomes a member of a binary, the post-exchange semi-major axis of the new-binary is larger than the initial, and it is not obvious that mass transfer begins. This is particularly critical when the companion is a white dwarf instead of a main sequence star: Roche lobe filling of a white dwarf requires periods less than ~ 0.1 day. Model calculations indicate a contribution of $\sim 50\%$ from exchanges, but the fraction of mass-transferring binaries is lower. A path that seems important to create a mass-transferring neutron star binary is through the physical collision of a neutron star with a red giant. The collisional cross section scales in this case as the stellar radius of the red giant $\sigma_{coll} \sim 2\pi R_{RG} G(m_{ns}+m_{RG})/V^2$ resulting in a collision rate such that a few neutron stars can become member of a binary through a collision. The post-encounter

---

[1] As in the field, CE evolution of He stars is an important channel for the formation of close binaries with neutron stars. But in a typical GC, a binary that survived a CCS has a relative massive secondary that experiences either a second CCS or a phase of unstable mass transfer. Thus very little of these neutron star formed in binaries remain in binaries.

binary is a neutron star with a He core or a white dwarf as companion. The binaries have periods ranging from 0.03 to 100 days, and below ~ 1 day they can start mass transfer.[2]

4.2 Recycling

4.2.1 LMXBs

MSP formation is considered to be the outcome of neutron star recycling through an accretion disc. The co-presence of LMXBs and (b)MSPs in the GCs indicates unambiguously the evolutionary link between the number of observed MSPs and of observed (q)LMXBs. In order to establish the extent of this connection, it is necessary to implement a dynamical/evolution model with a recipe of recycling, i.e. a prescription of the mass transfer process under a variety of conditions. The formation rate of LMXBs and their presence at a given time in a GC depends on the detail of mass transfer and in turn on the life-time t_{LMXB} of the LMXB, for a given donor. Low mass ($< 0.6 M_\odot$), low metallicity main sequence donors are likely transient all times and are therefore seen as qLMXBs for more than 1 Gyr. Heavier ($> 0.6 M_\odot$), metal rich donors are persistent (bright LMXBs) for 5–50 per cent of t_{LMXB}, and transients for the rest. In the case of neutron star-white dwarf LMXBs (representing the population of ultra-compact binaries mentioned in Sect. 3.2.2) t_{LMXB} is a few Gyr, but the time interval where they are bright is limited to 10^7 and 10^8 yr, only. With these timescales a typical GC can contain up to two LMXBs with a main sequence companion and up to one (q)LMXB with a white dwarf in an ultra-compact binary. In a dense GC, these numbers can be as large as a few up to a dozen, in agreement with the constraints imposed by the observations (see [58]).

4.2.2 Millisecond Pulsars

Fig. 6, taken from Ivanova et al. [59], shows the outcome of several independent runs highlighting the existence of a population of bMSPs that comprises the entire set of binaries resulting from primordial binaries, and from dynamical processes, i.e. exchanges, physical collisions and tidal capture interactions.

Circles correspond to the observed bMSPs in Terzan 5 and 47 Tucanae. It is clear that current simulated models can account for the observed bMSPs, suggesting in addition that not all mass-gaining events lead to a recycled MSP (e.g. accretion during common envelope, and mass transfer in neutron star-white dwarf systems).

MSPs are more likely to be located in the core and those in the halo are likely to be primordial or recoiled instead of being formed dynamically in the halo itself. Indeed roughly half of the observed pulsars are found inside the

---

[2] Tidal capture has also been considered but gives a less than one per cent contribution to the population of (q)LMXBs in a GC.

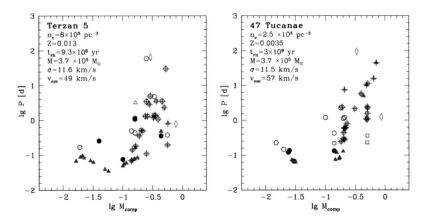

Fig. 7. bMSPs in simulated models from Ivanova et al. [59]. Left panel refers to Terzan 5, right panel to 47 Tucanae. The observed bMSPs are shown with circles. For the simulated population, triangles refer to bMSPs formed via binary exchanges; stars via a tidal capture, and squares via physical collisions. Diamonds refer to primordial binaries. Solid symbols mark binaries with a non-degenerate companion, or for the observed bMSPs, eclipsing systems.

core radius [8] and the radial distribution of most pulsars is that expected for a population that is produced in the core [35, 48].

4.3 Double Neutron Stars

Double neutron stars are considered to be the primary candidates for explaining the emission of Short Gamma Ray Bursts (SGRBs). The inspiral of the two neutron stars under the emission of gravitational waves leads to their coalescence and in this event a burst of gamma-rays and gravitational waves is emitted. Binary neutron stars are observed mostly in the field, but there is also one interesting candidate in the GC M15.

Recent statistical studies on the redshift distribution of SGRBs have highlighted the possibility that a considerable fraction of SGRBs at low redshift are occurring in GCs where they form dynamically [35, 97, 50]. In GCs, formation of double neutron stars from primordial binaries has been found [58] to be inefficient, given the little mass of the GC system compared to that of the Milky Way. Thus, double neutron stars should have mainly a dynamical origin, and again, a 3-body encounter is the favorite channel. A single neutron star can impinge on a binary hosting a low mass star and a neutron star (a bMSP or LMXB like system). The single neutron star exchanges off the lightest component, and an eccentric double neutron star system forms. Fig. 8 (from [35]) shows the eccentricities versus orbital periods of double neutron stars obtained simulating 3-body encounters with binaries with a neutron star.

If we approximate the post (pre)-exchange semi-major axis a (a_o) of the binary neutron star as $a \sim a_o(m_{\mathrm{ns}}/m_*)$, and take as $m_* = 0.9M_\odot$ the typical GC turn-off mass, we can compute a_o for Eq. (28) imposing $t_{\mathrm{GW}} = 11$ Gyr, for an eccentricity of 2/3. This gives $a_o = 5R_\odot$, and thus a $\Sigma = 2\pi a_o G(m_{\mathrm{ns}} + m_*)/V^2$. If the fraction of neutron stars in binaries for GC, $f_{\mathrm{ns-bin}}$ varies between 0.1% to 0.6% (as shown in [58]) one can infer a characteristic rate of 5×10^{12} yr, so that several hundred neutron stars are required to form a merging system per GC. Thus, if GCs are preferred sites for the dynamical formation of double neutron stars systems, SGRBs should be associated with "massive" GCs and in particular with those that underwent core collapse, as M15 where such a system is observed [53, 2].

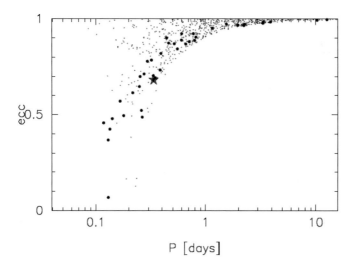

Fig. 8. Eccentricity versus binary period for double neutron star systems that form after the exchange of a neutron star off a (neutron star, star) binary, from [35]. The star denotes the observed double neutron star observed in M15. Big dots represent those simulated systems that are survived until now. Small dots represent those binaries that merged as result of gravitational waves emission.

5 Binary Black Holes in Globular Clusters

5.1 Observational Puzzles

The sample of galactic black hole X-ray binaries provides us with a unique opportunity of understanding the formation and evolution of black holes. At present, eighteen black hole X-ray binaries have been identified through dynamical mass measurements [83, 74]. The majority of these black hole candidates, with masses between 5 up to 15 $M_\odot$, are hosted in soft X-ray transients with a low mass donor whose non-persistent emission is currently interpreted in terms of a thermal instability growing in an irradiated accretion disc [26].

The important characteristic of all these black hole X-ray binaries is that they have only been found in the galactic disk. After more than 30 years of searches, there is no current observational evidence that the galactic GCs host any such system, in stark contrast to neutron star X-ray binaries that, as discussed in the previous sections, are detected in GCs in large numbers. All soft X-ray transients and qLMXBs in GCs are identified as accreting neutron stars, through their Type I bursts that unambiguously indicate the existence of a surface [68].

The lack of black hole X-ray binaries in GCs (down to a luminosity limit of $\sim 10^{30}$ erg s^{-1}) have been interpreted as evidence that stellar-mass black holes, resulting from the death of stars heavier than 25 $M_\odot$ and expected to be present in non-negligible numbers (as many as ~ 1000 in rich GCs from ordinary stellar evolution for a Salpeter IMF), evolve differently from neutron stars: either they are propelled out at the time of their birth, or evaporate during cluster evolution by some enhanced dynamical interaction [67, 98].

Natal kicks can be a source of black hole leakage from a GC. If black holes form in CCSe, again, asymmetries seeded in the massive collapsing core can be transmitted to the ejecta and the core recoils. This may occur if a hot proto-neutron star first forms that later collapses to a black hole due to fall-back material [40]. However if the collapse is prompt a much weaker kick is expected. If kicks scale approximately with the inverse of the mass of the compact object, and the mean three-dimensional kick velocity of neutron stars is 400 km s^{-1} [56], then kicks of black holes that formed through the proto-neutron stars stage would be ~ 40 km s^{-1}, so that many should remain in the parent GC. [3] If natal kick can not explain fully the paucity of GC black holes

---

[3] Proper motion have been measured for a handful of black hole X-ray binaries [78, 79] providing information on the 3-dimensional kinematic properties of these binaries. For a a few black hole candidates the inferred peculiar velocity is ~ 40 km s^{-1}, but two so called "runaway" black holes have been found with higher velocities. The first discovered is GRO J1655-40 [79] with an estimated peculiar velocity of 112±18 km s^{-1}. GRO J1655-40 is a source that shows α-elements in its optical spectrum indicating that a CCS explosion preceded the formation of the black hole [57]. In a binary, the sole mass loss following CCS can imprint a recoil velocity to the center of mass (in the plane of the binary) as large as ~ 100 km s^{-1}, but for GRO J1655-40 an additional component associated to a natal kick is likely to be present [111]. The second runaway black hole X-ray binary is the Nova XTE J1118+480 [78] whose orbit in the Milky Way resembles that of an halo object, suggesting that the black hole may have been ejected from a GC. However, stringent constraints on the age of the companion star from detailed stellar evolution calculations by [36] revealed that the age of the system

in low-mass X-ray binaries, the alternative expectation is that they undergo an evolution different from that experienced by neutron stars.

Black holes form in GCs within the first 10^7 yr, and once all massive stars have turned into relic objects, they become the most massive components in the GC. It is then clear that if present in a sufficient large number, they would follow a different evolution path, being unstable to mass segregation [101, 110, 71]. This instability leads to their thermal decoupling, and to their clustering at the center of the GC [98]. It is believed that in this dense core composed of black holes only, close dynamical interactions ultimately lead the system to evaporation. Evaporation may not be complete and one (a few) black hole(s), perhaps in a binary, is(are) left inside the GC.

Two puzzling observations seem to support this view, i.e. evaporation of black holes with some "left over". The first refers to the recent discovery by *Chandra* of a very luminous (4×10^{39} erg s^{-1}) and highly variable X-ray source located in a GC of the elliptical galaxy NGC 4472 [70]. Variability and softness of the spectrum indicate that the source is likely a black hole candidate. This detection provides the first clear evidence of a black hole "retained" in a GC. The second refers to the galactic GC NGC 6752. Two MSPs, PSR-B and PSR-E, have been discovered [18] showing unusual accelerations that, once ascribed to the overall effect of the cluster potential well, indicate the presence of > 1000 $M_\odot$ of under-luminous matter within the central 0.08 pc [25]. NGC 6752 is a very peculiar GC, since it also hosts two MSPs with unusual locations. PSR-A, a binary MSP with a white dwarf companion [18, 25, 3] and with a very small eccentricity ($\sim 10^{-5}$), holds the record of being the farthest MSP ever observed in a GC, at a distance of 3.3 half mass radii. PSR-C, an isolated MSP, ranks second in the list of the most off-set MSPs known, at a distance of 1.4 half mass radii from the gravitational center of the cluster [18, 17]. Colpi et al. [14] first explored the possibility that PSR-A descends from a primordial binary, either born in the halo or in the core (in the last case it would have been ejected as a consequence of a natal kick and/or symmetric mass loss in the binary). Their analysis however led them to discard both these hypotheses as well as that of a scattering or exchange event off a core star, given the constraint imposed by the binary nature of PSR-A (see [14] for detail). The distance of PSR-A was a severe constraint on the energetic for ejection from encounters with binary stars. Colpi et al. [14] thus conjectured that a *binary of two black holes* can be massive enough to provide the mechanism for propelling PSR-A into its current halo orbit at an acceptable event rate. Colpi et al. [15] later carried on an extensive analysis of binary–binary encounters to assess the viability of their scenario. They found that a 100 $M_\odot$ black hole with

---

is between 2 and 5 Gyr, too young to explain the GC origin of XTE J1118+480. After reconstructing the evolution history, from present state back to the time prior to the core collapse event, Gualandris et al. [36] found evidence that the black hole in XTE J1118+480 has received an asymmetric natal kick at birth of 93^{+55}_{-60} km s^{-1}, perpendicular to the orbital plane of the binary.

a companion less massive black hole of $10\,M_\odot$ would be the best target for imprinting the necessary thrust to PSR-A and at the same time to preserve it low eccentricity.[4]

These two observations leave open the possibility of probing the black hole content in GCs and in particular the formation of relativistic binaries with two black holes using either MSPs or X-ray sources.

5.2 Black Hole Evaporation?

In a GC, $\sim 10^{-4}$ up to 10^{-3} of the initial N stars (of mean mass $\langle m \rangle$) are expected to become stellar-mass black holes, depending on the shape of the initial mass function [98, 91]. Given the large black hole mass ($m_{\rm BH} \sim 10 M_\odot$) compared to the mean stellar mass $\langle m \rangle \sim 0.7\,M_\odot$ and short relaxation timescale $t_{\rm rel,BH} \sim t_{\rm rh}\,(\langle m \rangle/m_{\rm BH})$, the black holes are expected to assemble in $\sim 0.1 t_{\rm rh}$ at the center of the cluster in their attempt to reach equipartition.[5]

Unstable mass segregation sets in when W, given in Eq. (12), exceeds 0.32 [110]. For a total mass in black holes $M_{\rm GC}^{\rm BH}$, condition (12) implies

$$M_{\rm GC}^{\rm BH} > 5.4 \times 10^{-4} \left(\frac{10\,M_\odot}{m_{\rm BH}} \right)^{2.4} \left(\frac{\langle m \rangle}{0.7\,M_\odot} \right)^{2.4} N\langle m \rangle. \qquad (34)$$

Eq. (34) shows that under rather general conditions there is a sufficient number of black holes for the Spitzer instability to set in. Thus, in the first ~ 1 Gyr of the cluster lifetime, the black holes decouple dynamically from the rest of the stars and evolve as if in isolation. It is expected that with the development of mass segregation, the black holes exchange off binary stars and off binaries with black holes "building up a population of double black hole binaries". As a result of the high level of collisional relaxation present in this dense sub-cluster, three or four-body encounters among black holes lead to the evaporation of the sub-cluster as continued hardening and recoils eject them, either singly or in binaries.

In order to improve upon these simple considerations, a number of authors have studied the fate of black holes in GCs [91, 71, 37, 38, 76]. Portegies Zwart and McMillan [91] carried out a series of direct N-body simulations

---

[4] A larger black hole of $\sim 1000 M_\odot$ could provide the correct ejection velocity but would cause a dramatic growth of the eccentricity. In this case, PSR-A had to interact with the intermediate-mass black hole only before its recycling, which is a possibility. In this chapter we do not consider all physical issues related to the presence of a more massive, 1000–10000 $M_\odot$ black hole in a GC and refer the reader to [92] and Sect. 5.5 for a brief discussion.

[5] The spatial distribution of black holes at birth is not currently known. There are hints that they are clustered more than the main-sequence stars because of mass segregation of their progenitors [29] or because of selective formation of massive stars near the cluster center [80, 6]. Under these circumstances they inhabit preferentially the cluster core since birth.

with $N = 4096$ total stars and 1% fraction of equal-mass black holes (10 times heavier than the stars), and found that 90% of the black holes were ejected in the first Gyr, including $\sim 30\%$ of binaries.

A further step ahead in the theoretical study of black hole evaporation has been conducted by Miller and Hamilton [76] who highlighted the possibility that binary black holes could merge inside the cluster more rapidly than they can be ejected, opening the possibility of black hole mass growth through successive mergers and intermediate-mass black hole formation. Massive black hole binaries experience a lower recoil remaining more easily anchored inside the core. Their presence allows binary–binary encounters to produce long-lived hierarchical triple systems in which the outer black hole increases the inner binary's eccentricity via Kozai-type instabilities [77], thereby enhancing their merger rate via emission of gravitational waves before their next interaction.

To quantify the selective process of black hole retention and mass growth, consider a binary black hole of total mass $m_{T,BH}$ and a single black hole of mass $m_{BH,s}$. Hardening in three-body encounters varies the eccentricity and semi-major axis of the binary that can either coalesce under the emission of gravitational waves or be ejected out of the parent GC. Survival inside the cluster would win over ejection if the binary, at the separation a_{GW} (eq. [29]), has a binding energy smaller than the minimum binding energy for escape $E_{b,es}$ considering an encounter with a single black hole. Momentum conservation leads to a limiting

$$E_{b,es}(V_{es}) = \left(\frac{1}{2\xi}\right) m_{T,BH} \left(\frac{m_{T,BH}}{m_{BH,s}}\right)^2 V_{es}^2 \qquad (35)$$

where V_{es} is the escape speed from the GC and ξ. If $E_{b,esc} > |E_b|$ at a_{GW}, the kinetic energy of recoil is less than that of escape. In this case, the binary avoids ejection during its lifetime. The condition $E_{b,es} = |E_b(a_{GW})|$ selects the critical mass for the black holes in the binary above which ejection by dynamical recoil off a single black hole is avoided. Under rather general conditions Colpi et al. [15] have shown that binaries with masses $> 50 M_\odot$ can remain in the GC.

Gülteking et al. [37, 38] later followed the dynamics of an unequal mass black hole binary subject to a sequence of successive interactions with single black holes of $10\,M_\odot$ sampled from an isotropic background. The inclusion of gravitational wave back reaction along the course of binary evolution and the random changes in the eccentricity was found to enhance the rate of binary merger more than previously thought possible, indicating the possibility of black hole growth up to a few hundred $M_\odot$ when starting with a $50\,M_\odot$ black hole binary. They however noticed that mass growth is mainly limited here by the actual number of black holes at our disposal that are required to form the intermediate-mass black hole.

Recently, O'Leary et al. [82] tried to quantify the complex interaction of black holes studying their dynamics inside a homogeneous core of ~ 500

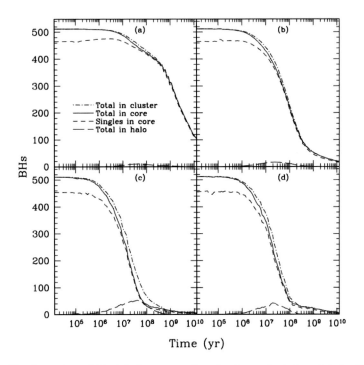

Fig. 9. Total number of black holes versus time for four selected models by O'Leary et al. [82]. Dashed-dotted lines refers to the total black holes population, solid line to the black holes in the core, long dashed line to black holes in the halo and. The dashed line refers to single black holes in the core.

single and binary stellar-mass black holes, the last with mass ratio q_{BH} sampled from a uniform distribution between 0.2 and 1, and with binary fraction $f_{bin} = 15\%$. Using a Monte Carlo technique to select interaction rates and a direct Few-body tool-kit to follow directly single-binary and binary–binary interactions, O'Leary et al. [82] confirmed that recoil and ejection cause a substantial depletion of the black holes on a timescale of ~ 1 Gyr, for typical GC parameters. Fig. 9 shows the progressive evaporation of the black holes for a series of simulated models (see details in [82]). It further indicates that the fraction of binary black holes drops and this is mainly due to strong binary–binary interactions. The holes that remain inside the cluster reach equipartition since their number becomes as low as < 40, (see Fig. 9) and it is this population that determine the content of black holes after 11 Gyr.

Fig. 10 shows the mass distribution of the remaining black holes at equipartition, from O'Leary et al. [82]. The distribution obtained indicates the genuine random behavior of this type of outcomes, and the potential of black hole growth up to a mass of $600 \, M_\odot$.

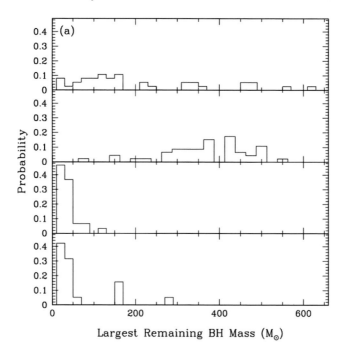

Fig. 10. Mass distribution of the largest remaining black holes at the time they reach equipartition and the bulk of has evaporated, for selected cluster models. From O'Leary et al. [82]

O'Leary et al. [82] have also included gravitational wave rocket recoil every time a binary black hole merger occurs. As shown in the chapter by Pretorius in this book, any asymmetry in the gravitational wave emission pattern at the time of coalescence leads to a linear recoil of the remnant black hole that may recede with a very large systemic speed. The introduction of gravitational wave recoil can change significantly the final distribution of the black hole masses that remain in the cluster, as black hole mass growth can be halted by the early escape of the first merged binaries. Even a small increase of the maximum gravitational wave recoil velocity above the escape threshold can lead to dramatic changes in the number of large black holes formed, cutting the probability of forming a $\sim 100\, M_\odot$ black hole in half.

In summary, there is now clear theoretical evidence that stellar-mass black holes in GCs undergo strong dynamical evolution and evaporate. At the end of this process, the number of black holes that remain is a random number. The "minimal" configuration that we can envisage is three remaining black holes in the form of a binary and a single black hole. The last binary–single interaction can result in three possible outcomes: all three black holes can receive sufficient recoil to escape, a binary can remain and eject the single black hole, or the single black hole can remain and eject the binary (depending on the relative masses).

What is the fate of the black holes that remain and those that escape GCs? In the next sections we explore their evolution highlighting the possibility to reveal those that remain through their free interaction with stars and MSPs, and those that escape and are in binaries through the emission of gravitational waves at the time of their coalescence.

5.3 The Fingerprint of Binary Black Holes in GCs

Binary black holes of masses $\sim 100\,M_\odot$ can affect the dynamics of stars and of MSPs that can be used as indirect probe of their existence in a GC. This has been explored in a number of papers [15, 73, 20].

5.3.1 Supra-Thermal Stars and Angular Momentum Alignment

A binary black hole acts as a source of heat and angular momentum for stars flying by, and can thus generate a family of supra-thermal stars, i.e. stars which remain bound to the GC but have a velocity higher than the velocity dispersion. Scattering experiments [73] have shown that typically a few hundred stars gain such excess energy, and a fraction of these tend also to align their angular momentum vector with that of the binary black hole. This alignment is relevant when the binary is sufficiently wide to sustain a momentum flow higher than that carried by the impinging stars.

Fig. 11 shows the post-encounter asymptotic velocity distribution of cluster stars interacting with a binary black hole of $(100\,M_\odot, 50M_\odot)$ and different semi-major axis. When the binary has $a > 10$ AU stars are heated to supra-thermal energies while for $a < 10$ AU, most of the stars are up-scattered to velocities well in excess of the escape velocity. The number of supra-thermal stars that a binary produces is a fraction f_{th} of the N_{int} interacting stars. The latter is given by

$$N_{\mathrm{int}} = \frac{m_{\mathrm{T,BH}}}{\langle m \rangle} \ln\left(\frac{a_0}{a_{\mathrm{st}}}\right) \tag{36}$$

where a_0 (set equal to 2000 AU) is the semi-major axis below which the binary is able to generate supra-thermal stars, and a_{st} (~ 0.1 AU) is the minimum semi-major axis below which the cross section for 3-body encounters become negligible. A typical binary produces up to 300 of supra-thermal stars over its lifetime.

Similarly, Fig. 12 illustrates the extent of the angular momentum alignment, showing the distribution of the cosine of the angle between the angular momentum of the binary black hole and that of stars prior (dashed lines) and after (solid lines) the encounter, for the same black hole binary mass as in Fig. 11, and for a semi-major axis of 100 AU. Stars initially are distributed isotropically in momentum space and so half are co-rotating (counter-rotating) with the binary, i.e. the projection of their angular momentum vector $\mathbf{J}_*$ on

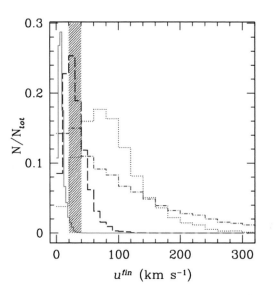

Fig. 11. Post-encounter asymptotic velocity distributions of cluster stars after interaction with a black hole binary of $(100\,M_\odot, 50\,M_\odot)$ with semi-major axis of 1 AU (dot-dashed line), 10 AU (dotted line), 100 AU (dashed line), and 1000 AU (solid line), respectively. The shaded area marks the window of supra-thermal stars.

$\mathbf{J}_{BH}$ is positive (negative). From the simulation it is found that at least 70% of the initially co-rotating stars remain co-rotating and that a high percentage ($\sim$ 70%) of the initially counter-rotating stars became co-rotating, flipping their angular momentum. The combination of these two tendencies (i.e. the tendency to remain co-rotating for initially co-rotating stars and the tendency to become co-rotating for initially counter-rotating stars) suggests the strongest evidence that a black hole binary can introduce a net rotation in a group of stars. This effect however is temporary as it requires a binary semi-major axis sufficiently large ($\sim$ 100 AU) for the angular momentum of the black hole binary to dominate over that of stars.

Fig. 13 gives the three timescales t_{rh}, t_{hard}, and t_{GW} (defined in Eqs. (2), (27), (28) respectively) as a function of the orbital semi-major axis (in AU) of the binary black hole, for a GC with stellar density $n = 10^5$ pc^{-3}, $\xi = 1$, and $\sigma = 8.5$ km s^{-1}. The shaded area refers to t_{GW} computed for black hole binaries with total mass between 60 and $210\,M_\odot$ and eccentricity 2/3. The figure illustrates (i) that a black hole binary hardens down to $\sim$ 1 AU in $\sim$ 5 Gyr due to gravitational encounters with the stars; (ii) that it further decays under the emission of gravitational waves; (iii) that the hardening time is smaller than the relaxation time as long as $a > 5$ AU. Since a binary black hole can sustain the formation of supra-thermal stars and angular momentum

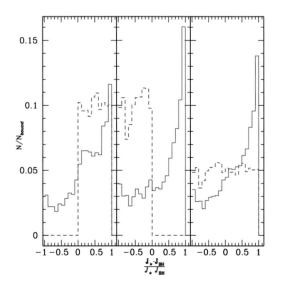

Fig. 12. The histograms show the distribution of the cosine of the angle between the angular momentum of the binary black-hole and that of the interacting star. In the three panels, the dashed line indicates the distribution before the encounter, the solid line that after the encounter. The left panel refers to stars which initially were co-rotating relative to the binary, the central panel refers to those stars which initially were counter-rotating. The right panel gives the sum of the pre- and post-encounter distributions.

anisotropies only in preferred windows of the binary semi-major axis, it can do so during its hardening and as long as $t_{hard} < t_{rh}$, since relaxation tends to erase the anisotropies induced by the black hole binary.

Supra-thermal stars tend to recover equilibrium, i.e. equipartition of kinetic energy with the other stars and thus it has become important to study their phase-space evolution after 11 Gyr in order to improve the chance of recognizing their origin. After following their dynamical evolution in the cluster potential taking into account relaxation and dynamical friction, Mapelli et al. [73] found that a few tens of supra-thermal stars are still present in the cluster within 6 core radii. Clearly the observation of a few tens of these stars with excess velocity would provide a key toward the discovery of these exotic binaries. Taking into account both projection and selection effects, only ~ 10 stars can be found in the supra-thermal tail of the Maxwellian within 6 core radii. The detectability appears to be problematic with present telescopes but in light of these findings the search of supra-thermal stars and of anisotropies in their angular momentum distribution can provide useful indications on the existence of binary black holes in GCs. It may become a feasible task for future instruments having much bigger collecting area and equipped with detectors

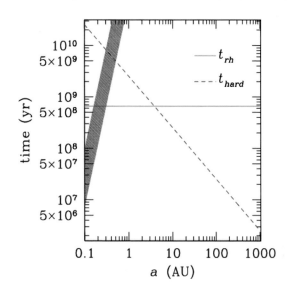

Fig. 13. Half-mass relaxation time t_{rh}, hardening time t_{hard}, as a function of the initial semi-major axis a. The shaded strip indicates the gravitational wave timescale t_{GW} computed for black hole binaries with total mass between 60 and 210 $M_\odot$ and $e = 2/3$.

having a much larger field of view, like the Thirty Meter Telescope or the Overwhelmingly Large Telescope.

5.3.2 Formation of a MSP- Intermediate-Mass Black Hole Binary

A MSP orbiting around a $\sim 100\,M_\odot$ black hole is a potential outcome of dynamical interactions in a GC, if the scenario of evaporation and mass growth through mergers is correct. The discovery of this exotic binary would have enormous impact on General Relativity and on our understanding of stellar evolution in GCs.

Devecchi et al. [20] explored the dynamical capture of a MSP by an intermediate-mass black hole (IMBH) of $\sim 100\,M_\odot$, either single or in a binary (with a star or another stellar-mass black hole). They carried out single–binary and binary–binary scattering experiments to compute cross section and rates considering the recycled MSP either single or as member of a binary with a white dwarf (i.e. a bMSP). For bMSPs, orbital periods and companion masses were extracted according to the period and mass distributions observed in the sample of bMSPs of the galactic GCs [8].

The main channel that dominates the formation rate is the scattering of a bMSP off the single $100\,M_\odot$ black hole. The binary (IMBH,MSP) that form has a distribution of semi-major axis and eccentricities such that it can become an important source of gravitational waves.

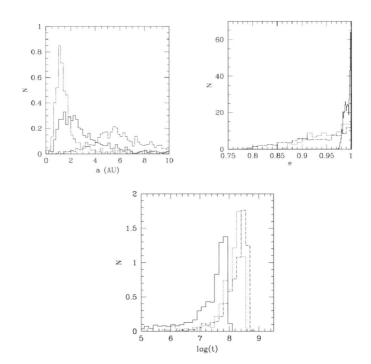

Fig. 14. Distribution of the semi-major axis, eccentricities and lifetimes of (IMBH,MSP) binaries formed in 3-body encounters between bMSPs (of different observed orbital periods; see [20] for details) impinging off a single IMBH of 100 $M_\odot$.

Fig. 14 shows the distribution of the semi-major axis $a_{\rm IMBH,MSP}$, eccentricies e and lifetimes of the (IMBH,MSP) binaries as inferred from a series of scattering experiments. The newly formed binary has a typical semi-major axis of a few AU. The peak of the distribution can be estimated analytically from the parameters of the initial bMSP, i.e. from the semi-major axis $a_{\rm MSP,WD}$ prior to the encounter:

$$a_{\rm IMBH,MSP} \sim \frac{a_{\rm MSP,WD}}{2\sqrt{2}} \frac{M_{\rm IMBH}}{m_{\rm WD}} \left(\frac{m_{\rm MSP} + m_{\rm WD}}{M_{\rm IMBH}} \right)^{1/3}. \tag{37}$$

As shown in Fig. 14, the distribution of the eccentricities is strongly peaked at very high values (> 0.9). This causes the newly formed binary to be relatively short-lived. Coalescence under the action of hardening off cluster stars and gravitational waves emission takes only 10^8–10^9 yr. For the current sample of MSPs in GCs, the formation rate was found to be a considerably low 10^{-11} yr^{-1}, given the short lifetime of the binary. However, next generation radio telescopes, like SKA, are expected to probe the faint-end of the MSP luminosity function and to detect at least one binary. A complete search for low-luminosity MSPs in the GCs of the Milky Way with SKA will thus have the potential of testing the hypothesis that black holes of $100\,M_\odot$ are hosted in GCs.

5.4 Binary Black Holes as Sources of Gravitational Waves

In the near future, GCs with their "ejection products" may be the first systems to have their population of relativistic binaries studied with the new window of gravitational radiation astronomy. The stellar-mass binary black holes ejected from the GCs in the first $\sim$ Gyr are sufficiently hard and eccentric, thanks to the close encounters experienced inside the GC core, that they are expected to coalesce under the emission of gravitational waves within a Hubble time [91]. Ground based laser interferometers as LIGO can just catch the instant at which their horizons touch when an intense burst is emitted.

As shown in Sect. 5.2 there is a limiting value of the binding energy $|E_{\mathrm{b}}^{\mathrm{crit}}|$ above which the black hole binary center of mass acquires a recoil velocity (eq. (35)) comparable to V_{esc}. This establishes a connection between the underlying GC potential and the orbital period of the binary below which escape is possible. For a King potential, the condition $(V_{\infty,\mathrm{rec}}^{\mathrm{bin}})^2 > V_{\mathrm{esc}}^2(0) = 2|\phi(0)|$, and $V_{\mathrm{esc}}^2 = 2W_o k_B T\langle m\rangle$ implies a lower limit on the binding energy for escape:

$$|E_{\mathrm{b}}^{\mathrm{crit}}| \sim \frac{8}{\xi}W_o\frac{m_{\mathrm{BH}}}{\langle m\rangle}k_B T. \tag{38}$$

For stellar-mass black holes to $10\,M_\odot$, $\langle m\rangle = 0.7\,M_\odot$, and $W_o \sim 5-10$, the critical binding energy $|E_{\mathrm{b}}^{\mathrm{crit}}| \sim 10^4 k_B T$ can greatly exceed the cluster thermal energy $k_b T$ measuring the kinetic energy of stars $\propto \sigma^2$ in the GC. As the cluster is in virial equilibrium, $k_B T = G\langle m\rangle^2 N/6R_{\mathrm{vir}}$ (where R_{vir} is the virial radius); combining all these relations, one can infer the limiting period

$$P_{\mathrm{lim}} \sim \left(\frac{R_{\mathrm{vir}}}{10W_o N\langle m\rangle}\right)^{3/2}\pi G^{-1/2}m_{\mathrm{BH}}: \tag{39}$$

black hole binaries with periods longer than P_{lim} are expected to be retained inside the GC, while those having periods much shorter than P_{lim} are expected to be ejected. For a typical GC, with $R_{\mathrm{vir}} \sim 1$ pc and a total GC mass of $10^6\,M_\odot$, $P_{\mathrm{lim}} \sim 1$ hr. The binaries ejected have such short periods that they can coalesce within a Hubble time under the emission of gravitational waves. Since during dynamical mass-segregation and rapid relaxation in the dynamically decoupled core, the black hole binaries are expected to establish a thermal distribution in their eccentricity, it is possible to infer their lifetime t_{GW} (eq. [28]), and thus determine a rate of coalescence.

Portegies Zwart and McMillan [91] (see also [92]) studied the characteristics of the black hole binary population ejected from the GC, from their direct N-body simulation, inferring a merger rate of about 1.6×10^{-7} yr^{-1} Mpc^{-3}.

For LIGO-I these authors expect one detection during the first two years of operation, but for the successor LIGO-II the rate may rise to one detection per day. For the black holes that remain inside the galactic GC, and that are expected to have orbital periods longer than a few hours, Benacquista et al [4] pointed out their potential contribution in low-frequency gravitational waves that could be detected by LISA. Assuming random GC hosts, Monte Carlo simulations have indicated that ∼ 10% of the galactic GCs may contain a black hole binary with sufficient signal strength in the higher harmonics to allow for detection and also for the angular position of the binary to be resolved inside the host GC.

These estimates suffer from severe uncertainties due to the random nature of the dynamical processes, but despite these limitation the identification and study of black hole binaries in GCs and in the Milky Way halo may provide invaluable information not only on the distribution of black hole masses (testing either stellar evolution or dynamical mass growth) but also on the history of the host GCs and of their stellar populations.

5.5 Central Intermediate Mass Black Hole in GCs ?

GCs have been considered to be prime sites for the formation of intermediate-mass black holes (IMBHs), with masses between 10^3–$10^4\ M_\odot$. In the Sect. 5.3, it was shown that a black hole as massive as $100\ M_\odot$ or more can form in an already evolved GC through the succession of close gravitational encounters among stellar-mass black holes. The growth to very large masses was however found to be limited by recoils. There is, however, a second root for the formation of an intermediate-mass black hole in a GC [22, 39, 93]. It has been suggested that inside young star clusters, stellar densities can be so large that the cluster itself can become vulnerable to unstable mass segregation and core collapse *before* the most massive stars explode as supernovae. A "runaway" collision among young massive stars may lead to the formation of a super-star which can then collapse to form an intermediate-mass black hole. Further interaction of the newly formed black hole with cluster stars over the GC lifetime can eventually lead to further growth [29, 93] and to the formation of a stellar cusp around the IMBH.

At present, there are observations hinting at the presence of an IMBH (see for example [105] for a review), but none is conclusive. Three are the most compelling. Gebhard, Rich and Ho [31] highlighted the possible presence of an intermediate-mass black hole of $2^{+1.4}_{-0.8} \times 10^4\ M_\odot$ in the cluster G1 of M31, on the basis of a joined analysis of photometric and spectroscopic measurements. Interestingly, this intermediate-mass black hole lies just on the low-end of the black hole mass versus line-of-sight velocity dispersion correlation observed in spheroids and bulges of nearby galaxies [24, 30].

In the galactic cluster M15, *HST* and ground-based observation of line-of-sight velocities and proper motions, indicate the presence of under-luminous matter of $500^{+2500}_{-500}\ M_\odot$ that could be ascribed to the presence of an IMBH [32].

By mapping the velocity field, van den Bosh found also evidence or ordered rotation in the central 4 arcsec of M15, hinting at a binary IMBH.

Recently Noyola et al. [81] measured the surface brightness profile from integrated light on an HST/ACS image of the center of the GC $\omega-$Cen, and found a central power-law cusp of logarithmic slope -0.08. From a kinematical study of the cluster they also found a clear increase in the velocity dispersion in the central region that could be imputed to the presence of an IMBH of $4.0^{+0.75}_{-1.0} \times 10^4 M_\odot$ for an isotropic, spherical dynamical model. In this case, the host of the IMBH is probably not a real GC but the stripped nucleus of a dwarf galaxy with an initial mass of $10^7 M_\odot$. Note that if that would be true, the inferred mass of the central IMBH would again lie on the M_{BH}-σ relation for spheroids.

At a theoretical level there is debate on the true interpretation of these data. Given all these uncertainties, we would like to point out the potential role played by MSPs to reveal a central IMBH in a GC of our Galaxy. Devecchi et al. [20] have shown that the interaction of an incoming bMSP off a black hole can lead to the formation of a new binary, i.e. a MSP orbiting around the black hole (see Sect. 5.3.2). In this exchange interaction, the binary that forms has a characteristic distribution of semi-major axis and eccentricities. According to [20] and Pfahl [85], the most likely end-state is a binary with a large eccentricity and semi-major axis that can be inferred by Eq. (37). For black hole masses as high as $10^4\ M_\odot$, a MSP can bind at a distance of 20-60 AU from the cluster core. The discovery of one of these systems would unambiguously prove the presence of an IMBH.

References

1. M. A. Alpar, A. F. Cheng, M. A. Ruderman, J. Shaham: A new class of radio pulsars, Nature **300**, 728 (1982)
2. Anderson, S. B.; Gorham, P. W.; Kulkarni, S. R.; Prince, T. A.; Wolszczan, A. Discovery of two radio pulsars in the globular cluster M15, Nature **346**, 42 (1990)
3. C. G. Bassa, F. Verbunt, M. H. van Kerkwijk, L. Homer: Optical identification of the companion to PSR J1911-5958A, the pulsar binary in the outskirts of NGC 6752, A&A textbf409, 31 (2003)
4. M. J. Benacquista: Gravitational radiation from black hole binaries in globular clusters, CQGra **19**, 1297 (2002)
5. J. Binney, S. Tremaine: *Galactic dynamics* (1987)
6. I. A. Bonnell, C. J. Clarke, M. R. Bate, J. E. Pringle: Accretion in stellar clusters and the initial mass function, MNRAS **324**, 573 (2001)
7. E. F. Brown, L. Bildsten: The Ocean and Crust of a Rapidly Accreting Neutron Star: Implications for Magnetic Field Evolution and Thermonuclear Flashes, ApJ **496**, 915 (1998)
8. F. Camilo, F. A. Rasio: Pulsars in Globular Clusters, in the Proceeding of the conference *Binary Radio Pulsars*, Edited by F. A. Rasio and I. H. Stairs (San Francisco: Astronomical Society of the Pacific, 2005) p.147

9. S. Campana, M. Colpi, S. Mereghetti, L. Stella, M. Tavani: The neutron stars of Soft X-ray Transients, A&ARv **8**, 279 (1998)

10. S. Chandrasekhar: Dynamical Friction. I. General Considerations: the Coefficient of Dynamical Friction, ApJ **97**, 255 (1943)

11. G. W. Clark: X-ray binaries in globular clusters, ApJ **199**, L143 (1975)

12. G. W. Clark, T. H. Markert, F. K. Li: Observations of variable X-ray sources in globular clusters, ApJ **199**, 93 (1975)

13. M. Colpi, U. Geppert, D. Page, A. Possenti: Charting the Temperature of the Hot Neutron Star in a Soft X-Ray Transient, ApJ **548**, L175 (2001)

14. M. Colpi, A. Possenti, A. Gualandris: The Case of PSR J1911-5958A in the Outskirts of NGC 6752: Signature of a Black Hole Binary in the Cluster Core?, ApJ **570**, 85 (2002)

15. M. Colpi, M. Mapelli, A. Possenti: Probing the Presence of a Single or Binary Black Hole in the Globular Cluster NGC 6752 with Pulsar Dynamics, ApJ **599**m 1260 (2003)

16. J. M. Cordes, R. W. Romani, S. C. Lundgren: The Guitar nebula – A bow shock from a slow-spin, high-velocity neutron star, Nature **362**, 133 (1993)

17. A. Corongiu, A. Possenti, A. G. Lyne, R. N. Manchester, F. Camilo, N. D'Amico, J. M. Sarkissian: Timing of Millisecond Pulsars in NGC 6752. II. Proper Motions of the Pulsars in the Cluster Outskirts, ApJ **653**, 1417 (2006)

18. N. D'Amico, A. Possenti, L. Fici, R. N. Manchester, A. G. Lyne, F. Camilo, J. Sarkissian: Timing of Millisecond Pulsars in NGC 6752: Evidence for a High Mass-to-Light Ratio in the Cluster Core, ApJ **570**, L570 (2002)

19. Dessart, L.; Burrows, A.; Ott, C. D.; Livne, E.; Yoon, S.-C.; Langer, N.: Multidimensional Simulations of the Accretion-induced Collapse of White Dwarfs to Neutron Stars, ApJ **644**,1063 (2006)

20. B. Devecchi, M. Colpi, M. Mapelli, A. Possenti: Millisecond pulsars around intermediate-mass black holes in globular clusters, MNRAS **380**, 691 (2007)

21. G. A. Drukier, C. D. Bailyn: Can High-Velocity Stars Reveal Black Holes in Globular Clusters?, ApJ **597**, L125 (2003)

22. Ebisuzaki, Toshikazu; Makino, Junichiro; Tsuru, Takeshi Go; Funato, Yoko; Portegies Zwart, Simon; Hut, Piet; McMillan, Steve; Matsushita, Satoki; Matsumoto, Hironori; Kawabe, Ryohei Missing Link Found? The Runaway Path to Supermassive Black Holes, ApJ **562**, 19L (2001)

23. Eggleton, Peter; Kiseleva, Ludmila An Empirical Condition for Stability of Hierarchical Triple Systems, ApJ **455**, 640 (1995)

24. Ferrarese, L., & Merritt, D.:A Fundamental Relation between Supermassive Black Holes and Their Host Galaxies, ApJ **539**, 9L (2000)

25. F. R. Ferraro, A. Possenti, E. Sabbi, P. Lagani, R. T. Rood, N. D'Amico, L. Origlia: The Puzzling Dynamical Status of the Core of the Globular Cluster NGC 6752, ApJ **595**, 179 (2003)

26. J. Frank, A. R. King, D. J. Raine: Accretion power in astrophysics

27. J. M. Fregeau: X-Ray Binaries and the Current Dynamical States of Galactic Globular Clusters, ApJ**673**, 25

28. M. Freitag, W. Benz: A new Monte Carlo code for star cluster simulations. I. Relaxation, A&A **375**, 711 (2001)

29. M. Freitag, M. A. Gürkan, F. A. Rasio: Runaway collisions in young star clusters II. Numerical results, MNRAS **368**, 141 (2006)

30. Gebhardt, Karl; Bender, Ralf; Bower, Gary; Dressler, Alan; Faber, S. M.;
Filippenko, Alexei V.; Green, Richard; Grillmair, Carl; Ho, Luis C.; Kormendy,
John; and 5 coauthors A Relationship between Nuclear Black Hole Mass and
Galaxy Velocity Dispersion, ApJ **539**, 13L (2000)

31. Gebhardt, Karl; Rich, R. M.; Ho, Luis C. A 20,000 $Msun$ Black Hole in the
Stellar Cluster G1, ApJ **578**, 41 (2002)

32. Gerssen, Joris; van der Marel, Roeland P.; Gebhardt, Karl; Guhathakurta,
Puragra; Peterson, Ruth C.; Pryor, Carlton Hubble Space Telescope Evidence
for an Intermediate-Mass Black Hole in the Globular Cluster M15. II. Kinematic Analysis and Dynamical Modeling, AJ **124**, 3270 (2002)

33. R. Giacconi, S. Murray, H. Gursky, E. Kellogg, E. Schreier, H. Tananbaum:
The Uhuru catalog of X-ray sources, ApJ **178**, 281 (1972)

34. R. Giacconi, S. Murray, H. Gursky, E. Kellogg, E. Schreier, T. Matilsky, D.
Koch, H. Tananbaum: The Third UHURU Catalog of X-Ray Sources, ApJ **27**,
37 (1974)

35. J. Grindlay, S. Portegies Zwart, S. McMillan: Short gamma-ray bursts from
binary neutron star mergers in globular clusters, Nature Physics, **2**, 116-119
(2006).

36. A. Gualandris, M. Colpi, S. Portegies Zwart, A. Possenti: Has the Black Hole
in XTE J1118+480 Experienced an Asymmetric Natal Kick?, ApJ **618**, 845
(2005)

37. K. Gultekin, M. C. Miller, D. P. Hamilton: Growth of Intermediate-Mass Black
Holes in Globular Clusters, ApJ **616**, 221 (2004)

38. K. Gultekin, M. C. Miller, D. P. Hamilton: Three-Body Dynamics with Gravitational Wave Emission, ApJ **640**, 156 (2006)

39. M. A. Gurkan, M. Freitag, F. Rasio: Formation of Massive Black Holes in
Dense Star Clusters. I. Mass Segregation and Core Collapse, ApJ **604**, 632
(2004)

40. A. Heger, C. L. Fryer, S. E. Woosley, N. Langer, D. H. Hartmann: How Massive
Single Stars End Their Life, ApJ **591**, 288 (2003)

41. D. C. Heggie: Binary evolution in stellar dynamics, MNRAS **173**, 729 (1975)

42. Heggie, Douglas C.; Hut, Piet Binary-single-star scattering. IV – Analytic
approximations and fitting formulae for cross sections and reaction rates, ApJ
85, 347 (1993)

43. Heggie, Douglas C.; Rasio, Frederic A. The Effect of Encounters on the Eccentricity of Binaries in Clusters, MNRAS **282**, 1064 (1996)

44. Heggie D. C.; Ramamani, N. Evolution of star clusters after core collapse,
MNRAS **237**, 757 (1989)

45. D. C. Heggie, P. Hut, S. L. W. McMillan: Binary–Single-Star Scattering. VII.
Hard Binary Exchange Cross Sections for Arbitrary Mass Ratios: Numerical
Results and Semianalytic FITS, ApJ **467**, 359 (1996)

46. D. C. Heggie, P. Hut: The Gravitational Million-Body Problem: A Multidisci-
plinary Approach to Star Cluster Dynamics, by Douglas Heggie and Piet Hut.
Cambridge University Press (2003)

47. C. O. Heinke, J. E. Grindlay, P. M. Lugger, H. N. Cohn, P. D. Edmonds,
D. A. Lloyd, A. M. Cool: Analysis of the Quiescent Low-Mass X-Ray Binary
Population in Galactic Globular Clusters, ApJ **598**, 501 (2003)

48. C. O. Heinke, J. E. Grindlay, P. D. Edmonds, H. N. Cohn, P. M. Lugger, F.
Camilo, S. Bogdanov, P. C. Freire: A Deep Chandra Survey of the Globular
Cluster 47 Tucanae: Catalog of Point Sources, ApJ **625**, 796 (2005)

49. J. G. Hills: Encounters between single and binary stars – The effect of intruder mass on the maximum impact velocity for which the mean change in binding energy is positive, AJ **99**, 979 (1990)

50. C. Hopman, D. Guetta, E. Waxman, S. Portegies Zwart: The Redshift Distribution of Short Gamma-Ray Bursts from Dynamically Formed Neutron Star Binaries, ApJ **643**, 91 (2006)

51. P. Hut: Dense Stellar Systems as Laboratories for Fundamental Physics, Review paper presented at the conference in Honor of van den Heuvel), astro-ph/0601232

52. P. Hut, J. N. Bahcall: Binary-single star scattering. I – Numerical experiments for equal masses, ApJ **268**, 319 (1983)

53. B. A. Jacoby, P. B. Cameron, F. A. Jenet, S. B. Anderson, R. N. Murty, S. R. Kulkarni: Measurement of Orbital Decay in the Double Neutron Star Binary PSR B2127+11C, ApJ **644**, 113 (2006)

54. J. I. Katz: Two kinds of stellar collapse, Nature **253**, 698 (1975)

55. T. T. Hamilton, D. J. Helfand, R. H. Becker: A search for millisecond pulsars in globular clusters, AJ **90**, 606 (1985)

56. G. Hobbs, D. R. Lorimer, A. G. Lyne, M. Kramer: A statistical study of 233 pulsar proper motions, MNRAS **360**, 974 (2005)

57. G. Israelian, R. Rebolo, G. Basri, J. Casares, E. L. Martín: Evidence of a supernova origin for the black hole in the system GRO J1655-40, Nature **401**, 142 (1999)

58. N. Ivanova, C. Heinke, F. A. Rasio, K. Belczynski, J. Fregeau: Formation and evolution of compact binaries in globular clusters: II. Binaries with neutron stars, MNRAS **386**, 553

59. N. Ivanova, C. Heinke, F. A. Rasio: Formation of Millisecond Pulsars in Globular Clusters, astro-ph/0711.3001, 40 YEARS OF PULSARS: Millisecond Pulsars, Magnetars and More. AIP Conference Proceedings, Volume 983, pp. 442-447 (2008).

60. N. Ivanova, C. O. Heinke, F. Rasio: Neutron Stars in Globular Clusters, astro-ph/0711.318

61. V. Kalogera, A. R. King, F. A. Rasio: Could Black Hole X-Ray Binaries Be Detected in Globular Clusters?, ApJ **601**, 171 (2004)

62. E. Khalisi, P. Amaro-Seoane, R. Spurzem: A comprehensive NBODY study of mass segregation in star clusters: energy equipartition and escape, MNRAS **374**, 703 (2007)

63. I. R. King: The structure of star clusters. III. Some simple dynamical models, AJ **71**, 64 (1966)

64. I. R. King, C. Sosin, A. M. Cool: Mass Segregation in the Globular Cluster NGC 6397, ApJ **452**, 33 (1995)

65. Y. Kozai: Secular Perturbations of Asteroids with High Inclination and Eccentricity, AJ **67**, 579 (1962)

66. P. Kroupa: The birth and early evolution of star clusters, astro-ph/0609370, To appear in "Mass loss from stars and the evolution of stellar clusters". Proc. of a workshop held in honour of H.J.G.L.M. Lamers, Lunteren, The Netherlands. Eds. A. de Koter, L. Smith and R. Waters (San Francisco: ASP)

67. S. P. Kulkarni, P. Hut, S. McMillan: Stellar black holes in globular clusters, Nature **364**, 421 (1993)

68. Kuulkers, E.; den Hartog, P. R.; in't Zand, J. J. M.; Verbunt, F. W. M.; Harris, W. E.; Cocchi, M. Photospheric radius expansion X-ray bursts as standard candles, A&A **399**, 663 (2003)
69. D. Lynden-Bell: Statistical mechanics of violent relaxation in stellar systems, MNRAS **136**, L101 (1967)
70. T. J. Maccarone, G. Bergond, A. Kundu, K. L. Rhode, J. J. Salzer, I. C. Shih, S. E. Zepf: An X-ray emitting black hole in a globular cluster, astro-ph/0710.3381
71. A. D. Mackey, M. I. Wilkinson, M. B. Davies, G. F. Gilmore: The effect of stellar-mass black holes on the structural evolution of massive star clusters, MNRAS **379** 40 (2007)
72. Mardling, Rosemary A.; Aarseth, Sverre J.: Tidal interactions in star cluster simulations, MNRAS **321**, 398 (2001)
73. M. Mapelli, M. Colpi, A. Possenti, S. Sigurdsson: The fingerprint of binary intermediate-mass black holes in globular clusters: suprathermal stars and angular momentum alignment, MNRAS **364**, 1315 (2005)
74. J. E. McClintock, R. A. Remillard: Black hole binaries, In: *Compact stellar X-ray sources*, ed. by W. Lewin & M. van der Klis (Cambridge Astrophysics Series, No. 39. Cambridge, UK: Cambridge University Press 2006) p. 157 - 213
75. M. C. Miller: Gravitational Radiation from Intermediate-Mass Black Holes, ApJ **581**, 438 (2002)
76. M. C. Miller, D. P. Hamilton: Production of intermediate-mass black holes in globular clusters, MNRAS **330**, 232 (2002)
77. M. C. Miller, D. P. Hamilton: Four-Body Effects in Globular Cluster Black Hole Coalescence, ApJ **576**, 894 (2002)
78. I. F. Mirabel, V. Dhawan, R. P. Mignani, I. Rodrigues, F. Guglielmetti: A high-velocity black hole on a Galactic-halo orbit in the solar neighbourhood, Nature *413*, 139 (2001)
79. I. F. Mirabel, R. Mignani, I. Rodrigues, J. A. Combi, L. F. Rodríguez, F. Guglielmetti: The runaway black hole GRO J1655-40, A&A **395**, 595 (2002)
80. S. D. Murray,D. N. C. Lin, Douglas: Coalescence, Star Formation, and the Cluster Initial Mass Function, ApJ **467**, 728 (1996)
81. E. Noyola, K. Gebhardt, M. Bergmann: Gemini and Hubble Space Telescope Evidence for an Intermediate Mass Black Hole in omega Centauri, astro-ph/0801.2782; New Horizons in Astronomy: Frank N. Bash Symposium ASP Conference Series, Vol. 352, Proceedings of the Conference Held 16-18 October, 2005 at The University of Texas, Austin, Texas, USA, p.269
82. R. O'Leary, F. Rasio, J Fregeau, N. Ivanova, R. O'Shaughnessy: Binary Mergers and Growth of Black Holes in Dense Star Clusters, ApJ **637** 937 (2006)
83. J. Orosz: Black Holes in the Milky Way, presented at the KITP: Astrophysics Seminars, 2004
84. P. C. Peters, J. Mathews: Gravitational Radiation from Point Masses in a Keplerian Orbit, PhRv **131**, 435 (1963)
85. Pfahl, E.: Binary Disruption by Massive Black Holes in Globular Clusters ApJ **626**, 849 (2005)
86. E. S. Phinney, S. Sigurdsson: Ejection of pulsars and binaries to the outskirts of globular clusters, Nature **349**, 220 (1991)
87. H. C. Plummer, H. C. On the problem of distribution in globular star clusters, MNRAS **71**, 460 (1911)

88. Ph. Podsiadlowski, N. Langer, A. J. T. Poelarends, S. Rappaport, A. Heger, E. Pfahl: The Effects of Binary Evolution on the Dynamics of Core Collapse and Neutron Star Kicks, ApJ **612**, 1044 (2004)

89. D. Pooley: X-ray binary systems in globular clusters. In: Proceedings of the "The X-ray Universe 2005", vol 1, ed. by A. Wilson, pp 125–130

90. D. Pooley,W. H. G. Lewin,S. F. Anderson, H. Baumgardt, A. V. Filippenko, B. M. Gaensler, L. Homer, P. Hut, V. M. Kaspi, J. Makino, and 5 coauthors: Dynamical Formation of Close Binary Systems in Globular Clusters, ApJ **591**, 131, 2003

91. S. Portegies Zwart, S. L. W. McMillan: Black Hole Mergers in the Universe, ApJ **528**, 17 (2000)

92. S. Portegies Zwart: The Ecology of Black Holes in Star Clusters in *Joint Evolution of Black Holes and Galaxies*, eds M. Colpi, V.Gorini, F.Haardt and U.Moschella Series in High Energy Physics, Cosmology and Gravitation. IOP Publishing, Bristol and Philadelphia (2005)

93. Portegies Zwart, Simon F. McMillan, Stephen L. W. The Runaway Growth of Intermediate-Mass Black Holes in Dense Star Clusters, ApJ **576**, 899 (2002)

94. G. D. Quinlan: The dynamical evolution of massive black hole binaries I. Hardening in a fixed stellar background, New Astronomy, vol 1, 35, 1996

95. S. M. Ransom, J. W. T. Hessels, I. H. Stairs, P. C. Freire, F. Camilo, V. M. Kaspi, D. L. Kaplan: Twenty-One Millisecond Pulsars in Terzan 5 Using the Green Bank Telescope, Science, Volume 307, Issue 5711, pp. 892-896 (2005).

96. Rasio, F. A. et al.: Neutron Stars and Black Holes in Star Clusters, astro-ph/0611615, Highlights of Astronomy, Volume 14, p. 215-243

97. R. Salvaterra, A. Cerutti, G. Chincarini, M. Colpi, C. Guidozzi, P. Romano: Short Gamma Ray Bursts: a bimodal origin?, MNRAS **388**, L6

98. S. Sigurdsson, L. Hernquist: Primordial black holes in globular clusters, Nature **364**, 423 (1993)

99. S. Sigurdsson, E. S. Phinney: Binary–Single Star Interactions in Globular Clusters, ApJ **415**, 613 (1993)

100. S. Sigurdsson, E. S. Phinney : Dynamics and Interactions of Binaries and Neutron Stars in Globular Clusters, ApJ **99**, 609 (1995)

101. L. Spitzer:Dynamical evolution of globular clusters, Princeton University Press (1987)

102. I. H. Stairs, S. Begin, S. Ransom, P. Freire, J. Hessels, J. Katz, V. Kaspi, F. Camilo: Pulsars in the Globular Cluster M28, 2007 AAS/AAPT Joint Meeting, American Astronomical Society Meeting 209, 159.02; Bulletin of the American Astronomical Society, Vol. 38, p.1118

103. Timmes, F. X.; Woosley, S. E.; Taam, Ronald E. The conductive propagation of nuclear flames. 2: Convectively bounded flames in C + O and O + NE + MG cores, ApJ **420**, 348 (1994)

104. van den Berk, J.; Portegies Zwart, S. F.; McMillan, S. L. W. :The formation of higher order hierarchical systems in star clusters, MNRAS **379**, 111 (2007)

105. van der Marel, R. P: Intermediate-mass Black Holes in the Universe: A Review of Formation Theories and Observational Constraints, in *Coevolution of Black Holes and Galaxies*, from the Carnegie Observatories Centennial Symposia. Published by Cambridge University Press, as part of the Carnegie Observatories Astrophysics Series. Edited by L. C. Ho, 2004, p. 37.

106. F. Verbunt, P. Hut, P.: The Globular Cluster Population of X-Ray Binaries, in *The Origin and Evolution of Neutron Stars*, ed. by D. J. Helfand and J.-H. Huang. Dordrecht, D. Reidel Publishing Co. (IAU Symposium, No. 125), 1987., p.187

107. F. Verbunt, W. h. G. Lewin: Globular Cluster X-ray Sources, astro-ph/0404136; "Compact Stellar X-ray Sources", eds. W.H.G. Lewin and M. van der Klis, Cambridge University Press

108. F. Verbunt, D. Pooley, C. Bassa: Observational evidence for the origin of X-ray sources in globular clusters. In *Dynamical evolution of dense stellar systems*, ed. by E. Vesperini

109. Walker, I. W.; Roy, A. E. Stability criteria in many-body systems. II - On a sufficient condition for the stability of coplanar hierarchical three-body systems, CeMec **24**, 195 (1981)

110. W. A. Watters, K. J. Joshi, F.A. Rasio: Thermal and Dynamical Equilibrium in Two-Component Star Clusters, ApJ **539**, 331 (2000)

111. B. Willems, M. Henninger, T. Levin, N. Ivanova, V. Kalogera, K. McGhee, F. X. Timmes, C. L. Fryer: Understanding Compact Object Formation and Natal Kicks. I. Calculation Methods and the Case of GRO J1655-40 , ApJ **625**, 324 (2005)

112. A. Wolszczan, S. R. Kulkarni, J. Middleditch, D. C. Backer, A. S. Fruchter, R. J. Dewey: A 110-ms pulsar, with negative period derivative, in the globular cluster M15, Nature **337**, 531 (1989)

Short Gamma Ray Bursts: Marking the Birth of Black Holes from Coalescing Compact Binaries

Davide Lazzati and Rosalba Perna

JILA, University of Colorado, 440 UCB, Boulder, CO 80309-0440, USA
lazzati,rosalba@colorado.edu

1 Introduction

As soon as the catalog of gamma-ray bursts (GRBs) detected by the BATSE (Burst And Transient Source Experiment) had enough events to allow a statistical study, it was discovered that GRB light curves could be separated into two families [45]. Long GRBs are characterized by a duration of more than 2 seconds and a somewhat soft spectrum. Short GRBs, on the other hand, are characterized by a duration of less than 2 seconds and a harder spectrum.

Not much more could be said in the BATSE era, due to the lack of precise localizations and impossibility of long wavelength follow-up that plagued both the long and short GRB populations. With the launch of the Italian-Dutch satellite BeppoSAX, the situation for long GRBs changed dramatically. X-ray, optical and radio afterglows were discovered [16, 88, 83] to follow the prompt phase. Spectroscopy revealed that the GRBs lie at cosmological distances and that they involve explosion energies similar to core collapse supernovae [55, 46]. Evidence of beaming and association to massive stars emerged [73, 78] leading to the now widely accepted scenario of long GRBs as collimated relativistic outflows associated to Type Ib/c supernova explosions [80, 38]. Unfortunately BeppoSAX was non-optimally designed to detect short GRBs and none of the above information was available for the short bursts that remained elusive and mysterious. The only advance came from the discovery that short GRBs also have longer wavelength emission on longer timescale [47]. This discovery was however made on a stacked light curve from past events, and did not allow for any follow-up observation. A general consensus was reached in those years that short GRBs could be associated with the merger of compact binary systems [22], based on a theoretical desire more than on any robust evidence.

More recently, thanks to the HETE-2 satellite and to Swift, afterglow observations have been performed also for the class of short GRBs, measuring their redshift and energetics, and finally giving some observational corroboration to the idea that they originate from binary mergers [89, 29]. Not everything has been clarified, though, the main problem being now the one of defining what is a long and what is a short burst, since the two classes seem to have a gray area between them with bursts sharing a complex set of properties. In this review, we critically present the new discoveries made with HETE-2 and Swift, the theoretical advances that were made possible by those discoveries and the still debated issues and future perspectives. This Chapter is divided in two main sections, the first observational and the second theoretical.

2 Observations

2.1 Pre-Swift Era

Observations performed in the pre-Swift (and HETE-2) era are mainly those performed with BATSE and Konus. Fig. 1 shows a histogram of the T_{90} distribution of 2041 GRBs detected by BATSE. The T_{90} is the time interval during which the GRB emits 90 per cent of the total fluence. The solid and dashed lines show the Gaussian fit to the distribution for the short and long bursts, respectively. Even though the bimodality is clear, it is also clear that the two populations are not entirely separated, since a considerable tail of short burst population (about 25 per cent) extends beyond $T_{90} > 2$ s, the traditional dividing line.

Additional separation between the two classes is provided by the spectral analysis of the light curves. Fig. 2 shows in greytones the two dimensional distribution of BATSE GRBs in the hardness-duration plane. The hardness is defined as the ratio of counts in the high energy channels (3 and 4) over the counts in the low energy channels (1 and 2). Even though short bursts appear to have a systematically harder spectrum, the two populations have a sizable overlap.

The origin of the spectral difference has been analyzed in detail by Ghirlanda et al. [30]. They performed spectroscopy on a sample of 36 bright short bursts comparing the results of several spectral models. They conclude that the spectra of short bursts are successfully fitted by a single power-law with an exponential high energy cut-off. They also find that short GRBs have harder spectra due to steeper low energy slopes of the power-law rather than due to a larger peak frequency. It is worth reminding that long GRBs are usually fitted with a smoothly broken power-law model or Band function [3].

Nakar and Piran [57] analyzed the light curves of a sample of short GRBs from BATSE. They find that, when high resolution data are available, short GRB light curves can be resolved into the superposition of many pulses, with statistical properties analogous to those of the long GRBs. This may indicate that the same dynamical and dissipation processes are powering the light curves, even though such processes are still far from being understood.

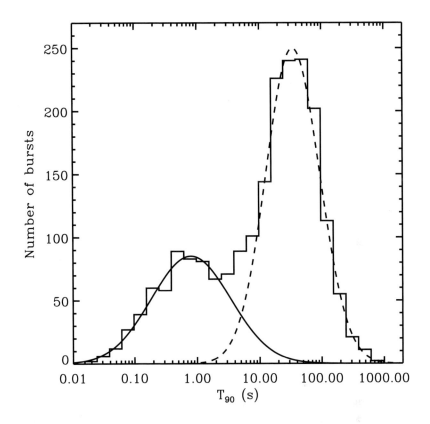

Fig. 1. Distribution of the durations of 2041 BATSE GRBs. Gaussian fits for the short (solid line) and long (dashed line) populations are overlaid.

The quest for short GRB afterglows went on for all the BATSE and BeppoSAX era, with very limited results. No afterglow of an individual short GRB was ever found. Lazzati et al. [47] stacked the background subtracted light curve of the 76 brightest short BATSE GRBs looking for an evidence of afterglow in the hard X-rays. They found (see Fig. 3) that there is indeed an excess soft component following the prompt emission of short GRBs lasting approximately 100 seconds. They showed that this component is consistent with being of afterglow origin. As we will see below, it later emerged from Swift observations that this component is more likely residual activity from the central engine.

2.2 The Swift Era

Many of the riddles of short GRB astrophysics have been solved in the Swift era, with a noticeable contribution from the HETE-2 satellite. Due to the

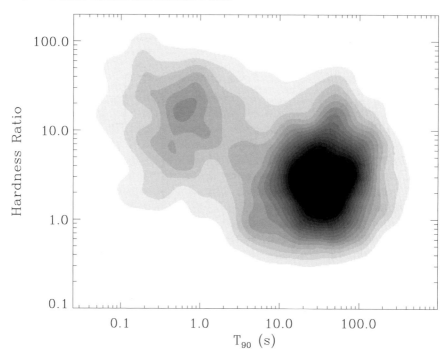

Fig. 2. Hardness-duration distribution of BATSE GRBs.

characteristics of the BAT detector, Swift observations did not increase our understanding of the prompt emission.

In the spring and summer of 2005, after years of struggle, afterglows of short GRBs were finally detected in X-rays, optical and radio wavelengths [29, 89, 14, 24, 39, 17, 6], bringing the short bursts in the afterglow era.

2.2.1 Afterglows

At first sight, afterglows of short GRBs are similar to the afterglows of long GRBs. They are fainter, but qualitatively analogous. Fig. 4 shows the X-ray afterglow of GRB 050724 as observed by the Swift XRT [13]. The afterglow shows an initially very bright phase, followed by a sharp decline, a possible t^{-1} power-law decay and a flare at late times. The afterglow is overall very faint. This is supposed to be due to a combination of an isotropic equivalent energy smaller than that of long bursts and of a low density interstellar medium, down to $n \sim 10^{-5}$ cm^{-3} [59]. Such low densities are thought to be an indication of the binary merger origin of short bursts, since they are of the order of magnitude of what is expected in the intergalactic medium.

The initial bright phase is likely to be the one detected in BATSE data [47] and was initially interpreted as the regular afterglow. It is now believed to be a sign of continued activity of the central engine (see below). X-ray

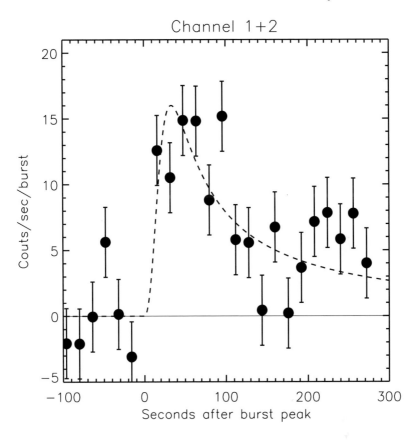

Fig. 3. Light curve of the excess emission found by Lazzati et al, (2001) in the composite light curve of 76 bright BATSE short GRBs. The dashed line is a best fit afterglow model.

flares are also common. Different from the X-ray flares detected on top of long GRB afterglows, the flares so far detected on short GRBs have long time scales ($\delta t/t \sim 1$) [50]. Their origin is not clear; a possible detection of rapid variation was reported for GRB 050709 [24], but is inconclusive. Since rapid variations are associated with late time activity of the engine [50], their confirmation would be of great relevance for our understanding of the short GRB engine physics.

It is still debated whether short GRB afterglows show conclusively the presence of beaming of the short GRB fireballs. The question is relevant since the total energy of the explosion depends on the beaming factor. In few cases, a jet break has been claimed. A good example is that of GRB 051221 [79] where a simultaneous break in the optical and X-ray light curves was detected approximately 5 days after the explosion. This is however an isolated case and

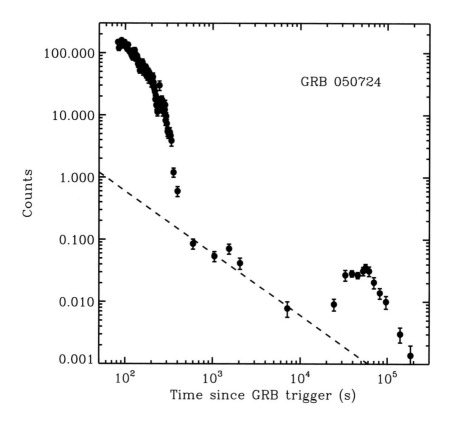

Fig. 4. Swift X-ray afterglow of GRB 050724 [13]. The dashed line shows a t^{-1} power-law for comparison purposes.

in most cases only a lower limit to the beaming can be obtained [36]. It seems fair to conclude that, even if beamed, short GRBs are less beamed than long GRBs, for which opening angles of a few degrees are routinely measured [31]. These observations support the idea that short GRBs are associated with the merging of compact objects, for which hydrodynamical collimation has a much lesser role than in massive stars [48].

Intensive searches for a supernova component have been performed in several short GRBs. Due to their moderate redshift, the searches are very sensitive, being able to detect the presence of a supernova bump even if the supernova was 10 times fainter than the faintest observed supernova [24, 9, 17, 79]. The lack of any detection strongly favors a different progenitor for short and long GRBs.

2.2.2 Host Galaxies

A debated and important aspect of short GRB research has been that of the identification of the host galaxy. It is important for at least two reasons. First, the identification of a host galaxy is usually the only way to obtain a redshift for short GRBs. Second, host galaxies can provide important clues to the nature of the progenitor of the short bursts and on the physics of their afterglow emission.

Unlike long GRBs, short bursts provided a harder challenge to astronomers for the identification of the host galaxy. While long bursts explode on top of the brightest region of galaxies, making the identification unquestionable [23], short GRBs explode in anonymous regions in the outskirts of galaxies, sometimes having more than one galaxy within their error circles.

The first well-localized short GRB offers a good example of the situation [40, 66, 29]. The X-ray error box was large enough to contain a low-redshift elliptical galaxy, member of a cluster, and several high redshift faint objects, analogous to the host galaxies of long GRBs. The implications of the choice of host were relevant. If the nearby elliptical was the host, short GRBs would be associated to non star forming objects, they would explode in the outskirts of the host – if not outside of them – and would be low redshift events, involving several orders of magnitude less energy than long events. On the other hand, if the progenitor was one of the high redshift objects, short GRBs could be analogous to the long ones. Circumstantial evidence favoring a low redshift origin of short bursts was emerging in the meantime, showing that the location of BATSE short GRBs correlates with local bright galaxies [82] or with clusters [32].

With more short GRB localizations and improvement in the error boxes there is now very little doubt that the population of short GRB host galaxies is markedly different from those of long bursts. Short GRB hosts seem to be a far less homogeneous sample than those of long ones. With the caveat that there may be some misidentifications, given that short GRBs do not explode in the brightest parts of their hosts but rather in their outskirts. Long GRB hosts are usually dwarf starbursting galaxies at moderate to high redshift, always found in the field. Morphologically, short GRB hosts can be of any kind, from early type ellipticals to late type spirals, and are found both in the field and inside clusters. In most cases, short GRB hosts have low star formation, $< 1 M_\odot/\mathrm{y}^{-1}(L_\star/L)$, or about one hundredth of the star formation rate in long GRB hosts [15]. The two populations are different at a very high confidence level [35] and at least in some cases, evidence of an old population of stars was found in the host [17, 79]. Approximately 20 per cent of Swift short bursts are associated to clusters of galaxies [7], in agreement with the fraction of stellar mass contained in such systems. Short GRBs seem therefore to be better unbiased samples of the stellar population in the low-moderate redshift universe than long ones.

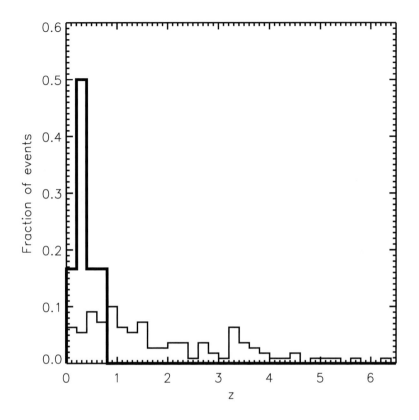

Fig. 5. Redshift distribution of short (thick line) and long (thin line) GRBs.

2.2.3 Redshift Distribution

The redshift of short GRBs is always measured as the redshift of the putative host galaxy and is therefore less robust than the long GRB redshifts, that are often measured both in absorption on the afterglow spectrum and in emission in the host galaxy. Fig. 5 shows the short GRB redshift distribution with a thick line histogram. Even though some of the values can be debated, it is clear that the redshift distribution of short GRBs is centered at much smaller redshifts than that of long GRBs. The long GRB redshift distribution is shown with a thin histogram for comparison and extends beyond redshift 6 [81, 37, 43].

2.2.4 Short or Not Short?

We have discussed so far a consistent body of observational evidence coherently supporting the idea that short and long GRBs have different physical origin and are two well separated populations. Long GRBs are long in γ-rays and soft, have bright afterglows with jet breaks and supernova bumps, they explode in star-forming regions inside high-redshift dwarf irregular star-forming galaxies. Short bursts, on the other hand, are short and hard in γ-rays, have dim afterglows and lack any evidence of an associated supernova, explode in low-intermediate redshift galaxies of all kind, with little sign of star formation.

Such an idyllic interpretation framework was suddenly shaken last summer, when two puzzling bursts were detected: GRB 060505 and GRB 060614 [20, 27, 28]. They appeared to be long in the γ-ray properties, but had faint afterglows and no sign of an associated supernova, down to very stringent limits. Possible interpretations range from short bursts with a particularly bright X-ray early afterglow [90] to long bursts with a non-massive star progenitor [28], or associated to a SN explosion that did not produce large quantities of ^{56}Ni [86]. The puzzle becomes even more intricate if secondary indicators are used. GRB 060614 is found to be consistent with the Amati correlation [1], like a long duration GRB, but to share the short GRB properties in terms of spectral lags (no significant lag was detected) [62].

Even though it is definitely premature to give up a classification successful for decades based on only two events, such observations should clearly ring a warning bell. It is obvious that if short GRBs are followed by relatively bright early X-ray afterglows [47, 13], the distinction between short and long events based only on their duration is band dependent and doomed to fail. However, a new robust scheme has not appeared, so far. Spectral lags are a promising tool, but they are still theoretically unexplained and may well not be associated to the physics of the progenitor, something that seem to be clearly different between the two classes. A classification based on the host galaxy properties is also a possibility that could be considered, with the problem that not all GRBs have host galaxy detections and so many events would become unclassified. It is clear that more events of difficult classification have to be detected before we can establish a new, successful, classification scheme.

2.3 SGRs

Considerable excitement was caused by the detection of a giant flare from the soft gamma repeater (SGR) SGR 1806-20 [65, 84, 12, 53]. The initial phase of the flare is a bright spike lasting less than one second, with a blackbody spectrum and enough energy to be detected by BATSE out to a distance of approximately 50 Mpc. It is followed by a pulsed X-ray tail that would be undetectable with any present instrumentation for an extragalactic flare. These properties suggested that a significant fraction of BATSE short GRBs could indeed be giant flares from SGRs in the local universe, out to the Virgo cluster of galaxies. A large effort was attempted to constrain this fraction and possibly identify the extragalactic SGRs in the BATSE sample. The search was based on positional coincidence with nearby galaxies [58, 63, 70], on the spectral properties of the prompt emission [49], and on the presence of an oscillating tail. None of these searches found any suitable candidate, implying a fraction of at most 15 per cent of BATSE short GRBs being SGR flares in the local universe. This limit is only marginally consistent with the galactic SGR rate, and requires the SGR1806-20 flare to be an exceptional event. Of course, caution should always be considered when statistics is based on a handful of events.

3 Theory

3.1 Binary Mergers as Progenitors of Short GRBs

The leading candidates as progenitors of short GRBs are mergers of NS–NS binaries [64, 33, 22, 61] and BH–NS binaries [61, 56]. Two compact objects in a binary are bound to eventually merge due to the emission of gravitational wave radiation that causes a loss of energy resulting in a gradual shrinking of the orbit. For typical binary parameters, binaries are expected to coalesce within a Hubble time, as discussed in more detail in the following. From an energetic point of view, there is enough gravitational binding energy that is liberated during the merger to power the GRB and its subsequent afterglow. The engine that helps channeling this energy into a relativistic flow is believed to be, like in the case of long GRBs, an hyper-accreting accretion disk. However, while in the case of long bursts (which are thought to be associated with the collapse of massive stars), the event duration is set by the collapse time of the star envelope (on the order of several tens of seconds) that keeps on feeding the disk, in the case of a binary merger the activity phase is set by the viscous timescale of the disk, which is a fraction of a second, the right order of magnitude needed for the power source of short bursts.

From an observational standpoint, double neutron stars are known to exist from direct observations in the Galaxy. Well-known examples are the Hulse-Taylor binary and the binary pulsar [11]. On the other hand, compact-object binaries where one of the components is a black hole have not been observed so far. However, theoretical modeling of binary evolution predicts that binaries with black holes are expected (e.g. [10, 71, 25, 4]). It should be noted that, even though BH-BH binaries are also predicted as a possible end state of binary evolution, however they are not expected to give rise to a GRB when they merge. This is because, in this scenario, there would be no fuel to power the accretion disk that could then provide the source of energy for the GRB (see Sect. 3.4).

In order for mergers of compact objects to be progenitors of short bursts, a fundamental condition is that the merger event rate be comparable to that of the bursts. With the growing number of short bursts with detected redshift, a comparison can be made not only for the overall (i.e. redshift-integrated) event rate, but also for its redshift evolution. As the sample grows, the calibration of the cosmic rate evolution, in combination with observations of the burst rate as a function of the galaxy type, can allow one to further constrain various types of binary models.

The main source of guidance for the expected characteristics of a population of bursts associated with the coalescence of two compact objects is provided by population synthesis calculations. In the following, we will describe the main assumptions, calculation methods, and results from these calculations (Sect. 3.2), with a special emphasis on their specific application to short bursts (Sect. 3.3). Finally (Sect. 3.4), we will discuss the highlights of numerical calculations that simulate the final moments of the binary, and the formation of the accretion disk that eventually leads to the GRB.

3.2 Binary Evolution – Theoretical Modeling

In the last few years, several groups [71, 4] have developed population synthesis calculations. These simulations, starting from some assumptions regarding the binary star population, track the evolution of stars both individually and in relation to the companion in the binary system; in output, they are able to predict the fraction of systems that end up with a certain type of compact objects, the merger time of each binary, and hence the cosmic event rate. In the following, we describe in more detail the highlights of the population synthesis calculations.

The evolution of each star in the binary system is computed given its zero-age main-sequence mass and its metallicity. The code tracks all the stages from the main sequence to the red giant branch to the core helium burning and the asymptotic giant branch. At each stage of evolution, the basic stellar parameters (radius, luminosity, stellar mass, core mass) are determined. While each star is evolved depending on its initial parameters, a number of effects that influence the binary orbit (e.g. mass and angular momentum losses due to stellar winds) are taken into account. At every evolutionary time step, the codes check for possible binary interactions. If any of the components fills its Roche lobe, then the resulting mass transfer is computed, and so the eventual resulting mass and angular momentum losses. If the binary survives the mass transfer event (i.e. both stellar components fit within their Roche lobe), then the evolution of the binary keeps on being followed. The calculation for each star ends at the formation of a stellar remnant: a white dwarf, a neutron star or a black hole. If the remnant is born with a supernova explosion, the effects of the supernova kicks and mass loss on the binary orbit are computed. Finally, once a binary consists of two remnants, its merger lifetime is calculated, i.e. the time until which the components merge due to emission of gravitational radiation and the consequent orbital decay.

Early population synthesis studies [71, 26] found that, for the NS–NS binary systems, merger times are generally long, $t_{\mathrm{merg}} > 0.1$–1 Gyr. Smaller timescales were found [87] as a result of the assumption that the secondary star, once it becomes a low-mass helium-rich star, can initiate an extra mass transfer phase. More recently, simulations [4] identified a new population of coalescing NS–NS binaries, which merge on a timescale which is much shorter than that of the "classical" population. These very short timescales, t_{merg} 0.001–1 Myr, are the result of allowing both the primary and the secondary star to initiate an extra mass transfer phase. Furthermore, the inclusion of natal kicks in the simulations [4] further contributes to decrease the merger timescales. Accounting for natal kicks, in fact, results in the disruption of the widest binaries and in an eccentricity gain for the remaining ones, which further reduces their lifetimes. This subpopulation of relatively short-lived NS–NS binaries was found to be the dominant channel, making up about 80% of the total population.

On the other hand, all population synthesis calculations agree on the typical distribution of merger times of the NS–BH systems, which is found to be similar to that of the "classical" population of NS–NS binaries, i.e. $t_{merg} > 0.1$–1 Gyr.

3.3 Theoretical Predictions for the Observational Properties of Short Bursts Due to Binary Mergers

a) Rates The distribution of merger times obtained from population synthesis calculations, combined with a prescription for the cosmic star formation rate, allows one to estimate the rate of binary mergers throughout the lifetime of the Universe. Since the simulations by Belczynski et al. [5] include the population of tight NS–NS binaries (i.e. the short-lived population), the predicted cosmic redshift rates are different for the population of NS–NS and of NS–BH binaries. An example is shown in Fig. 6, with data from the simulations by Belczynski et al. [5]. Those calculations used as star formation rate which peaks at a redshift of about 3 and has only a mild decline up to redshifts of about 5. Due to the merger time delays, however, the predicted binary merger rate peaks at a remarkably lower redshift. The larger the time delay, the larger the shift to lower redshifts of the binary merger event rate compared to that of the underlying star forming rate. This is especially emphasized in Fig. 6, where the rates due to the fraction of (both NS–NS and NS–BH) long-lived binaries ($t > 100$ Myr) are shown together with the overall rates. Whereas the assumed star-formation rate remains constant up to a redshift of ~ 5, the rate of the long-lived component drops substantially. For the overall binary merger rate, if the assumed SFR peaks at a redshift of ~ 3, the peak is found in the range ~ 1–2.

Another method of estimating the binary merger rates starts from a count of the observed number of NS–NS binaries in the Milky Way. This number is then corrected to account for the completeness of the survey and transformed into a local rate by weighing in the estimated lifetimes of the observed systems. Up to date, the latest calculations of the galactic merger rates [42] have yielded a value in the range 1.7×10^{-5} to 2.9×10^{-4} yr^{-1} at the 95% level.

b) Offsets from host galaxies, densities of the circumburst medium, afterglow brightnesses and galaxy types

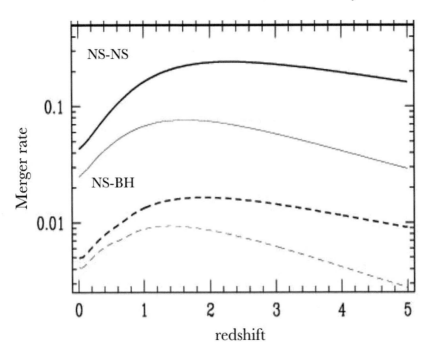

Fig. 6. *Thick lines*: the inferred merger rate as a function of redshift for the NS–NS and BH–NS mergers. *Thin lines*: the contribution from the long-lived ($t > 100$ Myr) population. Rates are in arbitrary units. Data from the simulations of Belczynski et al. [5].

Since the first studies on GRBs, it was clear that the location of a burst within a galaxy, and hence its environment, would provide important clues to the properties of the progenitor itself. In the past decade, a number of groups [26, 8, 67, 5] have performed statistical studies of the distribution of the binary merger sites in a galaxy. Generally speaking, progenitors with very short lifetimes are typically expected to produce GRBs close to their place of birth. However, if the progenitor is endowed with a kick velocity (such as in the case of binaries), then the merger sites could have a substantial offset from the place of birth. The amount of the offset will generally depend on a combination of the merger time and the velocity of the binary system, as well as the potential well in which the system moves. Kick velocities are estimated to be in the range of several tens to several hundreds km/s [41]. A binary with a velocity in the upper range of the distribution, and a long lifetime $\sim$ a few Gyr, can travel a substantial distance, on the order of the Mpc scale, before merging. On the other hand, for binaries with a short lifetime, on the order of a million years or less, the place of merger is always going to be close to the place of birth even for the fastest pulsars of the distribution. Therefore, it is clear that the expectations for the distribution of the merger sites are going to heavily depend on the expected distribution of merger times of the binary population under consideration.

Another element of consideration is the fact that different types of galaxies are expected to host different types of stellar populations, with, at one extreme, starburst galaxies which are dominated by very young stars, and at the other end, elliptical galaxies which have almost no star formation rate at all and hence are dominated by an old stellar population. These different types of galaxies, therefore, will select out different components of the binary mergers. Starburst galaxies are expected to be dominated by the short-lived binary population, while elliptical galaxies will mostly host the merger events of the long-lived population. Spiral galaxies are in between, since they host both young and hold star populations. As a result of these differences, merger events taking place in elliptical galaxies will generally occur at larger offsets (from the galaxy centers) with respect to merger events that occur in starburst and spiral galaxies. Furthermore, within a given morphological galaxy type, mergers in small galaxies will have substantially larger offsets than merger in large galaxies. This is due to the larger gravitational potential of the larger galaxies, which prevents the binaries from traveling too far.

The sites of the binary mergers are especially important in determining the observability of the GRB afterglows. Since the gas density declines with the distance from the galaxy center, mergers that occur in the outskirts of galaxies are expected to give rise to very dim afterglows. Based on the considerations above, short bursts occurring in large starburst galaxies are expected to be the brightest, while short bursts occurring in small ellipticals will be generally the dimmest. A fraction of these bursts are expected to be "naked", i.e. completely lacking any afterglow emission, especially at wavelengths longer than the X-ray band, where the effect of the density becomes more important.

As more data on short bursts gather, the information on the fraction of bursts as a function of the galaxy type will become an important element of study in order to establish whether there is a dominant channel of binaries (i.e. short-lived versus long-lived) that give rise to short bursts. However, if one wants to extract from these types of statistical studies meaningful physical information, one needs to keep in mind that there is a huge potential for selection biases. In fact, since bursts are localized through afterglow observations, and bursts occurring in elliptical galaxies are expected to be generally dimmer, there is going to be a selection effect toward the observation of a relatively larger fraction of the bursts occurring in starburst and spiral galaxies, i.e. the short-lived component of the binary population. Furthermore, since the mean redshift of merger events for the short-lived population is higher than that for the long-lived population, the selection effect described above will also result in a bias toward higher redshifts.

3.4 The Final Moments: From a Merging Binary to a Hyperaccreting Disk Around a Black Hole

Although there is no "direct" evidence for an accretion disk in GRBs, the GRB phenomenology provides strong hints in that direction. Firstly, accretion disks are a powerful way to tap gravitational energy and channel it into other sources. Second, the overall (short) burst durations, as discussed above, are naturally accounted by the viscous timescale of the disk, while the millisecond timescale variability (observed both in long and short bursts) is on the same order of the dynamical timescale of a compact disk accreting around a stellar-mass black hole. From an observational point of view, on the other hand, it is well known that other systems in nature believed to be associated with black holes accreting from a disk (i.e. active galactic nuclei, micro-quasars) are able to power relativistic jets, for which we have strong evidence also in GRBs.

In the last several years, a number of groups have devoted a substantial effort into simulations of NS–NS and NS–BH mergers, and the resulting structure of the hyperaccreting accretion disk. The early simulations [19] were Newtonian, used a polytropic equation of state, and did not include the effects of neutrinos. These, on the other hand, were shown to represent a substantial source of cooling in a number of semianalytical, 1D calculations of hyperaccreting disks around black holes [60, 21] . Neutrino effects, together with more realistic equations of state, were taken into account into later simulations [74, 75, 76, 77]. These simulations, independently of the numerical method used, found results in general agreement. Although details differ depending on whether the initial progenitor is a NS–NS or a NS–BH, some common features can be identified. As the binary members spiral in, within a few orbital periods the outcome is the formation of a dense and thick, hot torus, of mass on the order of $\sim 0.01 - 0.3 M_\odot$ that accretes onto a stellar mass black hole. In the case where the initial binary members are both neutron stars, the black hole will be formed as a result of the accretion of mass onto one of the NSs. Depending on the total initial mass of the binary system, the collapse of the hypermassive NS into a BH can occur promptly, or it can be delayed for an initial time during which the star is supported by differential rotation. The accreting material, on the other hand, is provided by the tidally disrupted debris of the NS.

The duration of the prompt GRB phase phase is set by the time during which efficient accretion occurs. Given the observed γ-ray luminosities ($\sim 10^{49}$–10^{50} erg/s assuming a beaming correction of a factor ~ 0.1), and taking an efficiency of conversion of accretion energy into γ-rays of a fraction of per cent to a per cent, the accretion rates of the disk must then be in the range ~ 0.01–$10 M_\odot$. The resulting hyperaccreting disk is very dense and hot, optically thick to photons, and cools mainly by neutrinos. In the upper end of the range of accretion rates, the density of the disk can however be so high that the innermost regions would become optically thick even to the neutrinos themselves.

An important component of the studies of GRB accretion disks deals with the processes by which the disk black-hole system is able to collimate and launch the relativistic jets known to power the GRBs. Two mechanisms have been suggested as being involved: neutrino–anti-neutrino annihilation, and magnetic fields. In the former process, suggested by a number of authors [34, 22, 61, 54, 56], the source of energy is provided by the neutrinos and anti-neutrinos emitted in the cooling disk which annihilate in a funnel above a disk (which has a lower density). Calculations [21] have estimated that the maximum efficiency by which the rest mass energy of the accreting material is converted into neutrino luminosity does not exceed a value of $\sim 10^{-4}$. For a disk mass of $\sim 0.1 M_\odot$, this would yield an energy output of about 10^{49} erg, therefore making this jet-production mechanisms viable only if there is a substantial degree of collimation in short bursts.

The second method of jet collimation and launch relies on the help of magnetic fields. A number of authors [85, 51] have suggested this possibility by noticing how, even if the magnetic field is initially low, it is likely to be amplified by the magneto-rotational instability within the disk [2]. Numerical simulations [52] find that magnetically driven jets, whose energy output increases with the spin of the BH, are generally more efficient than neutrino-powered jets. Magnetic fields are therefore considered to play an important role in collimating and driving the jets from the accretion disk.

Whereas substantial progress has been made in this area of research of GRB hyperaccreting accretion disks, recent, new observations with *Swift* have shown that the current picture is far from being complete. In particular, the detection of energetic X-ray flares superimposed on the smooth afterglow decay, with arrival times and durations from tens to tens of thousands of seconds, requires the presence of an engine with duration much longer than the fraction of a second that is sufficient to power the prompt phase of a short burst [50]. These observations have prompted a number of suggestions on how to make the lifetime of the accretion disk much longer than its viscous timescale. Perna, Armitage and Zhang ([68]; see also Piro and Phfal [69]) suggested that fragmentation of the outer parts of the accretion disk (which would then accrete at later times) could be responsible for creating a long-lived engine. Proga and Zhang [72], on the other hand, envisaged a scenario in which the accumulation of magnetic flux in the innermost parts of the accretion disk creates a barrier that can then produce intermittent accretion. Other suggestions, which do not involve accretion disks, include that of Dai et al. [18]. They showed that, if the NS–NS merger leads to a differentially rotating, millisecond pulsar, then the differential rotation can cause the interior magnetic field to wind up and break through the stellar surface, hence resulting in magnetic reconnection-driven explosive events. These events would be observed as X-ray flares.

4 Gravitational Waves from Short GRBs

Mergers of two compact objects have traditionally been of great interest as sources of gravitational waves. With the likely association of binary mergers with short GRBs, the interest of the gravitational wave community has been extended to that of short GRBs. The local rate of these sources, however, is not large enough to make a detection likely with a blind search with LIGO-I. However, the detection probability can be increased if the observations are made shortly after the γ-ray detection from the burst is detected. Estimates [44] suggest that in this case a positive detection in coincidence with a short GRB could be already made with current gravitational wave detectors. Clearly, such a signal, besides giving information on the last moments of the binary merger, would also provide the still needed conclusive evidence of the association of short bursts with mergers of compact objects.

References

1. Amati L., Della Valle M., Frontera F., Malesani D., Guidorzi C., Montanari E., Pian E., 2007, A&A, 463, 913
2. Balbus, S. A. & Hawley, J. F. 1991, ApJ, 376, 214
3. Band D., et al., 1993, ApJ, 413, 281
4. Belczynski, C., Bulik, T, & Kalogera, V. 2002, ApJ, 571, 147
5. Belczynski, K., Perna, R., Bulik, T., Kalogera, V., Ivanova, N. & Lamb, D. Q. 2006, ApJ, 648, 1116
6. Berger E., et al., 2005, Nature, 438, 988
7. Berger E., Shin M. S., Mulchaey J. S., Jeltema T. E., 2006, ApJ, 660, 496
8. Bloom, S. J., Sigurdsson, S. & Pols, O. 1999, MNRAS, 305, 763
9. Bloom J. S., et al., 2006, ApJ, 638, 354
10. Brown, G. E. 1995, ApJ, 440, 270
11. Burgay, M. et al. 2003, Nature, 426, 531
12. Cameron P. B., et al., 2005, Nature, 434, 1112
13. Campana S., et al., 2006, A&A, 454, 113
14. Castro-Tirado A. J., et al., 2005, A&A, 439, L15
15. Christensen L., Hjorth J., Gorosabel J., 2004, A&A, 425, 913
16. Costa E., et al., 1997, Nature, 387, 783
17. Covino S., et al., 2006, A&A, 447, L5
18. Dai, Z. G., Wang, Z. Y., Wu, X. F., Zhang, B. 2006, Science, 311, 1127
19. Davies, M. B., Benz, W., Piran, T., Thielemann, F. K., 1994, ApJ, 431, 742
20. Della Valle M., et al., 2006, Nature, 444, 1050
21. Di Matteo, T., Perna, R., & Narayan, R. 2002, ApJ, 579, 706
22. Eichler, D., Livio, M., Piran, T., Schramm, D. N. 1989, Nature, 340, 126
23. Fruchter A. S., et al., 2006, Nature, 441, 463
24. Fox D. B., et al., 2005, Nature, 437, 845
25. Fryer, C., Burrows, A. & Benz, W. 1998, ApJ, 496, 333
26. Fryer, C., Woosley, S. E.& Hartmann, D. H. 1999, ApJ, 526, 152
27. Fynbo J. P. U., et al., 2006, Nature, 444, 1047
28. Gal-Yam A., et al., 2006, Nature, 444, 1053

29. Gehrels N., et al., 2005, Nature, 437, 851
30. Ghirlanda G., Ghisellini G., Celotti A., 2004, A&A, 422, L55
31. Ghirlanda G., Ghisellini G., Lazzati D., 2004, ApJ, 616, 331
32. Ghirlanda G., Magliocchetti M., Ghisellini G., Guzzo L., 2006, MNRAS, 368, L20
33. Goodman, J. 1986, ApJ, 308, L47
34. Goodman, J., Dar, A. & Nussmov, S. 1987, ApJ, 314, L7
35. Gorosabel J., et al., 2006, A&A, 450, 87
36. Grupe, D., et al., 2007, ApJ, 653, 462
37. Haislip J. B., et al., 2006, Nature, 440, 181
38. Hjorth J., et al., 2003, Nature, 423, 847
39. Hjorth J., et al., 2005, Nature, 437, 859
40. Hjorth J., et al., 2005, ApJ, 630, L117
41. Hobbs, G., Lorimer, D. R., Lyne, A. G., Kramer, M. 2005, MNRAS, 360, 974
42. Kalogera, V. et al. 1994, ApJ, 601, L179
43. Kawai N., et al., 2006, Nature, 440, 184
44. Kochanek, C. S. & Piran, T. 1993, ApJ, 417, L17
45. Kouveliotou C., Meegan C. A., Fishman G. J., Bhat N. P., Briggs M. S., Koshut T. M., Paciesas W. S., Pendleton G. N., 1993, ApJ, 413, L101
46. Kulkarni S. R., et al., 1999, Nature, 398, 389
47. Lazzati D., Ramirez-Ruiz E., Ghisellini G., 2001, A&A, 379, L39
48. Lazzati D., Begelman M. C., 2005, ApJ, 629, 903
49. Lazzati D., Ghirlanda G., Ghisellini G., 2005, MNRAS, 362, L8
50. Lazzati, D. & Perna, R. 2007, MNRAS, 375, L46
51. Lyutikov, M. 2006, in AIP Conf. Proc. 838: Gamma-Ray Bursts in the Swift Era, ed. S. S. Holt, N. Gehrels and J. A. Nousek, 483
52. McKinney, J. C. & Gammie, C. F. 2004, ApJ, 611, 977
53. Mereghetti S., Götz D., von Kienlin A., Rau A., Lichti G., Weidenspointner G., Jean P., 2005, ApJ, 624, L105
54. Meszaros, P. & Rees, M.J. 1992, ApJ, 397, 570
55. Metzger M. R., Djorgovski S. G., Kulkarni S. R., Steidel C. C., Adelberger K. L., Frail D. A., Costa E., Frontera F., 1997, Nature, 387, 878
56. Mochkovitch, R., Hernanz, M., Isern, J., Loisean, S. 1995, A&A, 293, 803
57. Nakar E., Piran T., 2002, MNRAS, 330, 920
58. Nakar E., Gal-Yam A., Piran T., Fox D. B., 2006, ApJ, 640, 849
59. Nakar E., 2007, Physics Reports (astro-ph/0701748)
60. Narayan, R., Kumar, P. & Piran, T. 2001, ApJ, 557, 949
61. Narayan, R. Paczynski, B. & Piran, T. 1992, ApJ, 395, L83
62. Norris J. P., Marani G. F., Bonnell J. T., 2000, ApJ, 534, 248
63. Ofek, E., O., 2006, ApJ, 659, 339
64. Paczynski, B. 1986, ApJ, 308, L43
65. Palmer D. M., et al., 2005, Nature, 434, 1107
66. Pedersen K., et al., 2005, ApJ, 634, L17
67. Perna, R. & Belczynski, C. 2002, ApJ, 570, 252
68. Perna, R., Armitage, P. J. & Zhang, B. 2006, ApJL, 636L, 29
69. Piro, A. L. & Pfhal, E. 2006, ApJL, 658, 1173
70. Popov S. B., Stern B. E., 2006, MNRAS, 365, 885
71. Portegies Zwart, S., F. & Yungelson, L. R. 1998, A&A, 332, 173
72. Proga, D. & Zhang, B. 2006, MNRAS, 370, L61

73. Rhoads J. E., 1999, ApJ, 525, 737
74. Rosswog, S. et al. 1999, A&A, 341, 499
75. Rosswog, S. & Davies, M. B. 2002, MNRAS, 334, 481
76. Ruffert, M. & Janka, H.-T. 1998, A&A, 338, 535
77. Ruffert, M. & Janka, H.-T. 1999, A&A, 380, 544
78. Sari R., Piran T., Halpern J. P., 1999, ApJ, 519, L17
79. Soderberg A. M., et al., 2006, ApJ, 650, 261
80. Stanek K. Z., et al., 2003, ApJ, 591, L17
81. Tagliaferri G., et al., 2005, A&A, 443, L1
82. Tanvir N. R., Chapman R., Levan A. J., Priddey R. S., 2005, Nature, 438, 991
83. Taylor G. B., Frail D. A., Beasley A. J., Kulkarni S. R., 1997, Nature, 389, 263
84. Terasawa T., et al., 2005, Nature, 434, 1110
85. Thompson, C. 1994, MNRAS, 270, 480
86. Tominaga N., Maeda K., Umeda H., Nomoto K., Tanaka M., Iwamoto N., Suzuki T., Mazzali P. A., 2007, ApJ, 657, L77
87. Tutukov, A. V. & Yungelson, L. R. 1994, MNRAS, 268, 871
88. van Paradijs J., et al., 1997, Nature, 386, 686
89. Villasenor J. S., et al., 2005, Nature, 437, 855
90. Zhang B., Zhang B.-B., Liang E.-W., Gehrels N., Burrows D. N., Mészáros P., 2007, ApJ, 655, L25

Strong Gravitational Field Diagnostics in Binary Systems Containing a Compact Object

L. Stella

INAF - Osservatorio Astronomico di Roma, Via Frascati 33, 00040 Monteporzio Catone (Roma), Italy stella@mporzio.astro.it

This chapter provides a short survey of two powerful diagnostic tools that can probe *in situ* the very strong gravitational field in the vicinity of accreting neutron stars and black holes. These are the fast quasi-periodic signals that are sometimes present in the X-ray flux of accreting compact objects and the very broad profiles of Fe K-shell lines that are observed in their X-ray spectra in a number of cases.

1 Introduction

Relativistic binary pulsars yield some of the most accurate tests of gravity theory and general relativity. Their pulses are remarkably accurate clocks which, besides providing high precision information on the binary motion, directly probe the space time characteristics over distances of the order of the orbital separation of the two stars. These regions, however, are characterized by comparatively weak gravitational fields.

In situ probing of the strong gravitational field regions in the vicinity of collapsed objects, *i.e.* neutron stars and black holes, has so far relied upon a few diagnostics, the relevance of which has emerged clearly only in the last decade. The motion of matter accreting towards these objects is involved in all cases: the deeper the gravitational potential, the faster the motion of the inflowing matter. In these regions the gravitational energy of the accreting matter is usually converted into radiation (mainly X- and gamma-rays) with high efficiency. Since neutron stars and black holes are very compact objects, matter endowed with angular momentum (albeit small) will settle in a disk and gradually spiral toward the collapsed object, as a result of viscous stresses acting across layers of matter orbiting at different radii. Accretion disks can extend down to the radius of the marginally stable orbit ($r_{ms} = 6GM/c^2$ for a non-rotating black hole) or slightly inside that. The motion of matter in their innermost regions thus holds crucial information on the strong gravitational field characteristics in the close vicinity of neutron stars and black holes.

Two diagnostics have proven to be especially promising in this respect: (a) the quasi-periodic signals that are often present in the flux of a number of X-ray binary systems, which most likely reflect the fundamental frequencies of the motion of matter in the innermost disk regions (Sect. 2 and Sect. 3); (b) the very broad profile of the Fe Kα emission line that is produced by the combination of relativistic Doppler, gravitational and transverse shifts in the motion of disk matter in the vicinity of the compact object (Sect. 4 and Sect. 5).

2 Quasi-Periodic Oscillations

Old neutron stars (NSs) in low mass X-ray binaries (LMXRBs) display a variety of quasi-periodic oscillations (QPOs) in their X-ray flux. These QPOs manifest themselves only as broad peaks in the power spectra of the X-ray light curves, since the effective area of current instrumentation is not large enough to reveal individual QPO cycles (or even trains of cycles). The *low frequency* QPOs ($\sim$ 1–100 Hz) that were discovered and studied from high luminosity "Z-type" LMXRBs since the mid-eighties were classified into horizontal, normal and flaring branch oscillations (HBOs, NBOs and FBOs, respectively), depending on the the position occupied by a source as it moves around in the X-ray colour-colour diagram (for a review see [1]).

Much faster QPOs ($\sim$ 0.2 to $\sim$ 1.4 kHz) were detected and studied with *Rossi X-ray Timing Explorer* in a number of NS LMXRBs (see [2] and references therein); these "kHz QPOs" correspond to timescales comparable to the dynamical timescales close to the NS. A pair of kHz QPOs (centroid frequencies of ν_1 and ν_2) is often present simultaneously. In many sources, the kHz QPO frequency drifts in a way such that their separation $\Delta\nu \equiv \nu_2 - \nu_1 \approx 250$–360 Hz remains approximately constant. In some cases $\Delta\nu$ decreases by up to $\sim$ 100 Hz as ν_2 increases. These QPOs display remarkably similar properties across accreting NSs in LMXRBs that differ a great deal with respects to other properties (for instance the luminosity, which is on average a factor of $\sim$ 10 higher in Z-sources than in the so-called Atoll sources).

During type I bursts from about a dozen Atoll sources, a nearly coherent signal at a frequency of $\nu_{burst} \sim$ 250–600 Hz has also been detected (for a review see [3]). These "burst oscillations" are clearly associated to the spin frequency ν_s of the neutron star, as directly measured from the coherent pulsations observed in a few transient accreting millisecond pulsars (for reviews see [4] and [5])

HBOs are present in both Atoll and Z-sources. Their frequency, ν_{HBO} ($\sim$ 15 to $\sim$ 60 Hz) shows a nearly quadratic dependence ($\sim \nu_2^2$) on the higher kHz QPO frequency. The frequency changes of the kHz QPOs and HBOs in individual sources appear to be positively correlated with the instantaneous accretion rate, as inferred from several indicators. However, on timescales of days or longer, the QPO frequencies and other properties are not uniquely related to mass accretion rate, suggesting that there is a second parameter at work. HBOs show also a certain degree of multiplicity in their harmonic content, occasionally with up to 3 or 4 power spectrum peaks separated by a factor of 2 in frequency.

Two high frequency QPOs have also been detected in several black hole candidates. For example the X-ray transient GRO J1655-40 displayed ~ 300 Hz and ~ 450 Hz QPOs [6]. Similarly in XTE J1550-564 QPOs at $\nu_1 \simeq 187$ Hz were present simultaneously with the QPOs at $\nu_2 \simeq 268$ Hz during the April–May 2000 outburst of this galactic microquasar [7]. In these and other instances the frequency ratio of high frequency QPOs from BHCs was consistent with 3/2. Low frequency QPOs (tens of Hz) are often present simultaneously with the high frequency QPOs also in BHCs.

A remarkable correlation between the centroid frequencies of QPOs (or peaked noise components) from LMXRBs was discovered [8]. This "PBV" correlation extends over nearly 3 decades in frequency and encompasses both NSs and black hole candidate, BHC, systems (see the points in Fig. 2); an approximate linear relationship ($\nu_{HBO} \sim \nu_1^{0.95}$) holds. In kHz QPO NS systems, the relevant frequencies are the lower frequency kHz QPOs, ν_1, and the low frequency, HBO or HBO-like QPOs, ν_{HBO}.

It is worth noting that, besides the QPOs whose properties are summarized above, other QPOs and noise components were found and studied, especially in accreting neutron star systems (see [9] and references therein); these will not be discussed here. The frequencies of QPOs, in spite their limited coherence, provide some of the most accurately measured observables of LMXRBs. A primary goal is therefore to explain the frequency range and dependence of the different QPO types of these sources.

3 Qpo Models and Strong Field Gravity

QPO models can be broadly divided into two main classes: models that apply only to neutron stars *or* black holes, and models that apply to both neutron stars *and* black holes. Models that apply to neutron stars only include: the magnetospheric and sonic point beat frequency models [10, 11, 12, 13, 14], the two-oscillator disk model [15, 16], the photon bubble model [17] and neutron star oscillation models [18, 19]. This class of models regards similarities and correlations of QPOs across the two types of collapsed objects as coincidental.

QPO models that are applicable to both neutron stars and black holes involve: disk oscillations and trapped modes [20, 21, 22, 23] or disk warping and (nodal) precession [24, 25, 26, 27, 28] or the parametric epicyclic resonance [29, 30]. The relativistic (periastron and nodal) precession model [31, 32, 33, 34, 35, 36, 37] that is summarized below also belongs to the latter class of models.

A variety of models of both classes associate the QPO signals to the fundamental frequencies of particle motion in the immediate vicinity of the accreting compact object. In particular the higher frequency kHz QPOs at ν_2 are believed to correspond to the Keplerian frequency of matter motion at the innermost radius of the accretion disk that feeds matters to the compact object. The highest values of ν_2 measured to date provide limits on the mass-radius relation of accreting neutron stars and have been used to constrain the equation of state of ultradense matter [38].

The basic features of the relativistic precession model (RPM) are reviewed here.

For a circular geodesic in the equatorial plane of a Kerr black hole of mass M and specific angular momentum a, the coordinate ϕ-frequency measured by a static observer at infinity is

$$\nu_\phi = \pm M^{1/2} r^{-3/2} [2\pi (1 \pm a M^{1/2} r^{-3/2})]^{-1} \tag{1}$$

(we use $G = c = 1$ units; the upper sign refers to prograde orbits).

If we slightly perturb a circular orbit in the r and θ directions, the coordinate frequencies of the small amplitude oscillations within the plane (the epicyclic frequency ν_r) and in the perpendicular direction (the vertical frequency ν_θ) are given by

$$\nu_r^2 = \nu_\phi^2 (1 - 6Mr^{-1} \pm 8aM^{1/2}r^{-3/2} - 3a^2 r^{-2}), \tag{2}$$

$$\nu_\theta^2 = \nu_\phi^2 (1 \mp 4aM^{1/2}r^{-3/2} + 3a^2 r^{-2}). \tag{3}$$

In the Schwarzschild limit ($a = 0$) ν_θ is equal to ν_ϕ and thus the nodal precession frequency $\nu_{nod} \equiv \nu_\phi - \nu_\theta$ is identically zero. ν_r is lower than the other two frequencies, reaches a maximum for $r = 8M$ and goes to zero at $r_{ms} = 6M$ (the radius of the marginally stable orbit). This qualitative behaviour of ν_r is preserved in the Kerr field ($a \neq 0$). Therefore the periastron precession frequency $\nu_{per} \equiv \nu_\phi - \nu_r$ is dominated by a "Schwarzschild" term over a wide range of parameters.

In the RPM the higher and lower frequency kHz QPOs are identified with $\nu_2 = \nu_\phi$ and $\nu_1 = \nu_{per}$, respectively. Therefore $\Delta\nu \equiv \nu_2 - \nu_1 = \nu_\phi - (\nu_\phi - \nu_r) = \nu_r$. For $a = 0$, Equations (1) and (2) give

$$\nu_r = \nu_\phi (1 - 6M/r)^{1/2} = \nu_\phi [1 - 6(2\pi\nu_\phi M)^{2/3}]^{1/2}. \tag{4}$$

Fig. 1A shows different curves for ν_r versus ν_ϕ for $a = 0$ and selected values of M, the only free parameter in Eq. (4). The measured $\Delta\nu$ versus ν_2 for 11 NS LMXRBs is also plotted. It is apparent that for NS masses of $\sim 2\,M_\odot$, the simple model above is in qualitative agreement with measured values, including the decrease of $\Delta\nu$ for increasing ν_2 seen in Sco X-1, 4U1608-52, 4U1735-44 and 4U1728-34. Remarkably, most points are close to the maximum of the epicyclic frequency. The rising behaviour of ν_r at low frequencies, a clear prediction of the RPM, was recently confirmed through the QPO separation in Cir X-1 which was found to increase for increasing QPO frequencies [39] (see Fig. 2B).

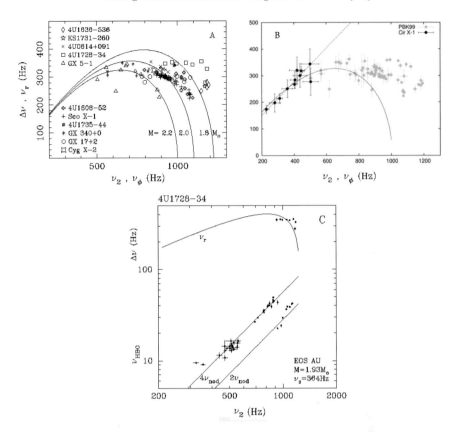

Fig. 1. (A) kHz QPO frequency difference $\Delta\nu$ versus higher QPO frequency ν_2 for 11 LMXRBs [32]. Error bars are not plotted for the sake of clarity. The curves give the r- and ϕ-frequencies of matter in nearly circular orbit around a non-rotating neutron star, of mass 2.2, 2.0 and 1.8 $M_\odot$. (B) $\Delta\nu$ versus ν_2 in Cir X-1 (black circles). The solid line shows the best fit RPM model. These data provide a confirmation of the low frequency behaviour of $\Delta\nu$ predicted by the RPM [39]. (C) kHz QPO frequency difference $\Delta\nu$ and (double-branched) HBO frequency versus higher QPO frequency ν_2 in 4U1728-34 [12, 45, 47]. The solid lines give the r-frequency and the 2nd and 4th harmonics of the nodal precession frequency ν_{nod} as a function of the ϕ-frequency for infinitesimally eccentric and tilted orbits in the spacetime of a 1.93 $M_\odot$ neutron star spinning at 364 Hz with EOS AU.

Analytical formulae to partly correct for the fact that the space-time around a rotating neutron star is different from Kerr's are given in [32]. The effects induced by the NS rotation on the $\nu_r - \nu_\phi$ relation are small, though non-negligible.

The behaviour of the curves in Fig. 1 (and thus the model's ability to match the observations), reflects the strong-field properties of the Schwarzschild metric; lower order expansions would not reproduce the observed frequencies.

Within the RPM, the maximum value of $\nu_r = \Delta\nu$ depends (mainly) on the mass of the compact object. The NS masses inferred from the curves in Figs. 1 and 2 are $\sim$ 1.8–2.0 $M_\odot$. While somewhat higher than most NS mass measurements, these values are compatible with masses obtained from spectro-photometric optical studies of a few X-ray binaries (e.g. Cyg X-2, $M = 1.78 \pm 0.23\,M_\odot$ [48] and Vela X-1, $M = 1.86 \pm 0.16\,M_\odot$ [49]).

In general within the RPM, $\Delta\nu$ should not be obviously related to the NS spin frequency ν_s (see [50] for a discussion of this issue). Besides those transient sources in which ν_s is measured directly from coherent pulsations, the RPM envisages that only the relatively stable frequency, ν_{burst}, seen during type I X-ray bursts (for a review see [40]) should be used to infer ν_s.

In the RPM the HBO frequency is related to the nodal precession frequency, ν_{nod}, at the same radius where the signals at ν_ϕ and ν_{per} are produced. From Eqs. (1) and (3) ν_{nod} is, in the slow rotation limit ($a/M \ll 1$),

$$\nu_{nod} \simeq 4\pi a\nu_\phi^2 \simeq 6.2 \times 10^{-5}(a/M)m\nu_\phi^2 \text{ Hz} \simeq 4.4 \times 10^{-8}\ I_{45}m^{-1}\nu_\phi^2\nu_s \text{ Hz}, \quad (5)$$

where $M = m\,M_\odot$. This is the well known Lense–Thirring formula. The latter equality refers to a rotating NS, where $aM = 2\pi\nu_s I$, with $I = 10^{45}I_{45}$ g cm^2 its moment of inertia. If ν_ϕ and ν_s are measured, the only free parameter in Eq. (5) is $I_{45}m^{-1}$; this can vary only over a limited range, $0.5 < I_{45}m^{-1} < 2$, for virtually any mass and EOS (see the rotating NS models of [41, 42]). The stellar oblateness induced by the star's rotation alters the nodal precession frequency somewhat (see [43] for a post-Newtonian formula). The Lense–Thirring term dominates over a wide range of parameters, such that ν_{nod} is expected to scale as $\sim \nu_\phi^2$. An approximately quadratic dependence of ν_{HBO} on the higher frequency kHz QPOs has been measured in a number of LMXRBs [31, 44, 45, 46] in agreement with one of the basic features of the RPM.

If in a source simultaneous kHz QPOs and HBOs are present and the NS spin frequency is measured, then the predictions of the RPM can be tested more directly. 4U1728-34 is one such source [12, 45, 47]. Its spin frequency is $\nu_s \simeq 364$ Hz, as inferred from burst oscillations. In the following application of the RPM we adopt a numerical approach in order to compute the spacetime metric and geodesics around the star [51].

Fig. 1C shows the measured values of $\Delta\nu$ and ν_{HBO} versus ν_2 in 4U1728-34. The solid lines represent the RPM frequencies for a 1.93 $M_\odot$ NS with EOS AU and $\nu_s = 364$ Hz. A good agreement is obtained if the HBO frequency, the lower of the two branches seen in 4U1728-34, is identified with the 2nd harmonics of ν_{nod} (i.e. $2\nu_{nod}$); correspondingly the upper HBO branch is well fit by $4\nu_{nod}$. The behaviour and values of the epicyclic frequency ν_r in this model are also in general agreement with the $\Delta\nu$ measurement.

In summary, the model presented here is capable of reproducing the salient features of both the $\Delta\nu$ versus ν_2 and ν_{HBO} versus ν_2 relationships in 4U1728-34. We note that only two free parameters are involved, M and the EOS, which can be varied over a very limited range. Concerning Z-sources and all other Atoll sources for which burst oscillations have not been detected yet, the NS spin should be regarded as a free parameter in applications of the RPM.

Another success of the RPM is that it reproduces accurately the PBV correlation, without resorting to any additional assumption [34]. Fig. 2A shows $2\nu_{nod}$ and ν_ϕ obtained from Eqs. (1) and (3) as a function of ν_{per} for co-rotating orbits and selected values of M and a/M. The high-frequency end of each line is dictated by the orbital radius reaching the marginally stable orbit. (For ν_{HBO} we use $2\nu_{nod}$ as in the case of 4U1728-34, see above). The different lines in Fig. 2A evidence that ν_{nod} depends weakly on M and strongly on a/M; the opposite holds for ν_ϕ, such that black hole masses and angular momentum can be inferred. We note that in the application of the RPM to BHC QPOs relatively small values of $a/M \sim 0.1$–0.3 are required.

For the case of rotating NSs we adopt the numerical approach outlined above, with a NS mass of $1.95\,M_\odot$, EOS AU and $\nu_s = 300, 600, 900$ and 1200 Hz. The measured QPO and peaked-noise frequencies giving rise to the PBV correlation are also plotted in Fig. 2B. The fact that the RPM dependence of ν_{nod} on ν_{per} matches nicely the observed $\nu_{HBO} - \nu_1$ correlation over ~ 3 decades in frequency (down to ν_1 of a few Hz) and encompasses both NS and BHC systems, provides additional support in favor of the RPM.

We have summarized above the salient characteristics and merits of the RPM and emphasized that the model holds the potential to address some key features of strong-field general relativity in the vicinity of collapsed objects and measure black hole masses and angular momenta.

It should be emphasized however that a number of other QPO models resort to the fundamental frequencies of motion in strong gravitational fields. Table 1 summarizes the frequency interpretation proposed in several classes of QPO models. We will not attempt to discuss the pros and cons of various models here; rather we simply stress again that QPOs hold great potential for probing "in situ" strong-field gravity effects. Further model developments and comparison with the properties of QPOs in accreting NSs and BHCs (including recent findings that challenge virtually all QPO models, see e.g. [52, 53]) will eventually tell us which QPO model is to be adopted in order to exploit this potential.

4 Relativistic Fe K-shell Lines from Accretion Disks

Fe K-shell emission lines at ~ 6–7 keV are observed in virtually all classes of accreting compact objects. They are likely driven by photoionization or Comp-

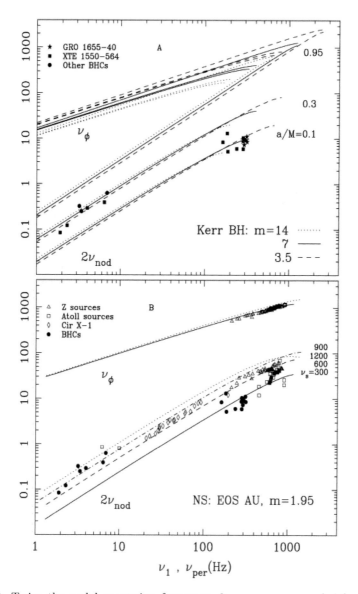

Fig. 2. Twice the nodal precession frequency, $2\nu_{nod} = \nu_{HBO}$, and ϕ-frequency, $\nu_\phi = \nu_2$, versus periastron precession frequency, $\nu_{per} = \nu_1$, for black hole candidates of various masses and angular momenta (panel A) and rotating neutron star models (EOS AU, $m = 1.95$) with selected spin frequencies (panel B) [34]. The measured QPO (or peaked noise) frequencies ν_1, ν_2 and ν_{HBO} giving rise to the PBV correlation are also shown in panel B for both BHC and NS LMXRBs and in panel A for BHC LMXRBs only; errors bars are not plotted (see [8] for a complete list of references). We included only those cases in which QPOs at ν_1 were unambiguously detected. NBO and FBO frequencies are not plotted.

Table 1. Approximate frequency identification in some QPO models applicable to neutron stars and black holes.

Model	ν_{HBO}	ν_1	ν_2
Generic	-	-	ν_ϕ
Disk oscillations	-	ν_{nod}	ν_r
Disk warping	-	-	ν_{nod}
Parametric epicyclic resonance	-	ν_r	ν_ϕ
Relativistic precession	ν_{nod}	ν_{per}	ν_ϕ

ton heating by high energy radiation emitted in the innermost regions of the accretion flows. Based on X-ray spectra of modest resolution ($E/\Delta E \sim 10$), it was realized in the mid-eighties that in some accreting black hole candidates and old neutron stars these lines are very broad and reach a full width at half maximum of 1 keV or more [54]. In Cyg X-1 and 4U1543-47 the line centroid energy is significantly lower than 6.4 keV, the lowest energy of Fe Kα transitions, indicating a substantial redshift [55, 56]. Blending of lines from different ionization stages of iron and thermal broadening can produce widths substantially smaller than observed. It is unlikely that the observed widths and redshifts arise from Comptonisation [57]. The iron K-line profile is most likely dominated by bulk plasma motions in the vicinity of the accreting collapsed object.

The first relativistic calculations of the profile from the innermost regions of an accretion disk as observed from infinity were carried out in the Schwarzschild geometry [58, 59]. Fig. 3 shows the results of some of these calculations. Different line profiles were obtained by stepping through a range of each parameter, the inner, r_i, and outer, r_o, radii of the line emitting region, the radial power law index of the line surface emissivity q and the inclination angle of disk to the line of sight i. At each step the other parameters are fixed at $r_i = 10\ r_s$, $r_o = 100\ r_s$, $i = 30$ deg and $q = -2$. Here $r_s = 2GM/c^2$ is the Schwarzschild radius.

In the first panel the inner radius is stepped: for decreasing r_i the red wing of the line extend to lower and lower energies, whereas the blue side of the line is virtually unaltered. This results from the increasing importance of gravitational and transverse redshifts for small radii. As the outer radius is increased the profile becomes narrower and the separation of the blue and red horn is reduced as a consequence of the lower velocities in the outer disk regions (see the second panel of Fig. 3).

Aberration, time dilation and blueshift, weighted by the relevant portion of the disk, make the blue horn brighter than the red horn. High inclinations make the effect more pronounced. For different values of q the importance of the line wings relative to the line core changes, due to a different contribution of the inner disk region to the integrated line profile (see the fourth panel of Fig. 3).

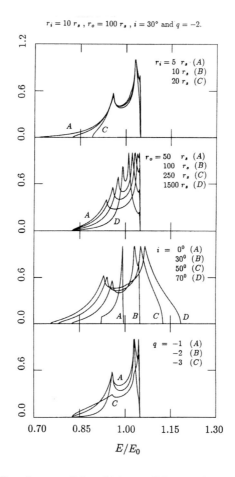

Fig. 3. Line profiles from a disk orbiting a Schwarzschild black hole. The line parameter listed at the top apply, except for the parameter that is stepped in each of the panels [59]. Radii are in units of the Schwarzschild radius $r_s = 2GM/c^2$

The properties of relativistic lines were shown to be fully compatible with the early measurements of the Fe K-line centroid and width. The line parameters of Cyg X-1, for example, could be reproduced for a line emitting disk extending down to $r_i = 10\ r_s$. Though based on X-ray spectra of modest resolution, these results evidenced the potential of Fe K-shell lines as a diagnostic of deep gravitational field regions. However it was clear that in order to study line profiles and fully exploit their potential, a higher spectral resolution, such as that of X-ray solid state spectrometers and CCDs ($E/\Delta E \sim 50$), would be required.

Line profiles in the Kerr metric were first calculated by [60]. The marginally stable orbit of prograde orbits around a rotating black hole is substantially smaller than that of a non-rotating black hole. Therefore an accretion disk around a Kerr black hole can extend deeper in the gravitational potential and give rise to broader and more redshifted lines than those for a Schwarzschild black hole.

Since these pioneering works, the study of Fe emission lines from the innermost regions of accretion flows towards collapsed objects has evolved into a major area of research, especially for the supermassive black holes that are hosted in active galactic nuclei, AGN (Fe Kα emission lines often have higher equivalent widths in AGNs than in X-ray binaries). By now relativistic Fe-line profiles have been observed in tens of accreting collapsed objects, especially AGNs and stellar mass black hole candidates, but also from several accreting weakly magnetic neutron stars. Three examples are shown in Fig. 4. These profiles provide a powerful diagnostic of the innermost accretion disk regions and, in some cases, black hole spin.

The need for a rotating black hole emerged clearly with the X-ray CCD observations of the Fe Kα line profile from MGC 6-30-15, a well studied Seyfert galaxy. The red wing from this line extended shortwards of the lowest energies than can be produced in a disk orbiting a Schwarzschild black hole, implying that the disk extended as far in as ~ 2–3 GM/c^2 (see Fig. 4A). This requires a Kerr black hole with $a/M \sim 0.93$ [61, 62, 63].

5 Relativistic Fe K-shell Line Variability

As most of the flux variability is produced in the close vicinity of the accreting collapsed object, variations in the central X-ray source are expected to induce changes in the disk line emissivity. Light travel time effects cause the innermost line emitting regions of the disk to respond first, while the outer regions respond at later times. For an observer at infinity any variation in the flux from the central source will be "reverberated" by a given front in the disk at any one time; such a front expands in time [66] (see Fig. 5). This translates into characteristic profile changes of the line, with a faster response of the line wings (the innermost disk regions have the highest projected velocities) than the core.

Fig. 5 shows an example of line profile and its response to a short-lived, step-like doubling of the flux from the central source. Once the "reverberating" front reaches the line-emitting portion of the disk, two features form in the wings of the profile. As time passes, the reverberation front expands outwards in the disk and the two features drift inward in the profile, due to lower and lower projected velocities. When the reverberation front propagates beyond the line emitting disk region, the line returns to its initial profile [67].

The drifting blue and red features provide a sensitive diagnostic of the inner disk regions, as their shift and intensity depends on the disk parameters and time. By measuring the observed position of the blue and red drifting features, the position of the reverberation front can be determined in units

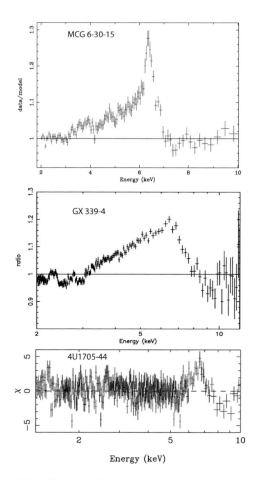

Fig. 4. Examples of Fe–K line profiles from accreting collapsed objects: (A) a massive black hole in an AGN [61, 62], (B) a stellar mass black hole [64] and (C) an old accreting neutron star [65].

of the Schwarzschild radius. Such a "relative" measurement does not provide information on the mass of the central object. However, by monitoring the blue or red drifting features at different times, the position of the reverberation front is evaluated also in absolute length units. The combination of the relative and absolute measurements provides the mass of the central object [67]. Extensive theoretical calculations were carried out in the Kerr metric for reverberating line emission from an accretion disk around a rotating black hole (see [68] and references therein). The variable line profiles arising from other effects, such as orbiting or radially falling blobs, were also calculated in detail.

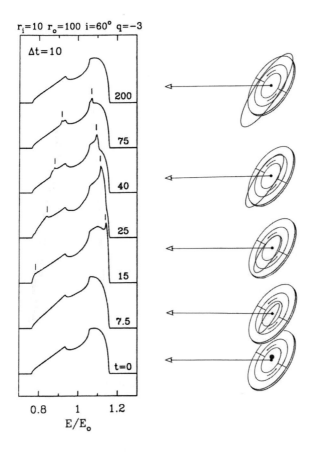

Fig. 5. Response of the line profile from a disk surrounding a Schwarzschild black hole to a short variation at $t = 0$ lasting ΔT. Times are in units of the Schwarzschild radius light crossing time r_s/c. [67]. The drawings on the right-hand side illustrate the expansion of the reverberation front.

AGNs offer the best prospects for observing the drifting features and other types of variability in the line intensity and profile. In fact the number of X-ray photons at the Earth per natural unit time (e.g. the Schwarzschild radius light crossing time or the Keplerian period at the marginally stable orbit) is a factor of $\geq 10^4$ higher for bright AGN than for X-ray binaries. Large area instruments and long observing times are required for this purpose.

Observations of relativistic line variability are in their infancy. Present generation X-ray satellites such as XMM/Newton and Chandra have already revealed different behaviours. Variability over some fraction of the line profile, possibly repeated over a few cycles, was observed in a few cases (NGC 3516 [69] and Mkn 766 [70]). In some other instances, red or blue-shifted transient features were detected in the line profiles (e.g. Mkn 509 [71]). Conspicuous short term flaring was observed in the photoionizing continuum (the power law spectral component) that was not mirrored in the intensity of the X-ray reflection component and the Fe-line (MCG 6-30-15 [72] and NGC 4051 [73]). The latter finding is contrary to the predictions of simple reverberation theory. Subsequent studies led to realization that, owing to strong field light deflection, changes in the position or geometry of the region emitting the ionizing continuum can results in large variations for an observer at infinity, whereas the illumination for the inner disk regions can be virtually unaffected [74, 75].

6 Conclusion

Fast quasi-periodic oscillations and very broad and redshifted Fe-line profiles offer to date the best prospects for probing the innermost regions of accretion flows toward black holes and neutron stars, down to radii comparable to the innermost stable orbit, where a variety of relativistic effects are important. Much remains to be done: on the one hand the interpretations that we have summarized above will need to be further developed and corroborated through additional observations and studies, before they can be considered unique. On the other hand higher throughput X-ray observations will be required. Ideally both diagnostics should be used simultaneously for the same source, in order to exploit all their information and cross check those parameters that can be determined independently through the two diagnostics. Very large area X-ray telescopes of the next generation (XEUS and Con-X) will likely be crucial in this area of study.

References

1. van der Klis, M.: In: X-ray Binaries, Eds. W. H. G. Lewin, J. van Paradijs & E. P. J. van den Heuvel (Cambridge University Press), p. 252 (1995)
2. van der Klis, M.: Ann. Rev. As. Ap., **38**, 717 (2000)
3. Strohmayer, T.E.: Adv. Sp. Res. **28**, 511 (2001)
4. Chakrabarty, D.: In: Interacting binaries: Accretion, Evolution, and Outcomes, AIP Conf Proc, **797**, p.71 (2005)
5. Di Salvo, T., et al: In: The Multicolored Landscape of Compact Objects, AIP Conf Proc, **924**, p. 613 (2007).
6. Remillard,R.A., Muno, M. P., McClintock, J. E., Orosz, J.A.: ApJ **580**, 1030 (2000)
7. Miller, J.M., et al: ApJ **563**, 928 (2000)
8. Psaltis, D., Belloni, T. & van der Klis, M.: ApJ **520**, 262 (1999)
9. van der Klis, M.: In: Compact stellar X-ray sources, Ed. W. Lewin & M. van der Klis. Cambridge Ap Ser, No. 39 Cambridge, UK: Cambridge University Press, p. 39 (2006)

10. Alpar, M. A., Shaham, J.: Nature **316**, 239 (1985)

11. Lamb, F. K., Shibazaki, N., Alpar, M. A., Shaham, J.: Nature **317**, 681 (1985)

12. Strohmayer, T.E.: ApJ **469**, L9 (1996)

13. Miller, M. C.; Lamb, F. K., Psaltis, D.: ApJ **508**, 791 (1998)

14. Lamb, F. K., Miller, M. C.: ApJ **554**, 1210 (2001)

15. Osherovich, V.; Titarchuk, L.: ApJ **518** L95 (1999)

16. Osherovich, V.; Titarchuk, L.: ApJ **522**, L113 (1999)

17. Klein, R.I. et al: ApJ **469**, L119 (1996)

18. Bildsten, L.: ApJ **501**, 89 (1998)

19. Bildsten, L., Cumming, A.: ApJ **506**, 842 (1998)

20. Nowak, M. A.; Wagoner, R. V.: ApJ **378**, 656 (1991)

21. Nowak, M. A.; Wagoner, R. V.: ApJ **393**, 697 (1992)

22. Nowak, M. A.; Wagoner, R. V.; Begelman, M. C.; Lehr, D. E.: ApJ **477**, L91 (1997)

23. Milsom, J. A.; Taam, R. E.: MNRAS **286**, 358 (1997)

24. Cui, W., Zhang, S. N., Chen, W.: ApJ **492**, L53 (1998)

25. Markovic , D. , Lamb, F. K.: ApJ **507**, 316 (1998)

26. Fragile, P. C., Mathews, G. J.; Wilson, J. R.: ApJ **553**, 955 (2001)

27. Merloni, A., Vietri, M.; Stella, L.; Bini, D.: MNRAS **304** 155 (1999)

28. Karas, V.: ApJ **526**, 953 (1999)

29. Abramowicz, M. A.; Kluzniak, W.: A&A **374**, L19 (2001)

30. Kluzniak, W. Abramowicz, M. A.: Ap&SS **300**, 143 (2005)

31. Stella, L. & Vietri, M.: ApJ **492**, L59 (1998)

32. Stella, L. & Vietri, M.: Phys. Rev. Lett **82**, 17 (1999)

33. Vietri, M. & Stella, L.: ApJ **503**, 350 (1999)

34. Stella, L., Vietri, M. & Morsink, S.M.: ApJ **524**, L63 (1999)

35. Psaltis, D. & Norman, C. 2000, astro-ph/0001391

36. Stella, L. & Vietri, M.: in "X-ray Astronomy 2000", ASP Conf Proc Vol **234** Ed. R. Giacconi, S. Serio, and L. Stella. San Francisco: As Soc Pac, p.213 (2001)

37. Psaltis, D Ad Sp Res **28** 481 (2001)

38. Miller, M. C., Lamb, F. K. & Cook, G. B.: ApJ **509**, 793 (1998)

39. Boutloukos, S. et al: ApJ **653**, 1435 (2006)

40. Strohmayer, T.E.: Ad Sp Res **28**, 511 (2001)

41. Friedman, J.L, Ipser, J.R. & Parker, L.: ApJ **304**, 115 (1986)

42. Cook, G.B., Shapiro, S.L., & Teukolsky, S.A.: ApJ **398**, 20 (1992)

43. Morsink, S. & Stella, L.: ApJ **513**, 827 (1999)

44. Vietri, M. & Stella, L.: In: The active X-ray sky, Nucl Phys B (Proc. Suppl.) **69**, 135 (1998)

45. Ford, E.C. & van der Klis, M.: ApJ **506**, L39 (1998)

46. Psaltis, D., et al: ApJ **520**, 763 (1999)

47. Mendez, M. & van der Klis, M.: ApJ **517**, L51 (1999)

48. Orosz, J.A. & Kuulkers, E.: MNRAS **305**, 132 (1999)

49. Barziv, O. et al: A&A **377**, 925 (2001)

50. Mendez, M. & Belloni, T.: MNRAS **381**, 790 (2007)

51. Stergioulas, N. & Friedman, J.L.: ApJ **444**, 306 (1995)

52. Linares, M., van der Klis, M., Altamirano, D. Markwardt, C. B.: ApJ **634**, 1250 (2005)

53. Linares, M., van der Klis, M., Wijnands, R.: ApJ **660**, 595 (2005)

54. White, N.E. et al: MNRAS **218**, 129 (1986)

55. Barr, P., White, N.E. & Page, C.G.: MNRAS **216**, 65 P (1985)
56. van der Woerd, H, White, N.E. & Kahn, S.M.: ApJ **344**, 320 (1989)
57. Kallman, T. & White, N.E.: ApJ **341**, 955 (1989)
58. Chen, K. & Halpern, J.P.: ApJ **344**, 115 (1989)
59. Fabian, A.C., Rees, M.J., Stella, L. & White, N.E.: MNRAS **238**, 729 (1989)
60. Laor, A.: ApJ **376**, 90 (1991)
61. Tanaka, Y., et al: Nature **375**, 659 (1995)
62. Fabian, A.C. et al: MNRAS **335**, L1 (2002)
63. Reynolds, C.S., Brenneman, L. W., Garofalo, D.: Ap&SS **300**, 7 (2005)
64. Miller, J.M. et al: ApJ **606**, L131 (2004)
65. Di Salvo, T. et al: ApJ **623**, L121 (2005)
66. Blandford, R. D., McKee, C. F.: ApJ **255**, 419 (1982)
67. Stella, L. Nature **344**, 747 (1990)
68. Reynolds, C.S.: In: X-ray Astronomy: Stellar Endpoints, AGN, and the Diffuse X-ray Background, Ed. N.E. White, G.Malaguti, and G. Palumbo AIP Conf Proc, **599**, 346 (2001)
69. Iwasawa, K., Miniutti, G., Fabian, A.C.: MNRAS **355**, 1073 (2004)
70. Turner, T.J., Miller, L.J., George, I.M., Reeves, J.M.: A&A **445**, 59 (2006)
71. Dadina, M. et al: A&A **442**, 461 (2005)
72. Fabian, A.C. et al: MNRAS **331**, L35 (2002)
73. Ponti, G. et al: MNRAS **282**, L53 (2006)
74. Fabian, A.C., Vaughan, S.: MNRAS **340**, L28 (2003)
75. Martocchia, A., Matt, G., Karas, V.: A&A **383**, L23 (2002)

White Dwarfs in Ultrashort Binary Systems

Gian Luca Israel and Simone Dall'Osso

INAF - Osservatorio Astronomico di Roma, Via Frascati 33, Monteporzio Catone, Italy gianluca,dallosso@mporzio.astro.it

1 Introduction

White dwarf binaries are thought to be the most common binaries in the Universe: in our Galaxy their number is estimated to be as high as 10^8. In addition, most stars are known to be part of binary systems, roughly half of which have orbital periods short enough that the evolution of the two stars is strongly influenced by the presence of a companion. Furthermore, it has become clear from observed close binaries, that a large fraction of binaries that interacted in the past must have lost considerable amounts of angular momentum, thus forming compact binaries, with compact stellar components. The details of the evolution leading to the loss of angular momentum are uncertain, but generally this is interpreted in the framework of the so-called "common-envelope evolution": the picture that in a mass-transfer phase between a giant and a more compact companion, the companion quickly ends up inside the giant's envelope, after which frictional processes slow down the companion and the core of the giant, causing the "common envelope" to be expelled, as well as the orbital separation to shrink dramatically [55].

Among the most compact binaries known, often called ultra-compact or ultra-short binaries, are those hosting two white dwarfs and classified into two types: *detached* binaries, in which the two components are relatively widely separated and *interacting* binaries, in which mass is transferred from one component to the other. In the latter class a white dwarf is accreting from a white dwarf like object (we often refer to them as AM CVn systems, after the prototype of the class, the variable star AM CVn; [56, 28]).

In the past many authors have emphasised the importance of studying white dwarfs in double-degenerate binaries (DDBs). In fact, the study of ultra-short white dwarf binaries is relevant to important astrophysical questions that have been outlined by several author. Recently, [32] listed the following ones:

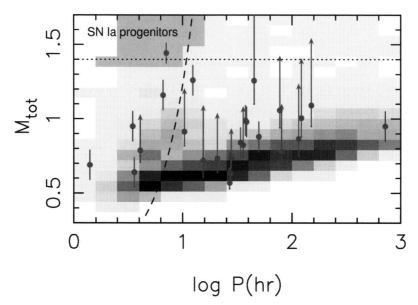

Fig. 1. Period versus total mass of double white dwarfs. The points and arrows are observed systems [29], the gray shade a model for the galactic population. Systems to the left of the dashed line will merge within a Hubble time, systems above the dotted line have a combined mass above the Chandrasekhar mass. The top left corner shows the region of possible type Ia supernova progenitors, where the gray shade has been darkened for better visibility (adapted from [32]).

- *Binary evolution* Double white dwarfs are excellent tests of binary evolution. In particular the orbital shrinkage during the common-envelope phase can be tested using double white dwarfs. The reason is that for giants there is a direct relation between the mass of the core (which becomes a white dwarf and so its mass is still measurable today) and the radius of the giant. The latter carries information about the (minimal) separation between the two components in the binary before the common envelope, while the separation after the common envelope can be estimated from the current orbital period. This enables a detailed reconstruction of the evolution leading from a binary consisting of two main sequence stars to a close double white dwarf [26]. The interesting conclusion of this exercise is that the standard schematic description of the common envelope – in which the envelope is expelled at the expense of the orbital energy – cannot be correct. An alternative scheme, based on the angular momentum, for the moment seems to be able to explain all the observations [30].
- *Type Ia supernovae* Type Ia supernovae have peak brightnesses that are well correlated with the shape of their light curve [35], making them ideal standard candles to determine distances. The measurement of the apparent brightness of far away supernovae as a function of redshift has led to the

conclusion that the expansion of the Universe is accelerating [34, 45]. This
depends on the reasonable assumption that these far-away (and thus old)
supernovae behave the same as their local cousins. However, one of the
problems is that we do not know what exactly explodes and why, so the
likelihood of this assumption is difficult to assess [37]. One of the proposed
models for the progenitors of type Ia supernovae are massive close double
white dwarfs that will explode when the two stars merge [16]. In Fig. 1
the observed double white dwarfs are compared to a model for the galactic
population of double white dwarfs [27], in which the merger rate of massive
double white dwarfs is similar to the type Ia supernova rate. The gray
shade in the relevant corner of the diagram is enhanced for visibility. The
discovery of at least one system in this box confirms the viability of this
model (in terms of event rates).

- *Accretion physics* The fact that in AM CVn systems the mass losing star
  is an evolved, hydrogen deficient star, gives rise to a unique astrophysical
  laboratory, in which accretion discs are made of almost pure helium [22,
  49, 13, 46, 57]. This opens the possibility to test the behaviour of accretion
  discs of different chemical composition.

- *Gravitational wave emission* Until recently the DDBs with two NSs were
  considered among the best sources to look for gravitational wave emission,
  mainly due to the relatively high chirp mass expected for these sources,
  In fact, simply inferring the strength of the gravitational wave amplitude
  expected from [11]

$$h = \left[\frac{16\pi G L_{GW}}{c^3 \omega_g^2 4\pi d^2} \right]^{1/2} = 10^{-21} \left(\frac{\mathcal{M}}{M_\odot} \right)^{5/3} \left(\frac{P_{orb}}{1\mathrm{hr}} \right)^{-2/3} \left(\frac{d}{1\mathrm{kpc}} \right)^{-1} \quad (1)$$

where

$$L_{GW} = \frac{32}{5} \frac{G^4}{c^5} \frac{M^2 m^2 (m + M)}{a^5}; \quad (2)$$

$$\mathcal{M} = \frac{(Mm)^{3/5}}{(M + m)^{1/5}} \quad (3)$$

where the frequency of the wave is given by $f = 2/P_{orb}$. It is evident that
the strain signal h from DDBs hosting neutron stars is a factor 5–20 higher
than in the case of DDBs with white dwarfs as far as the orbital period is
larger than approximatively 10–20 minutes. In recent years, AM CVns have
received great attention as they represent a large population of guaranteed
sources for the forthcoming *Laser Interferometer Space Antenna* [31, 51].
Double WD binaries enter the *LISA* observational window (0.1–100 mHz)
at an orbital period $\sim$ 5 hrs and, as they evolve secularly through GW
emission, they cross the whole *LISA* band. They are expected to be so
numerous ($\sim 10^3$–10^4 expected), close on average, and luminous in GWs
as to create a stochastic foreground that dominates the *LISA* observational
window up to $\approx$ 3 mHz [51]. Detailed knowledge of the characteristics of

their background signal would thus be needed to model it and study weaker background GW signals of cosmological origin.

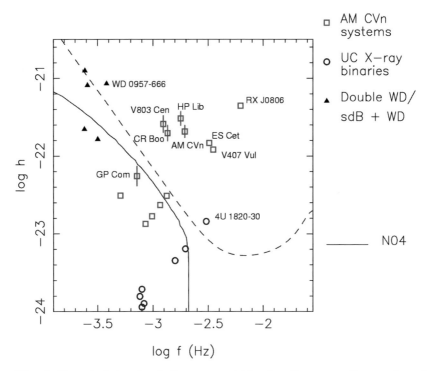

Fig. 2. Expected signals of ultra-compact binaries, the ones with error bars from (adapted from [46, 32].

A relatively small number of ultracompact DDBs systems is presently known. According to [42] there exist 17 confirmed objects with orbital periods in the 10–70 min in which a hydrogen-deficient mass donor, either a semi-degenerate star or a WD itself, is present. These systems (AM CVn) are roughly characterized by optical emission modulated at the orbital period, X-ray emission showing no evidence for a significant modulation (from which a moderately magnetic primary is suggested, [43]) and, in the few cases where timing analyses could be carried out, orbital spin-down consistent with GW emission-driven mass transfer.

In addition there exist two peculiar objects, sharing a number of observational properties that partially match those of the "standard" AM CVn's. They are RX J0806.3+1527 and RX J1914.4+2456, whose X-ray emission is ∼ 100% pulsed, with on-phase and off-phase of approximately equal duration. The single modulations found in their lightcurves, both in the optical and in X-rays, correspond to periods of, respectively, 321.5 and 569 s [21, 52] and were first interpreted as orbital periods. If so, these two objects are the binary systems with the shortest orbital period known and could belong to the AM CVn class. However, in addition to peculiar emission properties with respect to other AM CVn's, timing analyses carried out by the above cited authors demonstrate that, in this interpretation, these two objects have shrinking orbits. This is contrary to what expected in mass transferring double white dwarf systems (including AM CVn's systems) and suggests the possibility that the binary is detached, with the orbit shrinking because of GW emission. The electromagnetic emission would have in turn to be caused by some other kind of interaction.

Nonetheless, there are a number of alternative models to account for the observed properties, all of them based upon binary systems. The intermediate polar (IP) model [25, 17, 33] is the only one in which the pulsation periods are not assumed to be orbital. In this model, the pulsations are most likely due to the spin of a white dwarf accreting from non-degenerate secondary star. Moreover, due to geometrical constraints the orbital period is not expected to be detectable. The other two models assume a double white dwarf binaries in which the pulsation period is the orbital period. Each of them invoke a semi-detached, accreting double white dwarfs: one is magnetic, the double degenerate polar model [6, 39, 40, 18], while the other is non-magnetic, the direct impact model [27, 24, 39], in which, due to the compact dimensions of these systems, the mass transfer streams is forced to hit directly the accreting white dwarfs rather than to form an accretion disk.

After a brief presentation of the two X-ray selected double degenerate binary systems, we discuss the main scenario of this type, the Unipolar Inductor Model (UIM) introduced by [59] and further developed by [7, 8], and compare its predictions with the salient observed properties of these two sources.

1.1 RX J0806.3+1527

RX J0806.3+1527 was discovered in 1990 with the *ROSAT* satellite during the All-Sky Survey (RASS; [3]). However, it was only in 1999 that a periodic signal at 321 s was detected in its soft X-ray flux with the *ROSAT* HRI [17, 4]. Subsequent deeper optical studies allowed to unambiguously identify the optical counterpart of RX J0806.3+1527, a blue $V = 21.1$ ($B = 20.7$) star [18, 19]. B, V and R time-resolved photometry revealed the presence of a $\sim 15\%$ modulation at the ~ 321 s X-ray period [19, 39].

The VLT spectral study revealed a blue continuum with no intrinsic absorption lines [19]. Broad (FWHM ~ 1500 km s^{-1}), low equivalent width ($EW \sim 2$–6 Å) emission lines from the He II Pickering series (plus additional emission lines likely associated with He I, C III, N III, etc.; for a different interpretation see [44]) were instead detected [19]. These findings, together with the period stability and absence of any additional modulation in the 1 min– 5 hr period range, were interpreted in terms of a double degenerate He-rich binary (a subset of the AM CVn class; see [56]) with an orbital period of 321 s,

Table 1. Overview of observational properties of AM CVn stars (adapted from [28])

Name	P_{orb}^a (s)		P_{sh}^a (s)	Spectrum	Phot. var^b	dist (pc)	X-ray^c	UV^d
ES Cet	621	(p/s)		Em	orb	350	C^{3}X	GI
AM CVn	1029	(s/p)	1051	Abs	orb	606^{+135}_{-95}	RX	HI
HP Lib	1103	(p)	1119	Abs	orb	197^{+13}_{-12}	X	HI
CR Boo	1471	(p)	1487	Abs/Em?	OB/orb	337^{+43}_{-35}	ARX	I
KL Dra	1500	(p)	1530	Abs/Em?	OB/orb			
V803 Cen	1612	(p)	1618	Abs/Em?	OB/orb		Rx	FHI
SDSSJ0926+36	1698.6	(p)			orb			
CP Eri	1701	(p)	1716	Abs/Em	OB/orb			H
2003aw	?		2042	Em/Abs?	OB/orb			
SDSSJ1240-01	2242	(s)		Em	n			
GP Com	2794	(s)		Em	n	75±2	ARX	HI
CE315	3906	(s)		Em	n	77?	R(?)X	H
Candidates								
RXJ0806+15	321	(X/p)		He/H?[11]	"orb"		CRX	
V407 Vul	569	(X/p)		K-star[16]	"orb"		ARCRxX	

a orb = orbital, sh = superhump, periods from ww03, see references therein, (p)/(s)/(X) for photometric, spectroscopic, X-ray period.
b orb = orbital, OB = outburst
c A = ASCA, C = Chandra, R = ROSAT, Rx = RXTE, X = XMM-Newton kns+04
d F = FUSE, G = GALEX, H = HST, I = IUE

the shortest ever recorded. Moreover, RX J0806.3+1527 was noticed to have optical/X-ray properties similar to those of RX J1914.4+2456, a 569 s modulated soft X-ray source proposed as a double degenerate system [6, 38, 40].

In the past years the detection of spin–up was reported, at a rate of $\sim$ 6.2×10^{-11} s s^{-1}, for the 321 s orbital modulation, based on optical data taken from the Nordic Optical Telescope (NOT) and the VLT archive, and by using incoherent timing techniques [14, 15]. Similar results were also reported for the X-ray data (ROSAT and Chandra; [53]) of RX J0806.3+1527 spanning over 10 years of incoherent observations and based on the NOT results [14].

A Telescopio Nazionale Galileo (TNG) long-term project (started on 2000) devoted to the study of the long-term timing properties of RX J0806.3+1527 found a slightly energy–dependent pulse shape with the pulsed fraction increasing toward longer wavelengths, from $\sim$12% in the B-band to nearly 14% in the I-band (see lower right panel of Fig. 5; [21]). An additional variability, at a level of 4% of the optical pulse shape as a function of time (see upper right panel of Fig. 5 right) was detected. The first coherent timing solution was also inferred for this source, firmly assessing that the source was spinning-up: P = 321.53033(2) s, and $\dot{P}$ = $-3.67(1) \times 10^{-11}$ s s^{-1} (90% uncertainties are reported; [21]). Strohmayer [54] obtained independently a phase-coherent timing solutions for the orbital period of this source over a similar baseline, that is fully consistent with that of [21]. See [2] for a similar coherent timing solution also including the covariance terms of the fitted parameters.

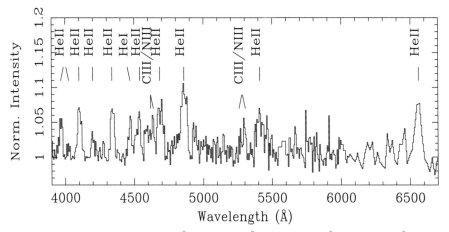

Fig. 3. VLT FORS1 medium (6Å; 3900–6000Å) and low (30Å; above 6000Å) resolution spectra obtained for the optical counterpart of RX J0806.3+1527. Numerous faint emission lines of HeI and HeII (blended with H) are labeled (adapted form [19]).

The relatively high accuracy obtained for the optical phase coherent P-Ṗ solution (in the January 2001 - May 2004 interval) was used to extend its validity backward to the ROSAT observations without loosing the phase coherency, i.e. only one possible period cycle consistent with our P-Ṗ solution. The best X-ray phase coherent solution is P = 321.53038(2) s, Ṗ = −3.661(5) × 10⁻¹¹ s s⁻¹ (for more details see [21]). Fig. 5 (left panel) shows the optical (2001–2004) and X-ray (1994–2002) light curves folded by using the above reported P-Ṗ coherent solution, confirming the amazing stability of the X-ray/optical anti-correlation first noted by ([20]; see inset of left panel of Fig. 5).

In 2001, a Chandra observation of RX J0806.3+1527 carried out in simultaneity with time resolved optical observation at the VLT, allowed for the first time to study the details of the X-ray emission and the phase-shift between X-rays and optical band. The X-ray spectrum is consistent with an occulting, as a function of modulation phase, black body with a temperature of ∼ 60 eV [20]. A 0.5 phase-shift was reported for the X-rays and the optical band [20]. More recently, a 0.2 phase-shift was reported by analysing the whole historical X-ray and optical dataset: this latter result is considered the correct one [2].

On November 1, 2002 a second deep X-ray observation was obtained with the *XMM–Newton* instrumentations for about 26000 s, providing an increased spectral accuracy (see left panel of Fig. 6). The *XMM–Newton* data show a lower value of the absorption column, a relatively constant black body temperature, a smaller black body size, and, correspondingly, a slightly lower flux. All these differences may be ascribed to the pile-up effect in the Chandra data, even though we can not completely rule out the presence of real spectral varia-

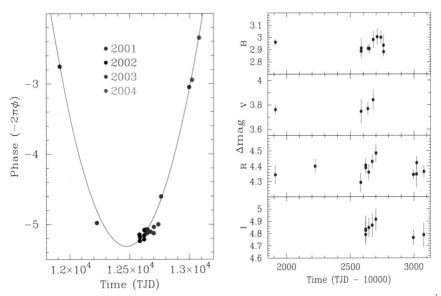

Fig. 4. Left panel: Results of the phase fitting technique used to infer the P-Ṗ coherent solution for RX J0806.3+1527: the linear term (P component) has been corrected, while the quadratic term (the Ṗ component) has been kept for clarity. The best Ṗ solution inferred for the optical band is marked by the solid fit line. Right panel: 2001–2004 optical flux measurements at different wavelengths.

tions as a function of time. In any case we note that this result is in agreement with the idea of a self-eclipsing (due only to a geometrical effect) small, hot and X-ray emitting region on the primary star. Timing analysis did not show any additional significant signal at periods longer or shorter than 321.5 s, (in the 5 hr–200 ms interval). By using the *XMM–Newton* OM a first look at the source in the UV band (see right panel of Fig. 7) was obtained confirming the presence of the blackbody component inferred from IR/optical bands.

Israel et al. [20] measured an on-phase X-ray luminosity (in the range 0.1–2.5 keV) $L_X = 8 \times 10^{31} (d/200 \text{ pc})^2$ erg s^{-1} for this source. These authors suggested that the bolometric luminosity might even be dominated by the (unseen) value of the UV flux, and reach values up to 5–6 times higher. The optical flux is only $\sim 15\%$ pulsed, indicating that most of it might not be associated to the same mechanism producing the pulsed X-ray emission (possibly the cooling luminosity of the WD plays a role). Given these uncertainties and, mainly, the uncertainty in the distance to the source, a luminosity $L_{\text{bol}} \simeq 10^{32} (d/200 \text{ pc})^2$ erg s^{-1} will be assumed as a reference value.

1.2 RX J1914.4+2456

The luminosity and distance of this source have been subject to much debate over the last years. Wu et al. [59] refer to earlier ASCA measurements that, for

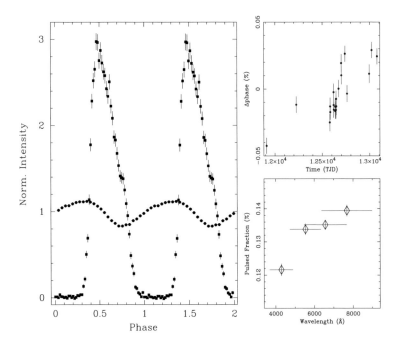

Fig. 5. Left panel: The 1994–2002 phase coherently connected X-ray folded light curves (filled squares; 100% pulsed fraction) of RX J0806.3+1527, together with the VLT-TNG 2001–2004 phase connected folded optical light curves (filled circles). Two orbital cycles are reported for clarity. A nearly anti-correlation was found. Right panels: Analysis of the phase variations induced by pulse shape changes in the optical band (upper panel), and the pulsed fraction as a function of optical wavelengths (lower panel).

a distance of 200–500 pc, corresponded to a luminosity in the range $(4 \times 10^{33}$–$2.5 \times 10^{34})$ erg s^{-1}. Ramsay et al. [41], based on more recent XMM-Newton observations and a standard blackbody fit to the X-ray spectrum, derived an X-ray luminosity of $\simeq 10^{35} d_{\mathrm{kpc}}^2$ erg s^{-1}, where d_{kpc} is the distance in kpc. The larger distance of ~ 1 kpc was based on a work by [50]. More recently, [42] find that an optically thin thermal emission spectrum, with an edge at 0.83 keV attributed to O VIII, gives a significantly better fit to the data than a blackbody model. The optically thin thermal plasma model implies a much lower bolometric luminosity of $L_{\mathrm{bol}} \simeq 10^{33}$ d$_{\mathrm{kpc}}^2$ erg s^{-1}.

Ramsay et al. [42] also note that the determination of a 1 kpc distance has large uncertainties and that a minimum distance of ~ 200 pc might still be possible: the latter leads to a minimum luminosity of $\sim 3 \times 10^{31}$ erg s^{-1}.

Given these large discrepancies, interpretation of this source's properties remains ambiguous and dependent on assumptions. In the following, we refer to the more recent assessment by [42] of a luminosity $L = 10^{33}$ erg s^{-1} for a 1 kpc distance.

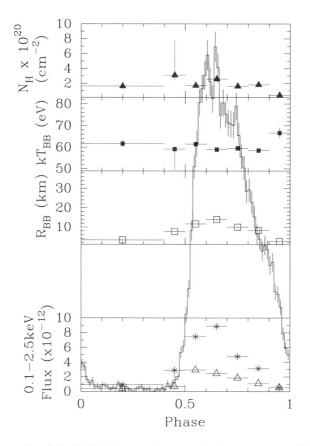

Fig. 6. The results of the *XMM–Newton* phase-resolved spectroscopy (PRS) analysis for the absorbed blackbody spectral parameters: absorption, blackbody temperature, blackbody radius (assuming a distance of 500 pc), and absorbed (triangles) and unabsorbed (asterisks) flux. Superposed is the folded X-ray light curve.

Ramsay et al. [42] also find possible evidence, at least in a few observations, of two secondary peaks in the power spectra. These are very close to ($\Delta\nu \simeq 5 \times 10^{-5}$ Hz) and symmetrically distributed around the strongest peak at $\sim 1.76 \times 10^{-3}$ Hz. Ramsay et al. [42] and Deloye & Taam [10] discuss the implications of this possible finding.

2 The Unipolar Inductor Model

The Unipolar Inductor Model (UIM) was originally proposed to explain the origin of bursts of decametric radiation received from Jupiter, whose prop-

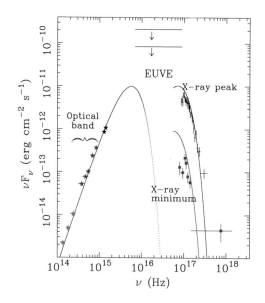

Fig. 7. Broad-band energy spectrum of RX J0806.3+1527 as inferred from the *Chandra*, *XMM–Newton*, VLT and TNG measurements and *EUVE* upper limits. The dotted line represents one of the possible fitting blackbody models for the IR/optical/UV bands.

erties appear to be strongly influenced by the orbital location of Jupiter's satellite Io [12, 36].

The model relies on Jupiter's spin being different from the system orbital period (Io spin is tidally locked to the orbit). Jupiter has a surface magnetic field ~ 10 G so that, given Io's good electrical conductivity (σ), the satellite experiences an electromagnetic field (e.m.f.) as it moves across the planet's field lines along the orbit. The e.m.f. accelerates free charges in the ambient medium, giving rise to a flow of current along the sides of the flux tube connecting the bodies. Flowing charges are accelerated to mildly relativistic energies and emit coherent cyclotron radiation through a loss cone instability (cf. [58] and references therein): this is the basic framework in which Jupiter decametric radiation and its modulation by Io's position are explained. Among the several confirmations of the UIM in this system, HST UV observations revealed the localized emission on Jupiter's surface due to flowing particles hitting the planet's surface - the so-called Io's footprint [5]. In recent years, the complex interaction between Io-related free charges (forming the Io torus) and Jupiter's magnetosphere has been understood in much greater detail [47, 48]. Despite these significant complications, the above scenario maintains its general validity, particularly in view of astrophysical applications.

Wu et al. [59] considered the UIM in the case of close white dwarf binaries. They assumed a moderately magnetized primary, whose spin is not synchronous with the orbit and a non-magnetic companion, whose spin is tidally locked. They particularly highlight the role of ohmic dissipation of currents flowing through the two WDs and show that this occurs essentially in the primary atmosphere. A small bundle of field lines leaving the primary surface thread the whole secondary. The orbital position of the latter is thus "mapped" to a small region onto the primary's surface; it is in this small region that ohmic dissipation – and the associated heating – mainly takes place. The resulting geometry, illustrated in Fig. 8, naturally leads to mainly thermal, strongly pulsed X-ray emission, as the secondary moves along the orbit. The source of the X-ray emission is ultimately represented by the relative mo-

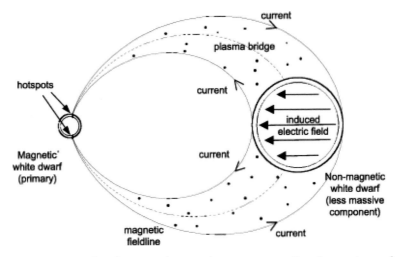

Fig. 8. Electric coupling between the asynchronous, magnetic primary star and the non-magnetic secondary, in the UIM (adapted from [59]).

tion between primary spin and orbit, that powers the electric circuit. Because of resistive dissipation of currents, the relative motion is eventually expected to be cancelled. This in turn requires angular momentum to be redistributed between spin and orbit in order to synchronize them. The necessary torque is provided by the Lorentz force on cross-field currents within the two stars.

Wu et al. [59] derived synchronization timescales $(\tau_\alpha) \sim$ few 10^3 yrs for both RX J1914.4+2456 and RX J0806.3+1527, less than 1% of their orbital evolutionary timescales. This would imply a much larger galactic population of such systems than predicted by population-synthesis models, a major difficulty of this version of the UIM. However, [7, 8] have shown that the electrically active phase is actually long-lived because perfect synchronism is never reached. In a perfectly synchronous system the electric circuit would be turned off, while GWs would still cause orbital spin-up. Orbital motion and primary spin would thus go out of synchronism, which in turn would switch the circuit on. The synchronizing (magnetic coupling) and de-synchronizing (GWs) torques are thus expected to reach an equilibrium state at a sufficiently small degree of asynchronism.

In the following we discuss in detail how the model works and how the major observed properties of RX J0806.3+1527 and RX J1914.4+2456 can be interpreted in the UIM framework. We refer to [1] for a possible criticism of the model based on the shape of the pulsed profiles of the two sources. Finally, we refer to [9, 10], who have recently proposed alternative mass transfer models that can also account for long-lasting episodes of spin-up in Double White Dwarf systems.

3 UIM in Double Degenerate Binaries

According to [59], we define the primary's asynchronism parameter $\alpha \equiv \omega_1/\omega_o$, where ω_1 and ω_o are the primary's spin and orbital frequencies. In a system with orbital separation a, the secondary star will move with the velocity $v = a(\omega_o - \omega_1) = [GM_1(1 + q)]^{1/3} \omega_o^{1/3}(1 - \alpha)$ relative to field lines, where G is the gravitational constant, M_1 the primary mass, $q = M_2/M_1$ the system mass-ratio. The electric field induced through the secondary is thus $\boldsymbol{E} = \frac{\boldsymbol{v} \times \boldsymbol{B_2}}{c}$, with an associated e.m.f. $\Phi = 2R_2E$, R_2 being the secondary's radius and $\boldsymbol{B_2}$ the primary magnetic field at the secondary's location. The internal (Lorentz) torque redistributes angular momentum between spin and orbit conserving their sum (see below), while GW-emission causes a net loss of orbital angular momentum. Therefore, as long as the primary spin is not efficiently affected by other forces, i.e. tidal forces (see Appendix A in [7]), it will lag behind the evolving orbital frequency, thus keeping electric coupling continuously active.

Since most of the power dissipation occurs at the primary atmosphere (see [59]), we slightly simplify our treatment assuming no dissipation at all at the secondary. In this case, the binary system is wholly analogous to the elementary circuit of Fig. 9. Given the e.m.f. (Φ) across the secondary star and the system's effective resistivity $\mathcal{R} \approx (2\sigma R_2)^{-1} (a/R_1)^{3/2}$, the dissipation rate of electric current (W) in the primary atmosphere is:

$$W = I^2\mathcal{R} = I\Phi = k\omega_o^{17/3}(1 - \alpha)^2 \tag{4}$$

where $k = 2(\mu_1/c)^2\sigma R_1^{3/2}R_2^3/[GM_1(1 + q)]^{11/6}$ is a constant of the system.

The Lorentz torque (N_L) has the following properties: i) it acts with the same magnitude and opposite signs on the primary star and the orbit, $N_L = N_L^{(1)} = -N_L^{(\mathrm{orb})}$. Therefore; ii) N_L conserves the total angular momentum in the system, transferring all that is extracted from one component to the other one; iii) N_L is simply related to the energy dissipation rate: $W = N_L\omega_o(1-\alpha)$.

From the above, the evolution equation for ω_1 is:

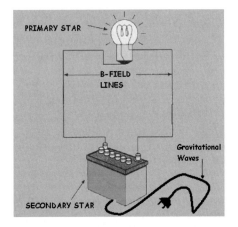

Fig. 9. Sketch of the elementary circuit envisaged in the UIM. The secondary star acts as the battery, the primary star represents a resistance connected to the battery by conducting "wires" (field lines.) Inclusion of the effect of GWs corresponds to connecting the battery to a plug, so that it is recharged at some given rate. Once the battery initial energy reservoir is consumed, the bulb will be powered just by the energy fed through the plug. This corresponds to the "steady-state" solution.

$$\dot{\omega}_1 = (N_L/I_1) = \frac{W}{I_1\omega_o(1-\alpha)} \tag{5}$$

The orbital angular momentum is $L_o = I_o\omega_o$, so that the orbital evolution equation is:

$$\dot{\omega}_o = -3(N_{\mathrm{gw}} + N_L^{(\mathrm{orb})})/I_o = -3(N_{\mathrm{gw}} - N_L)/I_o = -\frac{3}{I_o\omega_o}\left(\dot{E}_{\mathrm{gw}} - \frac{W}{1-\alpha}\right) \tag{6}$$

where $I_o = q(1+q)^{-1/3}G^{2/3}M_1^{5/3}\omega_o^{-4/3}$ is the orbital moment of inertia and $N_{\mathrm{gw}} = \dot{E}_{\mathrm{gw}}/\omega_o$ is the GW torque.

3.1 Energetics of the Electric Circuit

Let us focus on how energy is transferred and consumed by the electric circuit. We begin considering the rate of work done by N_L on the orbit

$$\dot{E}_L^{(\mathrm{orb})} = N_L^{(\mathrm{orb})}\omega_o = -N_L\omega_o = -\frac{W}{1-\alpha}, \tag{7}$$

and that done on the primary:

$$\dot{E}_{spin} = N_L\omega_1 = \frac{\alpha}{1-\alpha}W = -\alpha\dot{E}_L^{(\mathrm{orb})}. \tag{8}$$

The sum $\dot{E}_{spin} + \dot{E}_L^{(\mathrm{orb})} = -W$. Clearly, not all of the energy extracted from one component is transferred to the other one. The energy lost ohmic dissipation represents the energetic cost of spin–orbit coupling.

The above formulae allow to draw some further conclusions concerning the relation between α and the energetics of the electrical circuit. When $\alpha > 1$, the circuit is powered at the expenses of the primary's spin energy. A fraction α^{-1} of this energy is transferred to the orbit, the rest being lost to ohmic dissipation. When $\alpha < 1$, the circuit is powered at the expenses of the orbital energy and a fraction α of this energy is transferred to the primary spin. Therefore, *the parameter α represents a measure of the energy transfer efficiency of spin–orbit coupling*: the more asynchronous a system is, the less efficiently energy is transferred, most of it being dissipated as heat.

3.2 Stationary State: General Solution

As long as the asynchronism parameter is sufficiently far from unity, its evolution will be essentially determined by the strength of the synchronizing (Lorentz) torque, the GW torque being of minor relevance. The evolution in this case depends on the initial values of α and ω_o, and on stellar parameters. This evolutionary phase drives α towards unity, i.e. spin and orbit are driven towards synchronism. It is in this regime that the GW torque becomes important in determining the subsequent evolution of the system.

Once the condition $\alpha = 1$ is reached, indeed, GW emission drives a small angular momentum disequilibrium. The Lorentz torque is in turn switched on to transfer to the primary spin the amount of angular momentum required for it to keep up with the evolving orbital frequency. This translates to the requirement that $\dot{\omega}_1 = \dot{\omega}_o$. By use of Eqs. (5) and (6), it is found that this condition implies the following equilibrium value for α (we call it α_∞):

$$1 - \alpha_\infty = \frac{I_1}{k} \frac{\dot{\omega}_o/\omega_o}{\omega^{11/3}}. \tag{9}$$

This is greater than zero if the orbit is shrinking ($\dot{\omega}_o > 0$), which implies that $\alpha_\infty < 1$. For a widening orbit, on the other hand, $\alpha_\infty > 1$. However, this latter case does not correspond to a long-lived configuration. Indeed, define the electric energy reservoir as $E_{UIM} \equiv (1/2)I_1(\omega_1^2 - \omega_o^2)$, which is negative when $\alpha < 1$ and positive when $\alpha > 1$. Substituting Eq. (9) into this definition:

$$\dot{E}_{UIM} = -W. \tag{10}$$

If $\alpha = \alpha_\infty > 1$, energy is consumed at the rate W: the circuit will eventually switch off ($\alpha_\infty = 1$). At later times, the case $\alpha_\infty < 1$ applies.

If $\alpha = \alpha_\infty < 1$, condition (10) means that the battery recharges at the rate W at which electric currents dissipate energy: the electric energy reservoir is conserved as the binary evolves.

The latter conclusion can be reversed (see Fig. 9): in steady-state, the rate of energy dissipation (W) is fixed by the rate at which power is fed to the circuit by the plug ($\dot{E}_{UIM}$). The latter is determined by GW emission and the Lorentz torque and, therefore, by component masses, ω_o and μ_1.

Therefore the steady-state degree of asynchronism of a given binary system is uniquely determined, given ω_o. Since both ω_o and $\dot{\omega}_o$ evolve secularly, the equilibrium state will be "quasi-steady", α_∞ evolving secularly as well.

3.3 Model Application: Equations of Practical Use

We have discussed in previous sections the existence of an asymptotic regime in the evolution of binaries in the UIM framework. Given the definition of α_∞ and W (Eqs. (9) and (4), respectively), we have:

$$W = I_1\omega_o\dot{\omega}_o \frac{(1-\alpha)^2}{1-\alpha_\infty}. \tag{11}$$

The quantity $(1 - \alpha_\infty)$ represents the *actual* degree of asynchronism only for those systems that had enough time to evolve towards steady-state, *i.e* with sufficiently short orbital period. In this case, the steady-state source luminosity can thus be written as:

$$W_\infty = I_1\dot{\omega}_o\omega_o(1 - \alpha_\infty) \tag{12}$$

Therefore – under the assumption that a source is in steady-state – the quantity α_∞ can be determined from the measured values of W, ω_o, $\dot{\omega}_o$. Given its definition (Eq. (9)), this gives an estimate of k and, thus, μ_1.

The equation for the orbital evolution (Eq. (6)) provides a further relation between the three measured quantities, component masses and degree of asynchronism. This can be written as:

$$\dot{E}_{gr} + \frac{1}{3}I_o\omega_o^2(\dot{\omega}_o/\omega_o) = \frac{W}{(1-\alpha)}$$

that becomes, inserting the appropriate expressions for $\dot{E}_{gr}$ and I_o:

$$\frac{32}{5}\frac{G^{7/3}}{c^5}\omega_o^{10/3}X^2 - \frac{1}{3}G^{2/3}\frac{\dot{\omega}_o}{\omega_o^{1/3}}X + \frac{W}{1-\alpha} = 0, \tag{13}$$

where $X \equiv M_1^{5/3}q/(1+q)^{1/3} = \mathcal{M}^{5/3}$, $\mathcal{M}$ being the system's chirp mass.

4 RX J0806.3+1527

We assume here the values of ω_o, $\dot{\omega}_o$ and of the bolometric luminosity reported in Sect. 1 and refer to [8] for a complete discussion on how our conclusions depend on these assumptions.

In Fig. 10 (see caption for further details), the dashed line represents the locus of points in the M_2 versus M_1 plane, for which the measured ω_o and $\dot{\omega}_o$

are consistent with being due to GW emission only, i.e. if spin–orbit coupling was absent ($\alpha = 1$). This corresponds to a chirp mass $\mathcal{M} \simeq 0.3\,M_{\odot}$.

Inserting the measured quantities in Eq. (11) and assuming a reference value of $I_1 = 3 \times 10^{50}$ g cm^2, we obtain:

$$\frac{(1-\alpha)^2}{1-\alpha_\infty} \simeq \frac{10^{32}d_{200}^2}{1.3 \times 10^{34}} \simeq 8 \times 10^{-3}d_{200}^2. \tag{14}$$

In principle, the source may be in any regime, but our aim is to check whether it can be in steady-state, as to avoid the short timescale problem mentioned in Sect. 2. Indeed, the short orbital period strongly suggests it may have reached the asymptotic regime (see [8]). If we assume $\alpha = \alpha_\infty$, Eq. (14) implies $(1-\alpha_\infty) \simeq 8 \times 10^{-3}$.

Once UIM and spin–orbit coupling are introduced, the locus of allowed points in the M_2 versus M_1 plane is somewhat sensitive to the exact value of α: the solid curve of Fig. 10 was obtained, from Eq. (13), for $\alpha = \alpha_\infty = 0.992$.

From this we conclude that, if RX J0806.3+1527 is interpreted as being in the UIM steady-state, M_1 must be smaller than 1.1 $M_{\odot}$ in order for the secondary not to fill its Roche lobe, thus avoiding mass transfer. From $(1-\alpha_\infty) = 8 \times 10^{-3}$ and from Eq. (4), $k \simeq 7.7 \times 10^{45}$ (c.g.s.): from this, component masses and primary magnetic moment can be constrained. Indeed, $k = \hat{k}(\mu_1, M_1, q; \overline{\sigma})$ (Eq. (4)) and a further constraint derives from the fact that M_1 and q must lie along the solid curve of Fig. 10. Given the value of $\overline{\sigma}$, μ_1 is obtained for each point along the solid curve. We assume an electrical conductivity of $\overline{\sigma} = 3 \times 10^{13}$ e.s.u. [59, 8].

The values of μ_1 obtained in this way, and the corresponding field at the primary's surface, are plotted in Fig. 11, from which $\mu_1 \sim$ a few $\times 10^{30}$ G cm^3 results, somewhat sensitive to the primary mass.

We note further that, along the solid curve of Fig. 10, the chirp mass is slightly variable, being: $X \simeq (3.4 \div 4.5) \times 10^{54}$ g$^{5/3}$, which implies $\mathcal{M} \simeq (0.26–0.31)\,M_{\odot}$. More importantly, $\dot{E}_{\mathrm{gr}} \simeq (1.1–1.9) \times 10^{35}$ erg s^{-1} and, since $W/(1-\alpha_\infty) = \dot{E}_L^{(orb)} \simeq 1.25 \times 10^{34}$ erg s^{-1}, we have $\dot{E}_{\mathrm{gr}} \simeq (9–15)\,\dot{E}_L^{(orb)}$. Orbital spin-up is essentially driven by GW alone; indeed, the dashed and solid curves are very close in the M_2 versus M_1 plane.

Summarizing, the observational properties of RX J0806.3+1527 can be well interpreted in the UIM framework, assuming it is in steady-state. This requires the primary to have $\mu_1 \sim 10^{30}$ G cm^3 and a spin just slightly slower than the orbital motion (the difference being less than $\sim 1\%$).

The expected value of μ_1 can in principle be tested by future observations, through studies of polarized emission at optical and/or radio wavelengths [58].

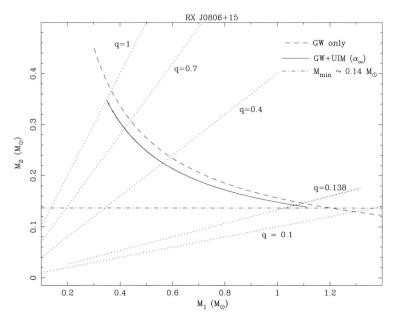

Fig. 10. M_2 versus M_1 plot based on the measured timing properties of RX J0806.3+1527. The dashed curve is the locus expected if orbital decay is driven by GW alone, with no spin–orbit coupling. The solid line describes the locus expected if the system is in a steady-state, with $(1 - \alpha) = (1 - \alpha_\infty) \simeq 8 \times 10^{-3}$. The horizontal dot-dashed line represents the minimum mass for a degenerate secondary not to fill its Roche-lobe at an orbital period of 321.5 s. Dotted lines are the loci of fixed mass ratio.

5 RX J1914.4+2456

As for the case of RX J0806.3+1527 , we adopt the values discussed in Sect. 1 and refer to [8] for a discussion of all the uncertainties on these values and their implications for the model.

Application of the scheme used for RX J0806.3+1527 to this source is not as straightforward. The inferred luminosity of this source seems inconsistent with steady-state. With the measured values of ω_o and $\dot{\omega}_o$, again assuming $I_1 = 3 \times 10^{50}$ g cm^2, the system steady-state luminosity should be $< 2 \times 10^{32}$ erg s^{-1} (Eq. (12)). This is hardly consistent even with the smallest possible luminosity referred to in Sect. 1, unless allowing for a large value of $(1 - \alpha_\infty \geq 0.15)$.

From Eq. (11) a relatively high ratio between the actual asynchronism parameter and its steady-state value appears unavoidable:

$$|1 - \alpha| \simeq 2.2(1 - \alpha_\infty)^{1/2}. \tag{15}$$

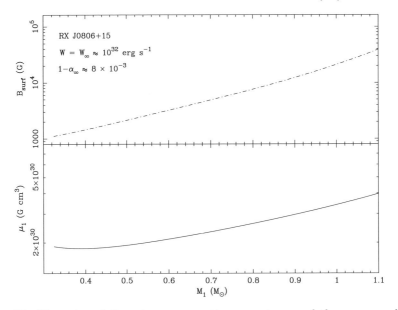

Fig. 11. The value of the primary magnetic moment μ_1, and the corresponding surface B-field, as a function of the primary mass M_1, for $(1 - \alpha) = (1 - \alpha_\infty) = 8 \times 10^{-3}$.

The Case for $\alpha > 1$

The low rate of orbital shrinking measured for this source and its relatively high X-ray luminosity put interesting constraints on the primary spin. Indeed, a high value of N_L is associated to $W \sim 10^{33}$ erg s^{-1}.

If $\alpha < 1$, this torque sums to the GW torque: the resulting orbital evolution would thus be faster than if driven by GW alone. In fact, for $\alpha < 1$, the smallest possible value of N_L obtains with $\alpha = 0$, from which $N_L^{(\min)} = 9 \times 10^{34}$ erg. This implies an absolute minimum to the rate of orbital shrinking (Eq. 6), $3\ N_L^{(\min)}/I_o$, so close to the measured one that unplausibly small component masses would be required for $\dot{E}_{\rm gr}$ to be negligible. We conclude that $\alpha < 1$ is essentially ruled out in the UIM discussed here.

If $\alpha > 1$ the primary spin is faster than the orbital motion and the situation is different. Spin–orbit coupling has an opposite sign with respect to the GW torque. The small torque on the orbit implied by the measured $\dot{\omega}_o$ could result from two larger torques of opposite signs partially cancelling each other.

This point has been overlooked by [23] who estimated the GW luminosity of the source from its measured timing parameters and, based on this estimate, claimed the failure of the UIM for RX J1914.4+2456. In discussing this and other misinterpretations of the UIM in the literature, [8] show that the argument by [23] actually leads to our same conclusion: in the UIM framework, the orbital evolution of this source must be affected significantly by spin–orbit coupling, being slowed down by the transfer of angular momentum and energy from the primary spin to the orbit. The source GW luminosity must accordingly be larger than indicated by its timing parameters.

5.1 Constraining the Asynchronous System

Given that the source is not compatible with steady-state, we constrain system parameters in order to match the measured values of W, ω_o and $\dot{\omega}_o$ and meet the requirement that the resulting state has a sufficiently long lifetime.

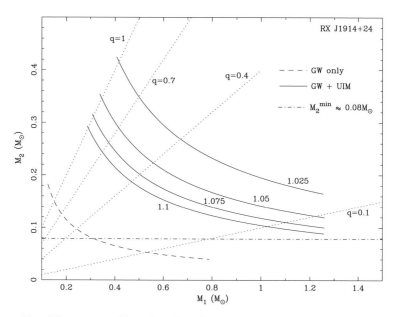

Fig. 12. M_2 versus M_1 plot based on measured timing properties of RX J1914.4+2456. The dot-dashed line corresponds to the minimum mass for a degenerate secondary not to fill its Roche-lobe. The dashed curve represents the locus expected if orbital decay was driven by GW alone, with no spin–orbit coupling. This curve is consistent with a detached system only for extremely low masses. The solid lines describe the loci expected if spin–orbit coupling is present (the secondary spin being always tidally locked) and gives a *negative* contribution to $\dot{\omega}_o$. The four curves are obtained for $W = 10^{33}$ erg s^{-1} and four different values of $\alpha = 1.025, 1.05, 1.075,$ and 1.1, respectively, from top to bottom, as reported in the plot.

Since system parameters cannot all be determined uniquely, we adopt the following scheme: given a value of α Eq. (13) allows us to determine, for each value of M_1, the corresponding value of M_2 (or q) that is compatible with the measured W, ω_o and $\dot{\omega}_o$. This yields the solid curves of Fig. 12.

As these curves show, the larger is α and the smaller the upward shift of the corresponding locus. This is not surprising, since these curves are obtained at fixed luminosity W and $\dot{\omega}_o$. Recalling that $(1/\alpha)$ gives the efficiency of energy transfer in systems with $\alpha > 1$ (see Sect. 3.1), a higher α at a given luminosity implies that less energy is being transferred to the orbit. Accordingly, GWs need being emitted at a smaller rate to match the measured $\dot{\omega}_o$.

The values of α in Fig. 12 were chosen arbitrarily and are just illustrative: note that the resulting curves are similar to those obtained for RX J0806.3+1527. Given α, one can also estimate k from the definition of W (Eq. (4)). The information given by the curves of Fig. 12 determines all quantities contained in k, apart from μ_1. Therefore, proceeding as in the previous section, we can determine the value of μ_1 along each of the four curves of Fig. 12. As in the case of RX J0806.3+1527, derived values are in the $\sim 10^{30}$ G cm^3 range. Plots and discussion of these results are reported by [8].

We finally note that the curves of Fig. 12 define the value of X for each (M_1, M_2), from which the system GW luminosity $\dot{E}_{\mathrm{gr}}$ can be calculated and its ratio to spin–orbit coupling. According to the above curves, the expected GW luminosity of this source is in the range $(4.6\text{--}1.4) \times 10^{34}$ erg s^{-1}. The corresponding ratios $\dot{E}_{\mathrm{gr}}/\dot{E}_L^{(orb)}$ are $1.15, 1.21, 1.29$ and 1.4, respectively, for $\alpha = 1.025, 1.05, 1.075$ and 1.1.

Since the system cannot be in steady-state a strong question on the duration of this transient phase arises. The synchronization timescale $\tau_\alpha = \alpha/\dot{\alpha}$

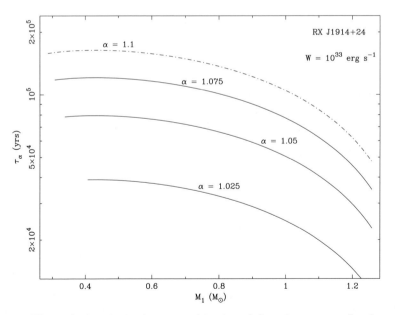

Fig. 13. The evolution timescale τ_α as a function of the primary mass for the same values of α used previously, reported on the curves. Given the luminosity $W \sim 10^{33}$ erg s^{-1} and a value of α, τ_α is calculated as a function of M_1.

302   Gian Luca Israel and Simone Dall'Osso

can be estimated combining Eqs. (5) and (6). With the measured values of W, ω_o and $\dot{\omega}_o$, τ_α can be calculated as a function of I_1 and, thus, of M_1, given a particular value of α. Fig. 13 shows results obtained for the same four values of α assumed previously. The resulting timescales range from a few $\times 10^4$ yrs to a few $\times 10^5$ yrs, tens to hundreds times longer than previously obtained and compatible with constraints derived from the expected population of such objects in the Galaxy. Dall'Osso et al. [8] discuss this point and its implications in more detail.

6 Conclusions

The observational properties of the two DDBs with the shortest orbital period known to date have been discussed in relation with their physical nature.

The Unipolar Inductor Model and its coupling to GW emission have been introduced to explain a number of puzzling features that these two sources have in common and that are difficult to reconcile with most, if not all, models of mass transfer in such systems.

Emphasis was put on the relevant new physical features that characterize the model. In particular, the role of spin–orbit coupling through the Lorentz torque and the role of GW emission in keeping the electric interaction active at all times has been thoroughly discussed in all their implications. It has been shown that the model does work over arbitrarily long timescales.

Application of the model to both RX J0806.3+1527 and RX J1914.4+2456 accounts in a natural way for their main observational properties. Constraints on physical parameters are derived in order for the model to work, and can be verified by future observations.

It is concluded that the components in these two binaries may be much more similar than it may appear from their timing properties and luminosities. The significant observational differences could essentially be due to the two systems being caught in different evolutionary stages. RX J1914.4+2456 would be in a luminous, transient phase that precedes its settling into the dimmer steady-state, a regime already reached by the shorter period RX J0806.3+1527. Although the more luminous phase is transient, its lifetime can be as long as $\sim 10^5$ yrs, one or two orders of magnitude longer than previously estimated.

The GW luminosity of RX J1914.4+2456 could be much larger than previously expected since its orbital evolution could be largely slowed down by an additional torque, apart from GW emission.

Finally, we stress that further developments and refinements of the model are required to address more specific observational issues and to assess the consequences that this new scenario might have on evolutionary scenarios and population synthesis models.

References

1. S. C. C. Barros, T. R. Marsh, P. Groot, et al.: MNRAS **357**, 1306 (2005)
2. S. C. C. Barros, et al.: MNRAS, **374**, 1334 (2007)
3. K. Beuermann, H.-C. Thomas, K. Reinsch, et al., Astr. & Astroph. **347**, 47 (1999)
4. V. Burwitz, and K. Reinsch: 2001, *X-ray astronomy: stellar endpoints, AGN, and the diffuse X-ray background*, Bologna, Italy, eds White, N. E., Malaguti, G., Palumbo, G., AIP conference proceedings, **599**, 522 (2001)
5. J. T. Clarke, et al.: Science, **274**, 404 (1996)
6. M. Cropper, M.K. Harrop-Allin, K.O. Mason, et al.: MNRAS, **293**, L57 (1998)
7. S. Dall'Osso, G. L. Israel & L. Stella: Astr. & Astroph. **447**, 785 (2006a)
8. S. Dall'Osso, G. L. Israel & L. Stella: astro-ph/0603795 (2006b)
9. F. D'Antona, P. Ventura, L. Burderi, & A. Teodorescu: Ap. J. **653**, 1429 (2006)
10. C. J. Deloye & R. E. Taam: Ap. J. Lett. **649**, L99 (2006)
11. C. R. Evans, I. Iben, and L. Smarr: Ap.J. **323**, 129 (1987)
12. P. Goldreich & D. Lynden-Bell: Ap. J. **156**, 59 (1969)1
13. P. J. Groot, G. Nelemans, D. Steeghs, and T. R. Marsh, Ap.J. Letters **558**, L123 (2001)
14. P. Hakala, G. Ramsay, K. Wu, L. Hjalmarsdotter, et al.: MNRAS, **343**, L10 (2003)
15. P. Hakala, G. Ramsay, and K. Byckling: MNRAS **353**, 453 (2004)
16. I. Iben, and A. V. Tutukov: Ap.J. Supplement, **54**, 355 (1984).
17. G.L. Israel, M.R. Panzera, S. Campana, et al.: Astr. & Astroph. **349**, L1 (1999)
18. G.L. Israel, L. Stella, W. Hummel, S. Covino and S. Campana, IAU Circ., **7835** (2002a)
19. G.L. Israel et al.: A&A, **386**, L13 (I02) (2002b)
20. G. L. Israel et al.: Ap. J. **598**, 492 (2003)
21. G. L. Israel, et al.: Memorie della Societa Astronomica Italiana Supplement **5**, 148 (2004)
22. T. R. Marsh, K. Horne, and S. Rosen: Ap.J. **366**, 535 (1991)
23. T.R. Marsh, and G. Nelemans: MNRAS **363**, 581 (2005)
24. T. R. Marsh, & D. Steeghs: MNRAS **331**, L7 (2002)
25. C. Motch, F. Haberl, P. Guillout, M. Pakull, et al.: A&A **307**, 459 (2006)
26. G. Nelemans, F. Verbunt, L. R. Yungelson, and S. F. Portegies Zwart: A&A **360**, 1011 (2000)
27. G. Nelemans, L. R. Yungelson, S. F. Portegies Zwart, and F. Verbunt: A&A **365**, 491 (2001)
28. G. Nelemans: ASP Conf. Ser. 330 – The Astrophysics of Cataclysmic Variables and Related Objects, **330** p. 27, (2005), astro-ph/0409676.
29. G. Nelemans , R. Napiwotzki , C. Karl , T. R. Marsh , et al. A&A **440**, 1087 (2005)
30. G. Nelemans, and C. A. Tout, MNRAS **356**, 753 (2005)
31. G. Nelemans, & P. G. Jonker: astro-ph/0605722 (2006); to appear in proceedings of "A life with stars" a conference in honour of Ed van den Heuvel's 60th birthday, New Ast. Rev
32. G. Nelemans: AIP Conf. ser. 873 (Ed. S.M. Merkowitz, J.C. LIvas), **873**, p. 397 (2007), astro-ph/0703292
33. A.J. Norton, C.A., Haswell, G.A. Wynn: A&A **419**, 1025 (2004)

34. S. Perlmutter, G. Aldering, M. della Valle, S. Deustua, et al.: Nature **391**, 51 (1998)
35. M. M. Phillips: Ap.J. **413**, L105 (1993)
36. J. H. Piddington: Moon **17**, 373 (1977)
37. P. Podsiadlowski , P. A. Mazzali , P. Lesaffre , C. Wolf , and F. Forster: astro-ph/0608324
38. G. Ramsay, M. Cropper, K. Wu, K.O. Mason, P. Hakala: MNRAS, **311**, 75 (2000)
39. G. Ramsay, P. Hakala, M. Cropper, et al.: MNRAS, **332**, L7 (2002a)
40. G. Ramsay, K. Wu, M. Cropper, et al.: MNRAS, **333**, 575 (2002b)
41. G. Ramsay, P. Hakala,K. Wu, et al.: MNRAS **357**, 49 (2005)
42. G. Ramsay, M. Cropper & P. Hakala: MNRAS **367**, L62 (2006a)
43. G. Ramsay et al.: astro-ph/0610357 (2006b), to appear in Proc 15th European White Dwarf Workshop, held in Leicester Aug 7-11th 2006, and to be published by ASP
44. K. Reinsch, V. Burwitz and R. Schwarz, Revista Mexicana de Astronomia y Astrofisica Conference Series, 2004, **20**, pp. 122, see astro-ph/0402458
45. A. G. Riess, L.-G. Strolger, J. Tonry, S. Casertano, et al.: Ap.J. **607**, 665 (2004)
46. G. H. A. Roelofs , P. J. Groot , T. R. Marsh , et al.: MNRAS **365**, 1109 (2006).
47. C. T. Russell et al: Planetary and Space Science **47**, 133 (1998)
48. C. T. Russell et al.: Advances in Space Research **34**, 2242 (2004)
49. N. S. Schulz, D. Chakrabarty, H. L. Marshall, et al.: Ap.J. **563**, 941 (2001)
50. D. Steeghs, T. R. Marsh, S. C. C. Barros et al.: Ap. J. **649**, 382 (2006)
51. A. Stroeer, A. Vecchio & G. Nelemans: Ap. J. Lett. **633**, L33 (2005)
52. T. E. Strohmayer: Ap. J. **581**, 577 (2002)
53. T. Strohmayer: Ap. J., **593**, L39 (2003)
54. T. E: Strohmayer: Ap. J. **627**, 920 (2005)
55. R. E. Taam, and E. L. Sandquist: ARAA **38**, 113 (2000)
56. B. Warner: Ap&SS, **225**, 249 (1995)
57. K. Werner, T. Nagel, T. Rauch, N. Hammer, and S. Dreizler: A&A **450**, 725 (2006)
58. A. J. Willes, and K. Wu: MNRAS, **348**, 285 (2004)
59. K. Wu, M. Cropper, G. Ramsay & K. Sekiguchi: MNRAS **331**, 221 (2002)

Binary Black Hole Coalescence

Frans Pretorius

Department of Physics, Princeton University, Princeton, NJ 08544, USA
`fpretori@princeton.edu`

1 Introduction

A black hole is one of the most fascinating and enigmatic predictions of Einstein's theory of general relativity. Its interior can have rich structure and is intrinsically dynamical, where space and time itself are inexorably led to a singular state. The exterior of an isolated black hole is, on the other hand, remarkably simple, described uniquely by the stationary Kerr solution. The dynamics of black holes are governed by laws analogous to the laws of thermodynamics, and indeed when quantum processes are included, emit Hawking radiation with a characteristic thermal spectrum. Most remarkable however, is that black holes, "discovered" purely through thought and the mathematical exploration of a theory far removed from every day experience, appear to be ubiquitous objects in our universe.

The evidence that black holes exist, though circumstantial, is quite strong [1]. The high luminosity of quasars and other active galactic nuclei (AGN) can be explained by gravitational binding energy released through gas accretion onto supermassive $(10^6$–$10^9 M_\odot)$ black holes at the centers of the galaxies [2, 3], several dozen X-ray binary systems discovered to date have compact members too massive to be neutron stars and exhibit phenomena consistent with matter interactions originating in the strong gravity regime of an inner accretion disk [4], and the dynamical motion of stars and gas about the centers of nearby galaxies and our Milky Way Galaxy infer the presence of very massive, compact objects there, the most plausible explanation being supermassive black holes [5, 6, 7].

In order to conclusively prove that black holes exist, one needs to "see" them, or conversely see the compact objects masquerading as black holes. The only direct way of observing black holes is via the gravitational waves they emit when interacting with other matter/energy (an isolated black hole does not radiate). The quadrupole formula says that the typical magnitude h of the gravitational waves emitted by a binary with reduced mass μ on a circular orbit measured a distance r from the source is (for a review of gravitational wave theory see [8])

$$h = \frac{16\mu v^2}{r}, \tag{1}$$

where v is the average tangential speed of the two members in the binary (and geometric units are used – Newton's constant $G = 1$ and the speed of light $c = 1$). This formula suggests that the strongest sources of gravitational waves are simply the most massive objects that move the fastest. To reach large velocities in orbit, the binary separation has to be small; black holes, being the most compact objects allowed in the theory, can reach the closest possible separations and hence largest orbital velocities. Therefore, modulo questions about source populations in the universe, a binary black hole interaction offers one of most promising venues of observing black holes through gravitational wave emission.

Joseph Weber pioneered the science of gravitational wave detection with the construction of resonant bar detectors. Weber claimed to have detected gravitational waves [9], though no similar detectors constructed following his claims were able to observe the putative (or any other) source, and the general consensus is that given the sensitivity of Weber's detector and expected strengths of sources it is very unlikely that it was a true detection [10]. Note that the *existence* of gravitational waves is not in doubt – the observed spin down rate of the Hulse–Taylor binary pulsar [11] and several others discovered since, is in complete accord with the general relativistic prediction of spin down via gravitational wave emission. Today a new generation of gravitational wave detectors are operational, including laser interferometers (LIGO [12], VIRGO [13], GEO600 [14], TAMA [15]) and resonant bar detectors (NAUTILUS [16], EXPLORER [17], AURIGA [18], ALLEGRO [19], NIOBE [20]). A future space-based observatory is planned (LISA [21]), and pulsar timing and cosmic microwave background polarization measurements also offer the promise of acting as gravitational wave "detectors" (for reviews see [22, 23]). The ultimate success of gravitational wave detectors, in particular with regard to using them as more that simply detectors, rather as tools to observe understand the universe, relies on source modeling to predict the structure of the waves emitted during some event. Even if an event is detected with a high signal-to-noise ratio (SNR), there simply is not enough information contained in such a one dimensional time series to "invert" it to reconstruct the event; rather template banks of theoretical waveforms from plausible sources need to be built and used to decode the signal. In rare cases an electromagnetic counterpart may be detected, for example during a binary neutron star merger if this is a source of short gamma ray bursts, which could identify the event without the need for a template. Though even in such a case, to extract information about the event, its environment, etc. requires source modeling.

Gravitational wave detectors have therefore provided much of the impetus for trying to understand the nature of binary black hole collisions, and the gravitational waves emitted during the process. However, from a theoretical perspective black hole collisions are fantastic probes of the dynamical, strong-field regime of general relativity. What is already know about this regime – the inevitability of spacetime singularities in gravitational collapse via the singularity theorems of Penrose and Hawking [24, 25]; the spacelike, chaotic "mixmaster" nature of these singularities conjectured by Belinsky, Khalat-nikov and Lifshitz (BKL) [26]; the null, mass-inflation singularity discovered by Poisson and Israel [27] that, together with regions of BKL singularities could generically describe the interiors of black hole; the rather surprising discovery of critical phenomena in gravitational collapse by Choptuik [28, 29]; etc. – together with the sparsity of solutions (exact, numerical or perturbative), suggests there is potentially a vast landscape of undiscovered phenomena. Of particular interest, and potential application to high energy particle collision experiments, are ultra-relativistic black hole collisions. It is beyond the scope of this article to delve much into these aspects of black hole coalescence, though a brief overview of this will be given in Sect. 5.3.

The two-body problem in general relativity, introduced in more detail in Sect. 2, is a very rich and complicated problem, with no known closed-form solution. Perturbative analytic techniques have been developed to deal with certain stages of the problem, in particular the inspiral prior to merger and ringdown after merger. Numerical solution of the full field equations are required during the merger, and this aspect of the problem is the main focus of this article. Much effort has been expended by the community over the past 15–20 years to numerically solve for merger spacetimes, and within the last two years an understanding of this phase of the two-body problem is finally being attained. Sect. 3 summarizes the difficulties in discretizing the field equations, and describes the methods known at present that work for black hole collisions, namely *generalized harmonic coordinates* and *BSSN* (Baumgarte-Shapiro-Shibata-Nakamura) with moving punctures. Preliminary results are discussed in Sect. 4, though given the rapid pace at which the field is developing much of this will probably be dated in short order. Sect. 5 concludes with a discussion of some astrophysical and other implications of the results.

2 The Two-Body Problem in General Relativity

Consider the classical two-body problem of finding the motion of two masses interacting only via the Newtonian gravitational force, and given the initial positions and velocities of the objects. The solution is well known – in the center of mass frame each body travels within the same plane along a conic section with a focus at the center of mass, and the type of conic (ellipse, hy-perbola, parabola) depends on the net energy of the system (bound, unbound, marginally unbound). In Newtonian gravity this setup is an idealization to the dynamics of two "real" objects in that one treats them as point sources without any internal structure. If one were to extend the problem to include the structure of the bodies there would be an infinite class of two-body problems depending on the nature of the material composition of the objects.

In general relativity (GR) the two-body problem is on one hand a significantly more challenging problem than its Newtonian counterpart due to the complexity of solving the Einstein field equations, yet if attention is restricted to the vacuum case, it is much simpler in that one can formulate the *exact problem without idealization*: given an initial spacelike slice of a vacuum spacetime containing two black holes, what is the subsequent evolution of the spacetime exterior to the event horizon?[1] If Penrose's cosmic censorship conjecture holds the solution will be unique and entirely independent of the interior structure of the black holes due to the causal structure of the spacetime. A wrinkle in this clean picture of the two-body problem in GR is that now, rather than a simple set of mass, position and velocity parameters, there are infinitely many degrees of freedom required to describe the initial conditions. These include the initial properties of each black hole and the gravitational wave content of the spacetime. To constrain the possibilities one could restrict the class of initial conditions to black holes that were, at some time in the past, sufficiently separated to each be well described by a Kerr metric with given mass and spin vector, require the initial spacetime slice to have "minimal" gravitational wave content and possess an asymptotic structure such that the black hole positions and relative velocities can unambiguously be defined. This class of initial conditions will cover the vast majority of conceivable astrophysical black hole binary configurations, and the black hole scattering problem setup discussed later on.

So what makes the two-body problem so interesting in GR? First of all, it almost goes without saying that as gravity is one of the fundamental forces influencing our existence and shaping the structure of the cosmos, and since GR is the best theory of gravity at our disposal, we want to understand all the details of of the more basic interactions in GR. A less prosaic reason to study this problem is the rich and fascinating phenomenology of solutions: what in Newtonian gravity is entirely describable by the mathematics of conic sections is now a problem that is unlikely to a have a closed form solution in any but the most trivial scenarios, featuring regimes with complicated orbital dynamics, and is accompanied by the emission of gravitational radiation. It is this latter feature which has the most profound implication: the two-body problem in GR for any bound system is *unstable*, and will eventually result in the decay of the orbit and collision of the two black holes. If cosmic censorship holds the collision will always result in a single black hole, and then, from the "no hair" theorems of Israel and Carter [30, 31, 32], one knows that the exterior

---

[1] This is not a technically precise definition, as the global structure of the spacetime is being ignored, and to capture the spirit of the two-body problem in a technically precise manner applicable to situations in our universe would probably require defining it using the concepts of isolated horizons [33], and furthermore restrict the solution to the future domain of dependence of an initial finite volume of the spacetime.

structure of the remnant black hole will eventually settle down to the Kerr solution. For an idea of how unstable orbits are to gravitational radiation, the time to merger t_m in units of the Hubble time t_H for an equal mass binary with each black hole having a mass M, initially separated by R_0 times the Schwarzschild radius $R_s = 2GM/c^2$, is roughly

$$\left(\frac{t_m}{t_H}\right) \approx \left(\frac{M}{M_\odot}\right)\left(\frac{R_0}{10^6 R_s}\right)^4. \tag{2}$$

For example, two solar mass black holes initially a million Schwarzschild radii, or $\approx 3 \times 10^6$ km apart, will merge within a Hubble time; two $10^9 M_\odot$ supermassive black holes need to be within $\approx 6 \times 10^3$ of their Schwarzschild radii, or roughly 1 parsec, to merge within the age of the universe.

In the following section the qualitative features of the two-body interaction in general relativity are described.

2.1 Stages of a Merger

This chapter is primarily concerned with numerical solution of the field equations as a tool to study the two-body problem. However, it is only in the final stages of coalescence where a full numerical solution is required to obtain an accurate depiction of the spacetime. This stage of a merger occurs on a very short time scale compared to other phases of the two-body interaction, which is fortunate, for due to the computational complexity of solving the field equations it is not feasible to evolve the spacetime for times much longer than this. In the following two sections a more detailed discussion of the various stages of a merger is given, in particular to set the scope for the remainder of the chapter and to highlight how much interesting phenomenology in the two-body problem is *not* addressed by full numerical simulation. We break the discussion up into two classes of merger, "astrophysical", and the black hole scattering problem. A merger scenario is considered astrophysical if, to some approximation and non-negligible likelihood the initial conditions could be realized by a binary system in our universe. The latter classification deals with the gedanken experiment of colliding two black holes with ultra-relativistic initial velocities and with an impact parameter of order the total energy of the system or less.

One reason for this classification is that we *might* expect very different qualitative behavior of the spacetime in these two cases. Consider two black holes of mass m_1 and m_2 with net ADM [34] energy E in the center of mass frame[2]. All astrophysical mergers are expected to take place in the rest-mass dominated regime where $(m_1 + m_2)/E \approx 1$, while in the scattering problem the kinetic energy of the black holes will dominate so that $(m_1 + m_2)/E \approx 0$. In the latter regime the geometry of each black hole also gets length contracted

---

[2] Here we consider m_1 and m_2 to be the total BH mass including spin energy, so not the irreducible mass.

into a pancake-like region, with the actual black holes occupying an ever smaller region of the non-trivial geometry as the boost factor increases. In fact, eventually it does not matter that it was a black hole that was boosted to large energies – any compact source will produce the same geometry in the limit. The ultra-relativistic limit might also be an interesting place to look for violations of cosmic censorship – the collision of plane-fronted gravitational waves generically leads to the formation of naked singularities [35], and though not exactly analogous to high energy black hole collisions, there are enough similarities that it would be worthwhile to explore this regime of the two-body problem in some detail. Note that the particular value of E is not relevant; there is no intrinsic length scale in the field equations of general relativity, and any solution with energy E_0 can trivially be re-scaled to a "new" solution with arbitrary energy E.

Astrophysical Binaries

Here the astrophysical merger scenario is broken down into four stages: *Newtonian, inspiral, plunge/merger* and *ringdown.*

Newtonian: In this stage the two black holes are sufficiently far apart that gravitational wave emission will be too weak to cause the binary to merge within a Hubble time t_H [2]. Thus, to have any hope of observing mergers of binaries formed in this stage, other "Newtonian" non-two-body processes need to operate. For example, in the stellar mass range, it is unlikely that a close black hole binary could be formed as the end point of the evolution of a massive binary star system. The reason is that at the requisite separations for a subsequent gravitational wave driven inspiral within t_H, the stars will most likely evolve through a common envelope phase, and recent results have suggested this will cause a merger of the stellar cores before a binary black hole system could be formed [36]. A likely mechanism then to produce hard binaries is through n-body interactions that occur in dense cluster environments [37, 38]. For supermassive black hole binaries, which are thought to form during galaxy mergers in the hierarchical structure formation scenario [39, 40], gas interactions, dynamical friction and other n-body processes are thought cable of driving most black holes close enough so that gravitational wave emission can take over and cause a merger [41, 42]. If such processes did not operate efficiently it would be in apparent contradiction with the observations that most galaxies harbor supermassive black holes at their centers [2, 3].

Inspiral: In the inspiral regime gravitational wave emission becomes the dominant process driving the black holes to closer separation, though the orbital time scale is still much shorter than the time scale over which orbital parameters change. The majority of non-extreme mass ratio binaries are expected to "form" with sufficiently large semi-major axis that the orbit will circularize via gravitational wave emission long before the binaries merge [43, 44]. Some exceptional cases might be stellar and intermediate mass binaries in dense star clusters, where numerous interactions with neighbors could frequently perturb the orbit, or triple systems where the Kozai mechanism operates [49, 45, 46, 47, 48, 50]. On the other hand, the majority of extreme mass ratio systems that will merge within the Hubble time are expected to have sizable eccentricities at merger [51, 52]. Note however that much of the theory behind the formation mechanisms and environments of binary black holes are not well known (indeed, no candidate binary black hole system has yet been observed), which offers gravitational wave detection a fantastic opportunity to help decipher some of these interesting questions.

The inspiral phase is well modeled by post-Newtonian (PN) methods [53]. For initially non-spinning, zero eccentricity binaries the higher order PN approximations and *effective one-body* (EOB) resummations [54] give waveforms that are surprisingly close to full numerical results even until very close to merger, well beyond when naive arguments suggest they should fail [55, 56, 57, 58, 59]; comparisons for more generic scenarios have yet to be made. Extreme mass ratio inspirals (EMRIs) can also be well described by geodesic motion in a black hole background together with prescriptions for computing the gravitational wave emission and effects of radiation reaction [64, 65, 66, 67, 68, 69, 70, 71]. Generic (non-equatorial) orbits about a Kerr black hole will not lie in a plane due to precession and frame-dragging effects, and thus during the lengthy course of an EMRI, which could be in LISA-band for thousands of cycles, the small black hole will "sample" much of the geometry of the background spacetime. The structure of the corresponding gravitational waves emitted will therefore contain a map of this geometry, and so EMRIs offer a remarkable opportunity to probe the geometry of a black hole, and will be able to confirm whether it is indeed described by the Kerr metric [72, 52].

Plunge/merger: Here, for non extreme-mass-ratio systems, gravitational wave emission becomes strong enough that the evolution of the orbit is no longer adiabatic, and the black holes plunge together to form a single black hole. Understanding this phase requires full numerical simulations, and it is only within the last couple of years that such simulations have become available. The interesting picture that is now emerging is that this phase is very short, lasting on the order of one to two gravitational wave cycles. To get an idea for the time scale of this regime, the Keplerian orbital angular frequency ω for an equal mass quasi-circular inspiral is

$$\frac{\omega}{2\pi} = \frac{1}{2\pi}\sqrt{\frac{M}{R^3}} \approx 11\,\text{kHz}\frac{M_\odot}{M}\left(\frac{R_s}{R}\right)^{3/2}, \tag{3}$$

where M is the total mass of the binary with corresponding Schwarzschild radius R_s, R is the separation, and the plunge/merger happens as $R_s/R \to 1$. Note that the frequency of the dominant quadrupolar ($\ell = 2, m = 2$) component of the gravitational wave that is emitted is twice the orbital frequency. This is the time of strongest gravitational wave emission, with the luminosity approaching on the order of one-hundredth of the Planck luminosity of 10^{59}

erg s^{-1}, making black hole mergers by far the most energetic events in the post-big-bang era of the universe. Furthermore, the frequency of the emitted wave rapidly grows to that of the dominant quasinormal mode frequency of the final black hole, causing the spectrum of the plunge/merger phase to occupy a large region of the frequency domain. For equal mass systems upwards of 3% of the rest mass energy of the system is radiated away here. A more detailed discussion of this phase is given in Sect. 4.

If cosmic censorship holds (and there are no signs that it is violated in any merger simulations to date), then Hawking's "no-bifurcation" theorem [73] states that a single black hole must result as the consequence of a merger. The uniqueness, or "no hair" theorems [30, 31, 32] further imply that the newly formed black hole must eventually settle down to the Kerr solution in the so-called ringdown phase.

Ringdown: The ringdown is the phase where the remnant black hole can be described as a perturbed Kerr spacetime. A more precise definition might be the time after which the gravitational waves emitted from the merger can, to good precision, be written entirely as a superposition of *quasi-normal modes* (QNMs) [74, 75, 76, 77, 78, 79] of the final black hole[3]. Given appropriate initial conditions, the ringdown phase could be calculated using perturbative techniques, in particular using the so-called *close limit* approximation [82]. As discussed more in Sect. 4, the early simulation results suggests this description is already adequate very shortly after formation of the common apparent horizon, which roughly coincides with the time of peak luminosity. Very shortly after ringdown begins, the waveform ($|h| \propto e^{-t/\tau_{22}} \sin(\omega_{22} t)$) is dominated by the least damped (fundamental harmonic) quadrupolar QNM, with angular frequency ω_{22} and decay time τ_{22}, given approximately by the following fitting formulas [80, 81]:

$$\frac{\omega_{22}}{2\pi} \approx \frac{1}{2\pi M} \left[1 - 0.63(1-j)^{0.3}\right] \approx 32\,\text{kHz}\frac{M_\odot}{M}\left[1 - 0.63(1-j)^{0.3}\right] \quad (4)$$

$$\tau_{22} \approx \frac{4M(1-j)^{-0.45}}{1 - 0.63(1-j)^{0.3}} \approx 20\,\mu\text{s}\frac{M}{M_\odot}\frac{(1-j)^{-0.45}}{1 - 0.63(1-j)^{0.3}}, \quad (5)$$

where M and $J = jM^2$ are the total mass and angular momentum of the final black hole (with $|j| \le 1$). The dominant ringdown frequency is several times higher than the orbital frequency in the last few inspiral cycles, and the decay time is quite short, so the majority of the energy lost during ringdown (1–2% of the rest mass) is emitted quite rapidly. Waves propagating in a curved spacetime like Kerr are back-scattered off the curvature, producing so-called *power law tails* [83]. They decay by integer powers of time, and so even though initially of much smaller amplitude than the ringdown waves, they will eventually dominate the late-time structure of the gravitational wave.

---

[3] The QNM spectrum is not complete, and so it is conceivable that they might not be able to *exactly* describe the wave structure.

Given their small amplitude it is unlikely that the tails could be detected by ground-based detectors.

The relative simplicity of the plunge/merger waveform, together with how short this phase is, suggests it may be possible to build effective analytic template banks of merger waveforms by stitching together PN inspiral waveforms with ringdown waveforms. Numerical simulations of the plunger/merger phase can provide instruction on exactly how this stitching should be performed, i.e., how long the transition region is, which set of quasi-normal modes are excited, how does the waveform interpolate between inspiral and ringdown modes, etc. In fact, this kind of prescription for constructing templates for mergers was already proposed several years ago by proponents of the effective-one-body (EOB) approach to binary dynamics [54, 61], and was recently demonstrated to work well for the extreme mass ratio problem [62] and a range of non-spinning comparable mass mergers [59]. Why might such a "simple" approach work so well for the merger phase, which was anticipated to be a showcase of the complexity and non-linearity of the field equations? First of all, recall that the PN approaches (including the EOB) are hardly simple, having required the dedicated effort of many researchers over a couple of decades to push to the orders presently known [53] – $(v/c)^7$ beyond Newtonian order for non-spinning binaries, and $(v/c)^5$ if spin is included. At such high orders in (v/c) it is not too surprising that much of the essential physics is already being captured, and the only question becomes how far the approach can be trusted. As the velocity of the black holes increases toward the merger one would expect the expansions to become increasingly inaccurate[4]. Though, at the same time the black holes are falling deeper into what is becoming the effective potential of the final black hole spacetime, and eventually details of the local dynamical geometry that may be poorly described by PN expansions will have little effect on the radiated gravitational wave structure. Also, a black hole by itself is not a simple, "linear" object, and thus perturbations thereof could also be expected to capture much of the late time physics of a merger.

The Black Hole Scattering Problem

Consider the collision of two black holes in the center of mass frame with masses m_1 and m_2, Kerr spin parameters a_1 and a_2, and initially moving toward each other with impact parameter b and (large) Lorentz γ-factors γ_1 and γ_2. At present very little is known about all the possible outcomes as a function of $(b, \gamma_{1,2}, m_{1,2}, a_{1,2})$, though one can speculate about several distinct stages, that will be classified here as *Lorentz, collision/ringdown, scatter* and *threshold*. Note that in contrast to the rest-mass dominated case, there is

---

[4] For the quasi-circular equal mass inspirals the coordinate velocities of the apparent horizons only approach around $v = 0.3$ prior to formation of the common horizon.

not necessarily such a straightforward progression through the phases. In particular, there could be a range of impact parameters where the black holes do not merge during the initial encounter, but have lost sufficient energy that they now form a bound system. Then subsequent evolution of the system will follow the stages of the astrophysical binaries outlined in the preceding section.

Lorentz: With sufficiently large γ factors the initial non-trivial geometry of each black hole is Lorentz contracted into a thin "pan-cake" (or plane-wave) transverse to the direction of propagation, and close to Minkowski spacetime on either side[5].

Collision/ringdown: As suggested by studies of colliding black holes in the infinite γ limit [84, 85, 86, 87, 88, 89, 90, 91, 92, 93], if the impact parameter is close to zero there will not be any phase analogous to inspiral; rather an encompassing apparent horizon forms at the moment of collision, and this will presumably settle down to a Kerr black hole. Estimates based on the size of the initial apparent horizon place an upper limit of 30% on the net energy of the spacetime that could be radiated in a head-on collision, though these limits weaken as the impact parameter increases. For the head-on collision case, perturbative studies suggest the energy radiated is close to 16% [88].

Scatter: For larger values of the impact parameter there will be a deflection of the two black hole trajectories, accompanied by a burst of radiation, after which they will move apart and the spacetime near each black hole will settle down to the *Lorentz* phase again. It has been suggested that there may even be a regime where a *third* or more black holes are formed during the interaction of the two black holes before they scatter, essentially due to the strong focusing of gravitational waves by the shock-fronts representing the boosted black holes [94]. This would be an astonishing addition to the phenomenology of the two-body problem in general relativity if the scenario can be realized.

Threshold: At intermediate values of the impact parameter there could be threshold-type behavior as seen when fine-tuning eccentric orbits in the rest-mass dominated regime [95, 96]. Namely, approaching a critical value $b = b^*$ of the impact parameter, the two black holes settle into the analogue of an unstable circular geodesic orbit, whirl around for an amount of time proportional to $-\ln|b - b^*|$, then either fly apart or plunge together (this is described in more detail in Sect. 4.4). During this phase copious amounts of energy could be radiated in gravitational waves; in fact, at threshold it is conceivable that essentially *all* of the kinetic energy of the system is radiated as gravitational waves in $O(10)$ orbits. If the black holes merge after the whirl phase, there will be a plunge/merger and ringdown phase similar to astrophys-

---

[5] In the the limit $\gamma \to \infty$ and $m \to 0$ with $m\gamma = E$ kept constant, one obtains the Aichelburg–Sexl solution [63], and the spacetime becomes exactly Minkowski on either side of a propagating C^0 kink in the geometry.

ical binaries. If they separate and have lost enough kinetic energy to form a bound system they will enter the *inspiral* phase of an astrophysical binary, otherwise they will fly apart as in the *scatter* phase above. It is tempting to speculate that exactly at threshold, $b = b^*$, the spacetime may approach a self-similar solution (see the discussion in Sect. 5.2).

3 Contemporary Successful Numerical Solution Methods

This section describes the two methods of formulating the field equations presently known that are amenable to stable numerical integration of binary black hole spacetimes[6], namely *generalized harmonic coordinates with constraint damping* (GHC) and the *Baumgarte–Shapiro–Shibata–Nakamura* (BSSN) formalism with moving punctures [97, 98, 99]. It is beyond the scope of this article to discuss either method in much detail, or all the variations and details of particular codes; rather the equations will be presented and briefly discussed to provide the reader with some appreciation for the similarities and differences between them. Note also that if a code produces an apparently stable, convergent solution, it is much more likely that the method actually *is* stable, compared to the opposite situation where a simulation "crashes" and from which one would like to conclude that the *method* is unstable. This is simply because bugs are easy to make, more difficult to find, and almost never "help" in any interpretation of the word. The point of this discussion is that there have been numerous good ideas and formulations of the field equations proposed over the past several years (see for example [100, 101]), and only in a few cases were they studied with sufficient detail to conclude that they were unstable; thus that we now know of two methods that are stable does not imply that all earlier methods are not. A case-in-point might be the $Z4$ formalism [103, 102] proposed several years ago, which is quite similar in some respects to generalized harmonic coordinates, and to which the same constraint damping mechanism can be applied [104]. On the other hand, the fact that "zeros" need to be added to the equations in just the right way to make things stable also tells us that the Einstein equations are even more subtle and intricate than previously thought.

3.1 Historical Background

The first attempt at a numerical solution to the field equations for a binary black hole spacetime was carried out by Hahn and Lindquist [105] in 1964. This was even before the word "black hole" had been coined by Wheeler, and they evolved what was then called the "worm hole" initial data of Misner [106]. They considered a time-symmetric scenario in axisymmetry, and reported a

---

[6] At least for the regions of parameter space studied to date, which are all in the rest-mass dominated regime.

run performed on an IBM 7090 computer, using a 51×151 mesh. It took about 4 hours to complete 50 time steps, after which they concluded that errors had grown too large to warrant further evolution. This corresponded to a time of $m/2$, with $m = \sqrt{A/16\pi}$, A being the area of the throat. Given the short run time, not much physics could be extracted from the simulation, yet even so there was no motivation to explore gravitational wave emission. In 1975 Smarr [107], and shortly afterwards Eppley [108], again simulated the head-on collision of two black holes, now with one of the primary goals being to compute the gravitational waves emitted in the process. Despite still being an axisymmetric simulation and almost a decade after Hahn and Lindquist, it was still beyond the capabilities of computers of the time to integrate the field equations with sufficient resolution to obtain very accurate results. Nevertheless, they were able to extract gravitational waveforms from the solutions, calculating (with uncertainties of a factor of 2) that upwards of 0.1% of the rest mass energy is released in gravitational waves in a time-symmetric case where the initial proper separation between the two throats is $9.6M$ [109]. Primarily because of the stringent computational requirements for numerical solution, no further work on the problem was carried out until the early 1990s, when prospects for the construction of LIGO became solid. LIGO was the impetus for returning to the two-body problem as it was realized early on [10] that given a practical design for the instrument, together with the estimated density of sources in our universe, matched-filtering would be an essential data analysis tool to allow a decent detection rate within a several year time-frame. Matched-filtering looks for known signals in a noisy data stream by convolving theoretical templates of the signals with the data. To be successful it is therefore imperative to understand the gravitational wave emission properties of the source with sufficient detail to construct the template libraries.

The early expectations following a revisit of the head-on collision case [110] were that although certain issues about the generic merger problem had yet to be fully addressed and could be complicated, such as having well behaved coordinates and providing astrophysically relevant initial conditions, a fair consensus was that the most significant hurdle to the problem was lack of computer power [111, 112]. Certainly a portion of the difficulties encountered may be traceable to attempting to find solutions with insufficient resolution, though it turns out that a host of additional issues had to be "discovered", understood and overcome to reach the state where the field is today.

The review of the history of the numerical two-body problem now continues, though switching to a non-traditional format: instead of trying to follow events in chronological order the ingredients needed for a successful simulation will be summarized, noting contributions that offered insights or solutions to the various problems.[7]

---

[7] And my apologies in advance to authors that I have missed here.

3.2 Historical Background Continued – Ingredients to Assemble a Successful Numerical Two-Body Code

The list of ingredients described below certainly "makes sense", and so one might wonder, why not satisfy all of them to begin with? First of all, many of the issues, such as choosing well behaved coordinates, are quite complicated, and one does not expect general solutions applicable to all spacetimes of interest. Second, it was perhaps not fully appreciated how vast the landscape of free-evolution schemes is, i.e. systems of equations that give solutions to the Einstein equations for only a restricted subset of initial and boundary conditions, and how important the behavior of these equations are for initial and boundary conditions that do *not* exactly satisfy those requirements. This is particularly so because it is numerical truncation error that sources "constraint violations", which is not *a priori* a problem as truncation error is a well understood, part-and-parcel component of any numerical solution. The surprising thing then is that, in dynamical systems language, it appears as if for the vast majority of free evolution formulations of the Einstein equations, trajectories through phase space denoting solutions to Einstein's equations form an *unstable* manifold in the space of all solution trajectories. A third reason is the ADM formulation, in the form popularized by York [113], certainly also "makes sense", and seems to be a very reasonable and intuitive approach to an initial boundary value formulation of the field equations. Furthermore, given the success of the ADM equations in early evolutions of many symmetry reduced spacetimes, there was not much reason to suspect problems with it.

3.2.1 Fix the Character of the Differential Equations

The Einstein field equations[8]

$$G_{ab} = 8\pi T_{ab}, \tag{6}$$

when expanded verbatim in terms of the metric g_{ab}

$$ds^2 = g_{ab}dx^a dx^b, \tag{7}$$

results in a coupled system of 10, quasilinear, second order partial differential equations for the 10 independent components of the metric, depending on the four spacetime coordinates x^a. However, these equations have *no* definite mathematical character – hyperbolic, parabolic or elliptic – and moreover, do not admit a well posed initial value problem. This in large part is due to the gauge invariance of the field equations: for a given, *unique* physical spacetime there are infinitely many *different* metric tensors describing it and all satisfying the same Eqs. (6). The first step towards obtaining a well posed system of

---

[8] Again, using *geometric units* where the speed of light $c = 1$ and Newton's constant $G = 1$.

equations is to specify enough of the gauge to fix the character of each of the four spacetime coordinates x^a. There are several possibilities, the most common being to choose one coordinate (t) to be timelike, and the remaining three (x, y, z) to be spacelike. After a bit more work, outlined in the following item, one comes up with a system of elliptic/hyperbolic equations. This "space-plus-time" (or 3+1) approach will be the exclusive focus of the remainder of the article, after briefly mentioned one alternative, *characteristic* or *null* coordinates (for more details, see for example [101, 114]). Here, one (single null) or two (double null) coordinates are chosen to be lightlike, and the rest of the coordinates spacelike. Single null evolution schemes have been very successful in evolving single black hole spacetimes[115]. Part of the reason for pursuing null evolution is that it is easy to extend the domain to future null infinity, which is ideally where one would want to measure gravitational waves. The difficulty with this approach is preventing or treating caustics that can form along the null coordinate in non-trivial, dynamical spacetimes, and no viable mechanism has yet been proposed that might be applied to a binary black hole spacetime. Hybrid null 3 + 1 schemes (often called *Cauchy-characteristic matching*) have been proposed (see [114] and the references therein), whereby a 3+1 scheme is used to evolve the spacetime in the vicinity of the binary, a characteristic scheme far from the binary, and the solutions mapped to one another in an intermediate zone where the coordinate systems overlap. The matching process is non-trivial, and to date the method has only been applied to single black hole spacetimes [116, 117, 118].

3.2.2 Find a Formulation Admitting a Numerically Well-Posed Initial Boundary Value Problem

To obtain a well-posed 3 + 1 formulation of the Einstein equations more work needs to be done than merely choosing one timelike and three spacelike coordinates. The choice of which set of fields to treat as the dependent variables of the system of PDEs, the gauge conditions that will be used, as well as modification of the equations by the addition of constraint terms (i.e. terms that are identically zero for any solution of the Einstein equations) all play an important role in determining the ultimate stability of the system. The "traditional ADM" approach, as outlined by York [113], is based on a Hamiltonian formulation of the field equations due to Arnowitt, Deser and Misner [34]. The end result is that the equations are rewritten in terms of quantities either *intrinsic* or *extrinsic* to $t =$ const. slices of the geometry. First, the four dimensional metric is decomposed as

$$ds^2 = -\alpha^2 dt^2 + h_{ij} \left(dx^i + \beta^i dt\right)\left(dx^j + \beta^j dt\right), \tag{8}$$

where h_{ij} is the spatial metric of the $t =$ const. hypersurface, α is the *lapse function* measuring the rate at which proper time flows relative to t for a hypersurface-normal observer, and β^i is the spatial *shift* vector describing how the spatial coordinate label for such an observer changes with time t. In other words, the time flow vector $(\partial/\partial t)^a$ is related to the unit hypersurface normal vector n^a by

$$\left(\frac{\partial}{\partial t}\right)^a = \alpha n^a + \beta^a \qquad (9)$$

In this way of describing the four-geometry the lapse and shift naturally represent the coordinate degrees of freedom in the theory. Second, the manner in which h_{ij} is embedded into the four-dimensional space is described by the *extrinsic curvature* tensor K_{ij}[9]

$$K_{ij} \equiv -h_i{}^a h_j{}^b \nabla_b n_a \qquad (10)$$

$$= -\frac{1}{2\alpha}\left(\frac{\partial h_{ij}}{\partial t} - \mathcal{L}_\beta h_{ij}\right). \qquad (11)$$

In terms of the variables $(h_{ij}, K_{ij}, \alpha, \beta^i)$ the field equations can be written as a set of 12 independent hyperbolic evolution equations for (h_{ij}, K_{ij}), 4 *constraint* equations that do not contain any time derivatives of K_{ij}, and need to be augmented with evolution equations for the gauge quantities (α, β^i) (see [101] for an overview of oft-used choices). A common way to proceed to solve these equations is by *free evolution* (see [119] for a discussion of the general alternatives): the constraints are only solved at the initial time, and the remaining equations are then used to evolve the variables with time. In a consistent discretization scheme the constraint equation will remain zero to within numerical truncation error, and hence, as mentioned before, that the constraints are not strictly enforced is not necessarily a problem. However, in general scenarios, i.e. when there are no symmetries that can be used to simplify the equations, it turns out that the standard ADM form of the equations just outlined is only *weakly hyperbolic*, and this implies that one cannot in general find a fully consistent, hence stable discretization scheme for the system [120]. This problem began to be appreciated by the numerical relativity community in the mid-90s, which sprouted a cottage industry of finding symmetric-hyperbolic reductions or various more "ad-hoc" modifications of the field equations [121, 123, 122, 124, 125, 126, 127, 128, 129, 130, 131, 132, 133, 134, 135, 136, 137, 138, 139, 140, 141, 142, 103, 143, 144, 145, 146, 147, 148]. Unfortunately, even though some of these methods were successfully applied to single black hole spacetimes, they offered only marginal improvements at best compared to ADM codes for the binary black hole problem [149, 150, 151, 152, 153, 154, 155, 156, 157, 158, 159, 160, 161, 163, 162, 164, 165, 166, 167, 168, 169, 170, 171], with the arguable exception of the BSSN formulation, which showed success in binary neutron star evolutions(see [172] and the works cited therein), and set the record for the longest binary black

---

[9] In the Hamiltonian picture the momentum π_{ij} canonically conjugate to h_{ij} is $\pi_{ij} = \sqrt{h}(K_{ij} - Kh_{ij})$, where h is the determinant of h_{ij} and K is the trace of K_{ij}.

hole evolution [173] prior to the breakthroughs of 2005 [97, 98, 99]. One reason why some of these methods, even though provably stable, can still "fail" in practice, is if the truncation error grows too rapidly with time. The truncation error $f_{te}(t, x^i)$ for a variable f will, to leading order in the mesh spacing h in an nth order discretization scheme, have the form $f_{te}(t, x^i) = e(t, x^i)h^n$. Formal stability only requires that the h-independent error term $e(t, x^i)$ does not grow *faster* than exponential, though with sufficiently rapid exponential growth it might be impractical to give high enough resolution (small h) to keep the error term small for the desired run-time. Another (and somewhat related) reason why stable codes could fail, and which actually appears to be *the* problem in most free evolution schemes, are "constraint violating modes" discussed next.

3.2.3 Curb Truncation-Error-Induced Growth of Constraints

Constraint violating modes (CVMs) are *continuum solutions* to the *subset* of Einstein equations that are evolved during a free evolution, but do not satisfy the constraint equations. These constraints could either be the Hamiltonian and momentum constraints inherent to the Einstein equations, or constraints arising from first order reductions or similar redefinitions of the underlying fields. Note that truncation error is not a CVM by this definition, however since in general truncation error will not satisfy the constraints it will be a source of CVMs in any free evolution scheme. For CVMs to be benign their growth rate must be sufficiently small to remain of comparable magnitude to truncation error during the evolution. At present only two formulations of the field equations appear to have this desired property "off the constraint manifold" for binary black hole evolutions – *generalized harmonic coordinates* with *constraint damping* [97], and variants of *BSSN* with appropriate gauge choices and methods for dealing with the black hole singularities [98, 99]. These two approaches will be described in more detail in Sect. 3.3 and Sect. 3.4. Constraint damping adds a particular function of the constraint equations to the Einstein equations to try to curb the growth of CVMs for solutions close to the desired one. This is not a very new idea [174, 121, 122, 177, 124, 126, 132, 175, 137, 163, 176, 125, 178, 102] though the particular method that works for harmonic evolution was only recently proposed by Gundlach et al. [104]. They were able to prove that their damping terms could curb all finite-wavelength constraint violating perturbations of Minkowski space. There is no mathematical proof that this should work for the binary black hole problem, and the evidence that the CVMs are adequately under control is entirely empirical. However, experience suggests there may never be a "black box" solver for the Einstein equations applicable to solving for arbitrary spacetimes; rather, schemes need to be tailored to the particular scenario, and constraint damping is probably no exception. The nature of the evolution of the constraints in BSSN is even less well understood.

Given all the problems with the constraints, an obvious alternative would be *constrained evolution*, whereby the constraints are solved at each time step in lieu of a subset of evolution equations. Such methods have worked very well in symmetry reduced situations, though with the exception of [179] have not yet been attempted in 3D. Part of the reason is that solving the constraints involves solving elliptic equations, which many people in the community have been reluctant to attempt. Also, it is not clear in a general 3D setting which degrees of freedom to constrain, and which to freely evolve. In [179], a spherical polar coordinate system is used, which does allow for a "natural" decomposition into free versus constrained variables; such a coordinate system is not well adaptive to studying binary black hole spacetimes. Several years ago Andersson and Moncrief [180] discussed an elliptic-hyperbolic formulation of the field equations that appears to be ideally suited for 3D constrained evolution, though to date no implementations of this system have been carried out. A related idea is *constraint projection* (similar to "divergence cleaning" in the solution of Maxwell's equations), whereby a free evolution system is used, then periodically the constraints are re-solved, modifying a subset of the variables accordingly. This technique was shown to have promise in a single black hole spacetime [168], though in that code excision boundary problems (apparently) prevented long-time stable evolution. In [181] a Langrange multiplier method was proposed to optimally project out the constraints; it was successfully implemented for scalar field evolution, though has not yet been applied to the full Einstein equations. One might think that another option for dealing with the constraints is at the numerical level via something akin to the *constrained transport* [182] scheme used in some magnetohydrodynamic codes, however Meier [183] showed that similar finite-difference based techniques will not work for the Einstein equations due to the non-linearity of the equations.

3.2.4 Provide Well Behaved Dynamical Coordinates Conditions

It almost goes without saying that covering the spacetime manifold with a well behaved, non-singular coordinate system is a necessary condition for stable evolution. The difficulty is that in a Cauchy evolution the dynamics of the fields describing the geometry are intimately linked to the coordinates, and thus when solving for a new spacetime where the future geometric structure is unknown, the future behavior of the coordinates is just as uncertain. A large number of analytic solutions discovered throughout the history of relativity have, in their original form, been riddled with coordinate pathologies (the most famous example of course being the event horizon of the Schwarzschild solution); dealing with them involved first understanding the nature of the pathology, then constructing a coordinate transformation to remove it. In *principle* this is an approach that could be applied in a numerical evolution: evolve to the point of a coordinate singularity, understand it, apply a coordinate transformation to remove it, and continue the evolution. However, given the nature of a numerical solution, i.e. discrete meshes of numbers representing either field values or coefficients of basis functions, this would be a very challenging endeavour in all but the simplest spacetimes. Thus, the universal approach in numerical relativity to try to avoid coordinate problems has been to devise coordinate *conditions* that typically either make the coordinates satisfy certain properties (e.g. constant mean curvature, or CMC slicing where $t =$ const. is a space of constant mean curvature, or harmonic coordinates described in Sect. 3.3), or conditions that force the variables to satisfy certain constraints (e.g. the unit-determinant condition on the conformal metric in the BSSN approach discussed in Sect. 3.4). Coordinate conditions usually take the form of algebraic or differential operators acting on the "gauge" fields of the formalism, which are most commonly the lapse and shift. It is beyond the scope of this article to describe the numerous coordinate conditions proposed over the years related to the binary hole problem (see [109, 113, 101] for more details), though in Sect. 3.3 and 3.4 we will described those that have been instrumental in the recent successful binary black hole simulations.

3.2.5 Specify Good Outer Boundary Conditions

"Good" outer boundary conditions for evolved fields in a simulation must have three properties: (1) be mathematically well posed, (2) be consistent with the constraints, and (3) be consistent with the physics being modeled, which here is asymptotic flatness[10] and no incoming gravitational radiation. The class of boundary conditions (3) will form a subset of (2), which in turn is a subset of conditions (1). A common approach is to apply either exact or some approximation to *maximally dissipative* boundary conditions, where the incoming characteristics of all fields normal to the boundary are set to zero. Though mathematically well-posed this in general is neither consistent with the constraints nor prevents outgoing waves of the solution from being reflected back. Much effort has been spent over the past several years devising constraint preserving boundary conditions (CPBCs) for various formulations [185, 186, 176, 187, 188, 189, 190, 191, 192, 193, 194, 195, 196, 197, 198, 199, 200, 201]. By themselves CPBCs do not alleviate the problem of spurious incoming radiation, and more recently research in CPBCs has begun to focus on subsets of CPBCs that do address this issue.

   An alternative approach to outer boundary conditions is to extend the computational domain to infinity, where the exact Minkowski spacetime boundary conditions can be placed. As mentioned in Sect. 3.2 Cauchy characteristic matching effectively extends the domain to future null infinity. The

---

[10] For the purposes of modeling the local geometry of a merger and extracting the resultant gravitational waves in the far-zone there is little practical distinction between an asymptotically flat versus Friedman–Robertson–Walker universe. The effects of a wave propagating across cosmological distances in an expanding universe can readily be accounted for analytically, as the wave amplitude decays as $1/D_L$ and its wavelength increases by a factor $1 + z$, where z is the redshift and D_L the luminosity distance to the source – see for example [184].

matching procedure is non-trivial however, and this technique has yet to be applied to a binary black hole merger scenario. Another option is to compactify the coordinates to spatial infinity, which is the approach used in the generalized harmonic evolutions in [97]. This rather straightforwardly solves all issues (1)–(3), though introduces potential numerical complications in that outgoing waves suffer an ever increasing blue shift as they travel toward the outer boundary [11]. Either increasing resolution and hence computational resources must be used to resolve the waves, or once they have passed the desired wave extraction radius be allowed to blue-shift to coarse resolution. With the latter option the numerical technique must therefore be robust to the introduction of high-frequency solution components; in the generalized harmonic evolution code this is achieved using Kreiss–Oliger style dissipation [202]. Note that this kind of dissipation is *not* akin to artificial viscosity sometimes used in hydrodynamical simulations, as Kreiss–Oliger dissipation modifies the difference equations at the level of the truncation error terms, and thus converges away in the continuum limit.

A couple of alternative methods of compactification include conformal compactification [140, 203, 204], and using asymptotically hyperboloidal or null slices [205, 206].

3.2.6 Deal with Black Hole Singularities

By the singularity theorems of Hawking and Penrose [24, 25] we know that all black holes contain true (geometric) singularities, which in a simulation will manifest as various field quantities diverging as the spacetime slice approaches the singularity. Infinities cannot be dealt with in a numerical code, and must be "regularized" in some fashion. The two contemporary successful approaches to deal with the singularities are *excision* and *punctures*. Both techniques rely on the causal property of a black hole spacetime that no information can flow out of the event horizon, and that cosmic censorship is valid, namely that all the geometric singularities that might exist in the spacetime are always inside the event horizon. If cosmic censorship were violated in a particular evolution, the codes would "crash"; thus a stable evolution is confirmation that in that instance cosmic censorship was not violated.

With excision, a 2-sphere *inside* the black hole and enclosing the singularity is designated as a boundary of the computational domain. By the assumed causal properties of the black hole there will always exist a class of such boundaries where all characteristics of the fields are directed *toward* the boundary, i.e. out of the computational domain. Mathematical theory (and common sense) says one is only allowed to place boundary conditions on the incoming components of fields satisfying hyperbolic equations. Thus *no* boundary conditions are specified on the excision surface; rather, the difference equations

---

[11] The blue shift is infinite in the limit, though it would take an infinite amount of time for the waves to reach the boundary.

are simply solved there[12]. The formal definition of a black hole event horizon is the boundary of the causal past of future null infinity, which is not a local property of spacetime and cannot be found during evolution. Instead, the *apparent horizon* – a marginally outer-trapped surface, which is a surface from which "outward" traveling photons have zero expansion – is used to determine where to excise. Excision surfaces on or inside the apparent horizon can also satisfy the requirement that all field-characteristics are directed outside of the domain, since, if cosmic censorship holds, the apparent horizon will always be *inside* the event horizon (see for example [207]).

Originally, a *puncture* was the singular point inside a maximally extended vacuum black hole spacetime representing the spatial infinity reached by a conformally mapped slice passing through an Einstein-Rosen bridge into a second asymptotically flat universe [208]. Therefore a puncture is a coordinate rather than geometric singularity. The manner in which the metric diverges approaching the puncture is known analytically, and can be factored out. Punctures were originally used to construct initial data for the binary black hole problem [208], though soon afterwards it was shown that punctures can be used in evolution [156]. The metric at the puncture was regularized by dividing out a time-independent conformal factor, however the extrinsic curvature was not regularized. Thus, to avoid problems this was anticipated to cause, the punctures were placed "between" grid points, and coordinate conditions were chosen to make the shift vector zero at the punctures so that derivatives of the extrinsic curvature across the punctures would not be needed. The vanishing of the shift vector implies the puncture locations are fixed in the grid. Maximal slicing was used for the lapse. The breakthrough in puncture evolutions discovered recently is to relax the condition on the shift vector, allowing the punctures to move through the domain. At the same time the slicing condition is altered to force the lapse to zero at the puncture, essentially "freezing" evolution there (see Sect. 3.4 for more on these coordinate conditions). This, remarkably, causes the codes to remain stable despite the irregular nature of the solution about the punctures. There have been several studies since attempting to understand geometrically what a moving puncture represents [209, 210, 211, 212, 213]; a couple of competing viewpoints at present are that 1) the puncture remains attached to spatial infinity of the alternate universe [213], and 2) the spacetime slice quickly evolves so that the alternate universe is "pinched-off", and the puncture effectively becomes a single excised point inside the black hole [209, 210, 212]. Though from the perspective of seeking solutions to the field equations exterior to the event horizons of the black holes, the question of what a puncture represents is academic.

---

[12] In a finite difference code this implies replacing centered derivative operators with "sideways" operators as appropriate to avoid referencing regions of the domain inside the excision boundary.

3.2.7 Provide Consistent and Relevant Initial Data

The initial data problem for binary black holes is not trivial. First, the initial conditions must satisfy the constraints, which typically involves solving systems of coupled, non-linear elliptic equations. Second, providing astrophysically relevant initial data is quite challenging, as for practical considerations the evolution must begin within several or tens of orbits before merger. This implies that there should already be a non-negligible amount of gravitational radiation from the prior inspiral of the black holes present in the initial data. Also, the closer the black holes are the more difficult it becomes to unambiguously associate relevant orbital parameters to the spacetime, for example the orbital eccentricity, binary separation, orbital frequency, etc. It is beyond the scope of this article to describe these issues – see [214, 215, 216] for review articles on contemporary initial data construction methods, and [217, 218, 219, 220] for suggestions to incorporate realistic initial conditions motivated by post-Newtonian expansions.

3.3 Generalized Harmonic Evolution

Generalized harmonic evolution, as its name implies, is an evolution scheme based on a generalization of *harmonic coordinates*. Harmonic coordinates are a set of gauge conditions that require each spacetime coordinate x^a to independently satisfy the covariant scalar wave equation:

$$\Box x^a = \frac{1}{\sqrt{-g}} \partial_b \left(\sqrt{-g} g^{bc} \partial_c x^a \right) \equiv 0, \tag{12}$$

where g is the determinant of the spacetime metric (7). The use of these coordinate conditions has a long and celebrated history in relativity, including DeDonder's analysis of the characteristic structure of general relativity [221], Fock's study of gravitational waves [222] and proofs of uniqueness and existence of solutions to the field equations by Choquet-Bruhat [223] and Fischer and Marsden [224]. In fact, as early as 1912 Einstein used harmonic coordinates, then known as isothermal coordinates, in his search for a relativistic theory of gravitation [225]. One of the key properties of harmonic coordinates that makes them so useful in these studies is that when (12) is substituted into the field equations, the principal part of the resultant equation for each metric element becomes a scalar wave equation for that particular metric element, with all non-linearities and couplings between the equations relegated to lower order terms. This has obvious benefits for formal analysis of the field equations, and is also a natural system to study the radiative degrees of freedom in the theory. Also, given that there are simple and effective numerical solution techniques available to solve wave equations, it would seem that harmonic coordinates would be a natural starting point for a numerical code.

However, only recently in numerical relativity have harmonic coordinates been used as the basis for numerical evolution schemes [226, 227, 228, 229, 230,

231], meaning discretizing the field equations *after* the harmonic conditions have been used to re-express the system as a set of wave-like equations. Prior to this harmonic coordinates had been advocated and used within the more traditional ADM space-plus-time formulation of the field equations [232, 233, 234, 112, 124, 235], where harmonic gauge (or variants of it) are imposed via conditions on the lapse function and shift vector. In such a decomposition the wave-like character of the field equations in not manifest, and the primary reason quoted for using harmonic gauge (in particular harmonic time slicing) was for its geometric "singularity avoiding" properties. However even within ADM evolutions harmonic coordinates were seldom used due to the notion that they would generically lead to the formation of "coordinate shocks" [235, 236, 237, 238, 239]. An in-principle solution to this problem noted by Garfinke [226] (and see an earlier discussion of this by Hern [237]) was to use *generalized harmonic coordinates* (GHC), first introduced by Friedrich [240]. Here, a set of arbitrary *source functions* are added to (12):

$$\Box x^a \equiv H^a. \tag{13}$$

To see that this can avoid coordinate pathologies, note that (13) can be regarded as a *definition* of the source functions; in other words, take *any* metric in *any* (well behaved) coordinate system, and (13) tells one what the corresponding source functions for the metric in GHC are. When imposing GHC in a Cauchy evolution, the H^a must be treated as independent functions to allow (13) to reduce the principal parts of the field equations to the desired wave-like equations. Thus additional evolution equations must be supplied for H^a to close the system, and so the issue of finding well-behaved coordinates for a particular dynamical spacetime becomes one of finding the appropriate evolution equations for H^a.

For concreteness, below an explicit form of the Einstein equations in GH form with constraint damping terms will be given, using the covariant metric elements and covariant source functions ($H_a = g_{ab}H^b$) as the fundamental variables. This is certainly not the only way to proceed – for a symmetric hyperbolic first order reduction see [229] (and see [241] for how the constraints introduced via auxiliary variables are kept under control), and versions using the densitized contravariant metric elements see [227, 230, 242]. Consider the Einstein equations in trace-reversed form

$$R_{ab} = 4\pi \left(2T_{ab} - g_{ab}T\right), \tag{14}$$

where R_{ab} is the Ricci tensor

$$R_{ab} = \Gamma^d_{ab,d} - \Gamma^d_{db,a} + \Gamma^e_{ab}\Gamma^d_{ed} - \Gamma^e_{db}\Gamma^d_{ea}, \tag{15}$$

Γ^g_{ab} are the Christoffel symbols of the second kind

$$\Gamma^g_{ab} = \frac{1}{2}g^{ge}\left[g_{ae,b} + g_{be,a} - g_{ab,e}\right] \tag{16}$$

T_{ab} is the stress energy tensor with trace T, and a comma is used to denote partial differentiation. Using the definition of GHC (13) and its first derivative (14) can be written out explicitly as

$$\frac{1}{2}g^{cd}g_{ab,cd} + \tag{17}$$

$$g^{cd}{}_{(,a}g_{b)d,c} + H_{(a,b)} - H_d\Gamma^d_{ab} + \Gamma^c_{bd}\Gamma^d_{ac} \tag{18}$$

$$+\kappa[n_{(a}C_{b)} - \frac{1}{2}g_{ab}n^dC_d] \tag{19}$$

$$= -8\pi\left(T_{ab} - \frac{1}{2}g_{ab}T\right). \tag{20}$$

Line (17) shows the principal, hyperbolic part of the equations, line (18) are the rest of the terms coming from (14) where all the couplings and non-linearities reside, line (19) are the constraint damping terms with adjustable parameter κ and unit timelike vector n^a normal to $t = \text{const.}$ hypersurfaces[13], and line (20) contains the coupling to matter. Here, the constraints are simply the definition of GHC

$$C_a \equiv g_{ab}\left(H^a - \Box x^a\right), \tag{21}$$

and are thus zero for any solution of the field equations. The relationship between the GH constraints and the more familiar form of the constraints of the Einstein equations, written as a one-form $\mathcal{M}_a$

$$\mathcal{M}_a \equiv (R_{ab} - \frac{1}{2}g_{ab}R - 8\pi T_{ab})n^b, \tag{22}$$

where the time-component $\mathcal{M}_a n^a$ is the Hamiltonian constraint and the momentum constraints are the components of the spatial projection $\mathcal{M}_a(\delta^a{}_b + n^a n_b)$, is [229]

$$\mathcal{M}_a = \nabla_{(a}C_{b)}n^b - \frac{1}{2}n_a\nabla_bC^b. \tag{23}$$

Furthermore, one can show that if the metric is evolved using (17-20), the constraints will satisfy the following evolution equation

$$\Box C^a = -R^a{}_bC^b + 2\kappa\nabla_b\left[n^{(b}C^{a)}\right]. \tag{24}$$

From this it is easy to see that, at the continuum level, a solution that initially satisfies the constraints will always do so if constraint-preserving boundary conditions are used during evolution. Part of the constraint damping modification in (24) – namely the term proportional to $n^b\nabla_bC^a$ – is a wave-equation damping term, so one might reasonably then expect that (24) will *not* admit exponentially growing solutions given small (truncation-error-sourced) initial

---

[13] In [104] it was suggested that an arbitrary timelike vector can be used in the constraint damping terms, though in all situations studied to date n^a has been chosen as the hypersurface normal unit timelike vector.

conditions. This "expectation" has been proven for small, finite-wavelength constraint-violating perturbations of Minkowski spacetime [104], though not yet for general spacetimes.

3.3.1 Source Function Evolution

To close the system of Eqs. (17)–(20), an additional set of evolution equations must be specified for the source functions, written schematically as

$$\mathcal{L}_a H_a = 0 \quad \text{(no summation)}. \tag{25}$$

$\mathcal{L}_a$ is a differential operator that in general is dependent upon the spacetime coordinates, the metric and its derivatives, and the source functions and their derivatives. The source functions directly encode the coordinate degrees of freedom of general relativity, as can be seen by writing the definition of GHC (13) in terms of ADM variables (8):

$$H_a \, n^a = -K - \partial_\nu (\ln \alpha) n^\nu \tag{26}$$

$$H_b \, h^{ab} = -\bar{\Gamma}^a_{jk} h^{jk} + \partial_j (\ln \alpha) h^{aj} + \frac{1}{\alpha} \partial_b \beta^a n^b, \tag{27}$$

where $\bar{\Gamma}^i_{jk}$ is the connection associated with spatial metric $h_{ij} \equiv g_{ij} + n_i n_j$. Thus, the temporal source function $H_a n^a$ is related to the time derivative of the lapse α, whereas a spatial source function $H_b h^{ab}$ is related to the time derivative of the corresponding component of the shift vector β^a. Not much research has been done on finding source function evolution equations to achieve a particular slicing or satisfy some coordinate conditions directly within the GH framework, though the above relationship between H^a and the lapse and shift allows many of the ideas developed over the years for ADM evolutions to be adopted in a GH evolution [228, 243]. We end this section by showing one example of a set of source evolution equations, used in [97]:

$$\Box H_t = -\xi_1 \frac{\alpha - 1}{\alpha^\eta} + \xi_2 H_{t,\nu} n^\nu, \quad H_i = 0. \tag{28}$$

This equation for H_t is a damped wave equation with a forcing function designed to prevent the lapse α from deviating too far from its Minkowski value of 1, which helps alleviate an apparent instability in the code of [228] that sets in when the lapse drops close to zero inside a black hole[14]. In (28) the parameter ξ_2 controls the damping term, and ξ_1, η regulate the forcing term. Ranges of useful parameter values are discussed in [95].

---

[14] Note that α is not an independent variable in the formalism – in the code it is replaced by its definition in terms of the metric g_{ab}.

3.4 BSSN with 'Moving Punctures'

The BSSN formulation of the field equations [244, 234, 177] begins with the
ADM (8) decomposition of spacetime, then continues by performing a York-
Lichnerowicz-like conformal decomposition of the spatial metric and extrinsic
curvature [245]. The conformal metric is defined via

$$\tilde{h}_{ij} \equiv e^{-4\phi} h_{ij} \tag{29}$$

and is *chosen* to have unit determinant, so that

$$e^{4\phi} = h^{1/3}, \tag{30}$$

where h is the determinant of h_{ij}. Continuing, the trace K, and conformal,
trace-free part of the extrinsic curvature (10)

$$\tilde{A}_{ij} \equiv e^{-4\phi}(K_{ij} - \frac{1}{3}h_{ij}K), \tag{31}$$

are treated as fundamental variables. The final ingredient in the BSSN formalism is to also evolve the conformal connection coefficients

$$\tilde{\Gamma}^i \equiv \tilde{h}^{jk} \tilde{\Gamma}^i_{jk} = -\tilde{h}^{ij}{}_{,j} \tag{32}$$

independently, where $\tilde{\Gamma}^i_{jk}$ is the Christoffel symbol of the conformal spatial
metric. In summary then, ϕ, $\tilde{h}_{ij}$, K, $\tilde{A}_{ij}$, Γ^i, α and β^i are the fundamental
variables of the BSSN formalism. The evolution equations for ϕ, $\tilde{h}_{ij}$ and Γ^i
derive from their definitions

$$\frac{d}{dt}\phi = -\frac{1}{6}\alpha K, \tag{33}$$

$$\frac{d}{dt}\tilde{h}_{ij} = -2\alpha\tilde{A}_{ij}, \tag{34}$$

$$\frac{\partial}{\partial t}\tilde{\Gamma}^i = -2\tilde{A}^{ij}\alpha_{,j} + 2\alpha\left(\tilde{\Gamma}^i_{jk}\tilde{A}^{kj} - \frac{2}{3}\tilde{h}^{ij}K_{,j} - \tilde{h}^{ij}S_j + 6\tilde{A}^{ij}\phi_{,j}\right)$$
$$+ \frac{\partial}{\partial x^j}\left(\beta^l\tilde{h}^{ij}{}_{,l} - 2\tilde{h}^{m(j}\beta^{i)}{}_{,m} + \frac{2}{3}\tilde{h}^{ij}\beta^l{}_{,l}\right). \tag{35}$$

and the evolution equations for K and $\tilde{A}_{ij}$ come from the Einstein equations

$$\frac{d}{dt}K = -h^{ij}D_jD_i\alpha + \alpha(\tilde{A}_{ij}\tilde{A}^{ij} + \frac{1}{3}K^2) + \frac{1}{2}\alpha(\rho + S), \tag{36}$$

$$\frac{d}{dt}\tilde{A}_{ij} = e^{-4\phi}\left(-(D_iD_j\alpha)^{TF} + \alpha(R_{ij}^{TF} - S_{ij}^{TF})\right) + \alpha(K\tilde{A}_{ij} - 2\tilde{A}_{il}\tilde{A}^l{}_j) \tag{37}$$

with

$$R_{ij} = \tilde{R}_{ij} + R_{ij}^{\phi}, \tag{38}$$

$$R_{ij}^{\phi} = -2\tilde{D}_i\tilde{D}_j\phi - 2\tilde{h}_{ij}\tilde{D}^l\tilde{D}_l\phi + 4(\tilde{D}_i\phi)(\tilde{D}_j\phi) - 4\tilde{h}_{ij}(\tilde{D}^l\phi)(\tilde{D}_l\phi), \tag{39}$$

$$\tilde{R}_{ij} = -\frac{1}{2}\tilde{h}^{lm}\tilde{h}_{ij,lm} + \tilde{h}_{k(i}\partial_{j)}\tilde{\Gamma}^k + \tilde{\Gamma}^k\tilde{\Gamma}_{(ij)k}$$
$$+ \tilde{h}^{lm}\left(2\tilde{\Gamma}^k_{l(i}\tilde{\Gamma}_{j)km} + \tilde{\Gamma}^k_{im}\tilde{\Gamma}_{klj}\right) \tag{40}$$

and matter projections

$$\rho = n_a n_b T^{ab}, \tag{41}$$

$$S_i = -h_{ia}n_b T^{ab}, \tag{42}$$

$$S_{ij} = h_{ia}h_{jb}T^{ab}. \tag{43}$$

The gauge variables α and β^i are freely specifiable. In the above the operator d/dt is defined to be

$$\frac{d}{dt} \equiv \frac{\partial}{\partial t} - \mathcal{L}_{\beta}, \tag{44}$$

where $\mathcal{L}_{\beta}$ is the Lie derivative with respect to the shift vector β^i (and note that $\tilde{h}_{ij}$ and $\tilde{A}_{ij}$ are tensor densities of weight $-2/3$), $D_i(\tilde{D}_i)$ is the covariant derivative operator with respect to $h_{ij}(\tilde{h}_{ij})$, and TF denotes the trace-free part of the expression. The BSSN equations listed above were taken from [177]; some of the actual implementations use slightly different variables (for example $\chi \equiv e^{-4\phi}$ is used instead of ϕ in [98]), differ in whether and/or how certain algebraic constraints in the formalism are enforced (such as the trace-free nature of $\tilde{A}_{ij}$ or that $\tilde{h}_{ij}$ has unit determinant), replace undifferentiated occurrences of $\tilde{\Gamma}^i$ with its definition (32), or add multiples of the constraints inferred by (32) to the evolution equation for $\tilde{\Gamma}^i$[246, 162, 247, 248].

There are several reasons often quoted as motivation behind the BSSN formalism. First, the conformal decomposition in part separates the extrinsic curvature into "radiative" versus "non-radiative" degrees of freedom (though within the York–Lichnerowicz formalism it is the *transverse* trace-free part of the extrinsic curvature that represents the radiative degrees of freedom). Second, the constraint equations are used to eliminate certain terms from the "bare" evolution equations (in particular the Hamiltonian constraint is used to eliminate a Ricci scalar term from the evolution equation for K, and the momentum constraints to eliminate a divergence of $\tilde{A}_{ij}$ term from the evolution equation of $\tilde{\Gamma}^i$), and so in a sense this is a partially constrained evolution system [249]. Third, with appropriate gauge conditions the BSSN system of equations is hyperbolic [250, 145, 251, 252, 253]. An important step in achieving hyperbolicity is treating the connection functions $\tilde{\Gamma}^i$ as independent quantities, which makes the principle part of the differential operator acting on the conformal metric in (40) elliptic. Incidentally, this is *exactly* what would be done if one were to express the spatial conformal metric in generalized harmonic form, with $\tilde{\Gamma}^i$ being the source functions.

Moving Punctures

An important element in achieving stable evolution of binary black hole space-times with the BSSN formulation is using coordinates that allow the punctures hiding the black hole singularities to move through the grid, yet do not allow any evolution *at* the puncture point itself (i.e., the lapse is forced to go zero at the puncture, though not the shift vector, hence the "frozen" puncture can be advected through the domain). The conditions that have so far proven successful are modifications to the so-called *1+log slicing* and *Gamma-driver shift* conditions [123, 254]:

$$\frac{d}{dt} = -2\alpha K \tag{45}$$

$$\partial_t \beta^i \equiv \xi B_i, \quad \partial_t B_i = \chi \partial_t \tilde{\Gamma}^i - \eta B^i - \zeta \beta^j \partial_j \tilde{\Gamma}^i. \tag{46}$$

In the above, ξ, χ, η and ζ are parameters (that are required to be within certain ranges for stable evolution, though do not require fine-tuning); a couple of examples for typical choices: $(\xi = 3\alpha/4, \chi = 1, \eta = 4, \zeta = 1)$ [255] and $(\xi = 1, \chi = 1, \eta = 1, \zeta = 0)$ [256]. Common initial conditions are $\beta^i = B^i = 0$, and $\alpha = 1/\psi_{BL}^2$, where $\psi_{BL} = 1 + \sum_i m_i/2|r - r_i|$ is the Brill–Linquist conformal factor for the initial data containing black holes with mass parameter m_i at coordinate location r_i.

It is uncertain exactly why these coordinate conditions work as well as they do (similar to why the rather ad-hoc equations used in the generalized harmonic scheme shown in (28) improve the evolution); an alternative way of phrasing this is that it is not known why *fixed* puncture evolutions are prone to instabilities. Recently in [239] it was suggested that 1+log slicing could generically lead to the formation of gauge shocks near the punctures as anticipated in [235], and that these have not yet been observed in current 3D simulations due to poor resolution about the punctures (though again, as long as stability can be maintained this in theory is not problematic for studying the geometry exterior to the horizon).

3.5 Comparison of the Two Techniques

After discussion of the two evolution formalism, generalized harmonic and BSSN, a "required" section deals with a comparison of the methods. That section is here, though there really is not much to say on the matter. Personal preferences and aesthetics aside, *both* methods are capable of finding discrete solutions describing similar physical processes within the context of the *same* theory – general relativity – and thus are equivalent from a scientific per-spective. In terms of technical issues, one could argue that moving punctures are much easier to get working than excision. However, this is more an issue of dealing with black hole singularities, and in principle either method could be implemented within either formalism. A technical issue of some relevance to numerical implementation is that there is (presently) no known fully first order, symmetric hyperbolic reduction of the BSSN equations, which would be a requirement for a spectral implementation using the methods of the Caltech/Cornell group.

3.6 Numerical Algorithms

It is beyond the scope of this article to delve into computational issues involved in solving the field equations – here a few references to related material in the literature are given. With the exception of the Caltech/Cornell pseudo-spectral code [257, 258], all contemporary binary black hole evolution codes use finite-difference methods (for a broader view of the use of spectral methods in relativity see [259]). The complexity of the field equations and the physical set-up of the binary black hole problem requires solution in a parallel computing environment, and adaptive mesh refinement (AMR) to adequately resolve all the relevant length scales (the only code at present not employing AMR is the *LazEv* code [260], however there a non-linear "fisheye" coordinate transformation is used to resolve the length scales in the vicinity of the binary). Some of the parallel/AMR software presently used is the Cactus Computational Toolkit [261] with the *Carpet* thorn for AMR [262], *paramesh* [263], *PAMR/AMRD* [264], *HAD* [265] and *BAM* [266]. Descriptions of some of the more computational aspects of the merger codes can be found in [268, 228, 260, 255, 231, 229, 266, 269].

4 Results from Black Hole Merger Simulations

In this section some results from recent merger simulations are discussed. This is a rapidly evolving field, and much of what is said could be dated in short-order. Also, in many respects this is still a very young field, and though there has been a flurry of early results, systematic, in-depth studies are sparse. We break the discussion up into the following classes of binary: a) equal mass, minimal eccentricity and spin, b) unequal mass, minimal eccentricity and spin, c) equal mass, non-negligible spin, minimal eccentricity, d) equal mass, large eccentricity, minimal spin, and e) generic.

4.1 Equal Mass, Minimal Eccentricity and Spin

The equal mass, minimal eccentricity and spin case is one of the simplest configurations in that there is no precession of the orbital plane, and no recoil imparted to the final black hole. Thus, the key parameters that characterize the merger are essentially only the final mass and spin of the remnant black hole. Recent simulations [98, 99, 270, 248, 271, 272, 58, 273] of this scenario have used either Cook–Pfeiffer [274] or Bowen–York [275] initial data. These results indicate that the energy emitted during the last orbit, plunge/merger and ringdown is 3.5%(±0.2%) of the total total energy of the system, resulting in a Kerr black hole with $a = 0.69$ (±0.02). Based on the binding energy of the initial data configurations, or PN/EOB estimates of the energy radiated during the inspiral (see for example [274, 54]), an additional 1.5% (±0.2%) of the available energy is radiated prior to this, implying that an equal mass, non-spinning inspiral beginning at infinite radial separation looses 5.0%(±0.4%) of its total rest-mass energy to gravitational waves during the entire merger event[15]. As mentioned in the discussion in Sect. 3.2, these families of initial data do not exactly capture the conditions of the equivalent astrophysical scenario, and though it is difficult at present to estimate precisely what the effects of this are, systematic studies suggest the artifacts are small. In particular, the initial data lack the correct initial gravitational radiation content, though within roughly an orbital light-crossing time this "junk" radiation leaves the vicinity of the orbit and is quickly replaced by radiation emitted by the binary motion. The energy content of the junk radiation also appears to be negligible to within the quoted uncertainties. Other noticeable "artifacts" in some of the cited simulation results is a small amount of orbital eccentricity (due to the choice of initial data having zero initial radial momentum), and the black holes are initially co-rotating for the Cook–Pfeiffer data presently in use; again, the effect on the waveforms appear to be small, and can also be removed without much effort [272, 273]. There have been several suggestions for how the correct radiation content can be inserted into initial fields [217, 218, 219, 220], though none of these methods has yet been implemented.

For illustrative purposes, Figs. 1 and 4 show some of the simulation results, taken from evolution of Cook–Pfeiffer initial data, described in detail in [55]. Fig. 1 is a plot of the orbital trajectory, Fig. 2 shows the real component of the Newman–Penrose scalar Ψ_4 in the orbital plane (which far from the source represents the second time derivative of the "plus" polarization of the usual gravitational wave strain), Fig. 3 shows the plus and cross polarizations of the waveform extracted on the axis normal to the orbital plane, and Fig. 4 shows the instantaneous gravitational wave frequency (divided by 2) and energy flux versus time together with labels depicting some phases of the merger.

Gravitational Wave Structure

Decomposed into a spin-weight 2 spherical harmonic basis, the waveform throughout the evolution is dominated by the quadrupole ($\ell = 2, |m| = 2$) component. The next leading order component ($\ell = 4, |m| = 4$) has an amplitude less than $1/10^{th}$ the quadrupole mode during the inspiral phase, growing

---

[15] The uncertainties reflect the authors best conservative "guess" based on the various results published in the literature to date; the uncertainty in the PN inspiral value is that the results usually quoted are for the integrated energy up to the ISCO (innermost stable circular orbit), which only approximately corresponds to the "last" orbit of the numerical results.

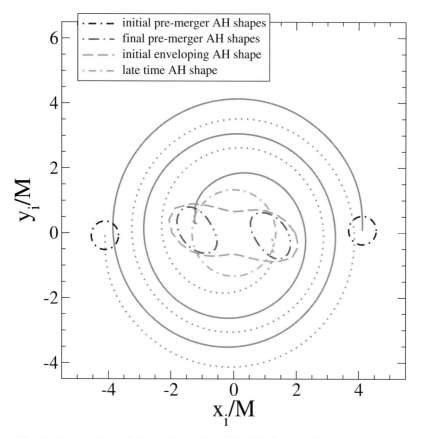

Fig. 1. A depiction of the trajectories of the black holes from a merger simulation (the "d=16" Cook–Pfeiffer case, from [55]). The green lines are the centers of the apparent horizons of each black hole. The trajectories end once a common horizon is found. Also shown are the coordinate shapes of the apparent horizons at several key moments. SEE COLOUR PLATES

briefly to $1/5^{th}$ of it near the peak of the emitted energy flux (and note that the energy content of a mode is proportional to the *square* of its amplitude) [55, 276]. Moreover, the *quadrupole formula* seems to describe the physics of gravitational wave *production* quite accurately throughout the orbital phase, in that the coordinate motion of the apparent horizons taken from the simulation and plugged verbatim into the quadrupole formula for two point-masses shows remarkable agreement with the full numerical waveform [55, 96]. Several studies have now also shown that the higher order PN and EOB methods can reproduce, to within the various uncertainties of the comparison (including numerical error, mapping parameters between the two descriptions, when to begin the comparison, etc.), up until close to merger [55, 56, 58, 57, 59]. The most accurate study to date [58] of exactly *when* the waveform from a

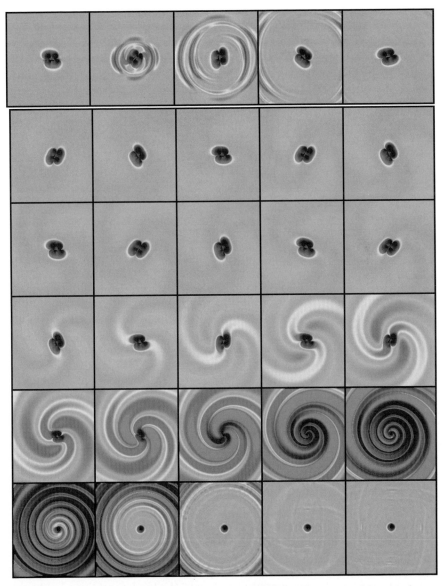

Fig. 2. A depiction of the gravitational waves emitted during the merger of two equal mass black holes (specifically "d=19" Cook–Pfeiffer initial data [55]). Shown is a color-map of the real component of the Newman–Penrose scalar Ψ_4 multiplied by r along a slice through the orbital plane, which far from the black holes is proportional to the second time derivative of the plus polarization (green is 0, toward violet (red) positive (negative)). The time sequence is from top to bottom, and left to right within each row. Each image is $25M$ apart, and a common apparent horizon is first detected at $t = 529M$ (i.e., the "merger"), which is a little after the frame in row 5, column 3. In the first several frames the spurious radiation associated with the initial data, and how quickly it leaves the domain, is clearly evident. The width/height of each box is around $100M$. SEE COLOUR PLATES

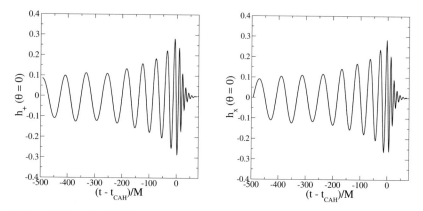

Fig. 3. The plus (left) and cross (right) polarizations of the waveform (multiplied by coordinate distance r from the source, and by the total mass M of the spacetime to non-dimensionalize) from the simulation shown in Fig. 1, though here measured along the axis normal to the orbital plane. t_{CAH} is the time when a common apparent horizon is first detected.

particular PN approach begins to deviate from the numerical signal to within the errors of the simulation – 0.3% in the phase and 1% in the amplitude over 18 cycles (9 orbits) of inspiral – showed that despite a relatively large amplitude disagreement of 7% between the restricted 3.5PN Taylor waveforms, the cumulative phase difference was 0.15 after 13 cycles, suggesting for the given numerical accuracy only the last 4.5 orbits of the inspiral would require numerical solution.

The transition from inspiral to ringdown does not last very long, only on the order of 10–$20M$. There is also no noticeable "plunge" in the orbital motion from the time there are two distinct black holes to a single one (see Fig. 1). However, in the Fourier transform of the waveform there seems to be a distinct change in the slope of the spectrum from the leading order PN prediction of $-7/6 \approx -1.2$ to somewhere between -0.6 and -0.8 before asymptoting to the dominant ringdown frequency [55, 60].

The ringdown portion of the waveform is dominated by the the fundamental harmonic ($n = 0$) of the quadrupole moment ($\ell = 2, m = 2$) of the final black hole's quasi-normal modes (QNMs) [55, 276]. The first two overtones of the quadrupole mode have amplitudes close to the fundamental mode, though they decay rapidly and are thus only discernible early on during the inspiral. Higher order multiple modes are also present, though as with the waveform itself at a much reduced amplitude compared to the quadrupole mode. An interesting property of the waveform is that from the moment of the peak in the flux onwards it can quite accurately be represented as a sum of QNMs. One reason why this is interesting is that here one would expect to be furthest into the regime where "non-linear effects" are most apparent, yet the wave can be described as coming from a linearly perturbed black hole. Proponents of

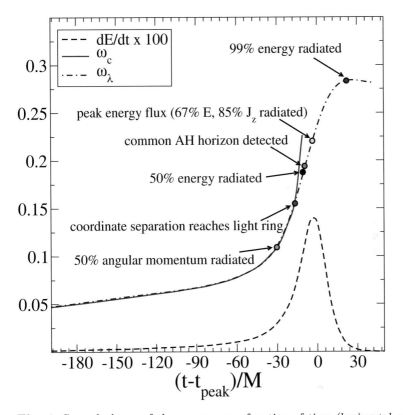

Fig. 4. Several phases of the merger as a function of time (horizontal axis) and orbital/wave angular frequency (vertical axis), from [55]. ω_c is the orbital angular frequency of the apparent horizons in coordinate space (multiplied by the total mass M to non-dimensionalize); this curve ends once a common apparent horizon forms. ω_λ is the instantaneous frequency of the emitted gravitational wave divided by 2 (and normalized by M again). dE/dt is the luminosity of the wave integrated over the wave extraction 2-sphere. J_z is the component of the angular momentum of the gravitational waves normal to the orbital plane. The "light ring" here is defined as the coordinate location of the unstable equatorial photon orbit of the final Kerr black hole. One cannot define this precisely or unambiguously in the binary spacetime, though it is interesting that the orbital and gravitational wave frequencies decouple roughly at this separation. This is also the time when the EOB approach advocates stitching together the inspiral waveform from resummed PN calculations to a ringdown signal.

the EOB approach predicted this behavior, and in fact have further suggested that with a sufficient number of QNM overtones and harmonics that the entire post-inspiral portion of the waveform may be described as a ringdown. This prescription has been carried out quite successfully for the extreme mass ratio case [62], and a range of non-spinning near equal mass mergers (with mass ratios from 1:1 to 4:1) [59], though may not be as straightforward (or possible at all) for general configurations with spin.

4.2 Unequal Mass, Minimal Eccentricity and Spin

Relaxing the condition of equal mass from the configuration discussed in Sect. 4.1, several qualitative features of the merger and corresponding waveform change [255, 277, 278, 276]. First, the equal mass case maximizes the total energy emitted and also maximizes the final spin of the remnant black hole. To a good approximation the total energy radiated decreases by a factor $(\eta/\eta_1)^2$, and the final spin decreases linearly in η via $a/M_f \approx 0.089 + 2.4\eta$, where the symmetric mass ratio $\eta = q/(1+q)^2$, $\eta_1 = 1/4$, $q = M/m$ with $q \geq 1$, and $M_f = m + M$ [278, 276]. The second difference is that although the quadrupole mode still dominates in the waveform, certain higher multipole modes, in particular the $\ell = 3, |m| = 3$ component, become non-negligible [276]. The simple explanation for this is quadrupole-formula physics again: the reduced quadrupole *moment* of the effective source energy distribution now has higher multipole *modes* when expressed in terms of a spherical harmonic decomposition, and this will be reflected in the structure of the gravitational waves emitted. The final significant difference is that there is an asymmetric beaming of the gravitational radiation in the orbital plane due to the mass difference. If not for the inspiral this would average to zero over a complete orbit, however the inspiral, combined with fact that the radiation eventually ceases due to merger, results in the asymmetry. This imparts a "kick", or recoil of the final black hole within the orbital plane to compensate for the net linear momentum carried away by the radiation. The dependence of recoil speed v on mass ratio can be approximated by the Fitchett formula [279] $v = A\eta^2\sqrt{1-4\eta}(1 + B\eta)$, with $A \approx 1.20 \times 10^4$ and $B \approx -0.93$ [278]. The maximum is upwards of $175 km/s$ achieved around a mass ratio $\approx 3{:}1$. Note that the direction of recoil, which in this case occurs within the orbital plane, depends on the "initial" phase of the orbit, and thus for astrophysical sources can be regarded as a uniform random variable.

4.3 Equal Mass, Non-negligible Spin, Minimal Eccentricity

Simulations of binary black hole spacetimes where the initial black holes have spin angular momentum have to date largely focused on equal mass black holes, and with the spin vectors having "non-generic" alignments [280, 281, 282, 283, 284, 285, 286, 287, 288, 289, 290, 291, 292]: either both black holes were given spins aligned and/or anti-aligned with the orbital angular momentum or the spin vectors were set equal in magnitude but opposite in direction and lying within the orbital plane. In all these configurations the net angular momentum vector is aligned with the orbital angular momentum, and thus there will be no precession of the orbital plane during evolution (this, ignoring radiation-reaction, is a consequence of conservation of angular momentum, which incidentally is also why precession *does* occur in cases where the orbital and net angular momentum are misaligned). A couple of exceptional studies examining more generic spin configurations have been presented in [282, 286].

There have not yet been the kinds of detailed or systematic studies of inspiral with spin as described for the non-spinning case in Sect. 4.1 in terms of understanding the multipole structure of the waves, comparison with PN inspiral waveforms, extraction of the QNMs, etc. Though at least one can describe certain qualitative features of the merger. Also, one of the more sought-after answers has been to the question of what the range of magnitudes of the recoil velocity are when spin is included, and many of the above cited papers have recently addressed this. In the next section we will outline the basics of what changes during merger with spin, and the section following that will describe the recoil results.

Qualitative Features of a Merger of Spinning Black Holes

When the black holes are given spin, several aspects of the merger are changed compared to the non-spinning, equal mass case. First, the net amount of energy/angular momentum radiated can change significantly, and consequently the final spin and mass of the remnant black hole. If the component of the net spin in the direction of the orbital angular momentum has the same (opposite) sign as the orbital angular momentum, then typically more (less) energy and angular momentum will be radiated compared to the non-spinning case. One explanation for this comes from the PN description of the spin–orbit interaction (see for example [293]), where in the aligned (anti-aligned) case this interaction term results in a repulsive (attractive) force between the black holes, thus causing them to orbit for a longer (shorter) amount of time emitting more (less) net radiation before merger. As an example, [280] (see also [285]) considered the merger of two equal mass black holes with spin parameters $a = 0.76$; for the case where the two spin vectors were aligned with the orbital angular momentum $\approx 6.7\%$ of the rest-mass energy was radiated, leaving a black hole with a spin of ≈ 0.89, whereas in the anti-aligned case $\approx 2.2\%$ energy was emitted, and the final black hole had a spin of only ≈ 0.44 (in the direction of the orbital angular momentum). Components of spin in the orbital plane have a much smaller effect on the dynamics of the orbit, and consequently the amount of energy emitted; for example, the configurations that result in the largest recoil velocities described in the next section are equal mass, have zero-net spin angular momentum with the spin vectors lying in the orbital plane, and in this case the total energy and angular momentum radiated is very close to the amount for the equivalent non-spinning case [283, 289].

A second significant effect of spin in a merger is that the spin vectors and orbital angular momentum vector will in general precess, and near the time of merger by potentially large enough amounts to cause spin and orbital plane "flips". In PN-terms this can be thought of as due to spin-spin and spin–orbit interactions [293]. A more Newtonian way of thinking about these interactions is that a spinning black hole effectively has a quadrupole moment, and thus the exterior gravitational field of the second black hole will in general exert a torque on the first black hole (and vice-versa), causing precession of the spins, and hence the orbital plane to conserve angular momentum (ignoring radiation). The only full numerical study to date of these effects were presented in [286, 282], though it will certainly not be long before more systematic studies are available from several groups.

Recoil Velocities

Any property of an orbit resulting in an asymmetric beaming pattern for the gravitational waves could, via conservation of linear momentum, impart a kick to the remnant black hole. As discussed in Sect. 4.2 unequal masses produce such an asymmetry, and so can individual black hole spins. An obvious example is the asymmetry that would be produced by precession of the orbital plane, and if the precession time scale is shorter than the orbital and inspiral time scales (which it is near merger) then there will not be enough time to average the momentum beamed in any one direction to zero before merger, thus resulting in a net momentum flux in some direction. For near-equal mass mergers this can produce larger kicks that an unequal mass ratio alone – typically around several hundred km s^{-1}. A less obvious source of asymmetry, though one resulting in the largest kicks of up to 4000 km s^{-1} [283, 289, 286, 288, 292], are equal mass black holes with equal but opposite spin vectors lying within the orbital plane. At a first glance this is a rather surprising configuration for producing a large recoil, as it is not obvious where the asymmetry in the energy emission is. Thus it will be instructive to spend a bit more time describing this configuration in the next couple of paragraphs, and see how the kick can be understood as a frame dragging (or gravitomagnetic) effect. More technical explanations of the source of the kick can be found in [292, 291]. A discussion of the astrophysical implications of such large recoil velocities is deferred to Sect. 5.

Consider the orbit depicted in Fig. 5, and imagine what the effect of rotation of black hole 2 on the motion of black hole 1 would be. The rotation of black hole 2 causes spacetime to be "dragged" about it following the right hand rule: grasp the spin vector with your right hand so that the thumb points in the direction of spin; then the direction in which the rest of your fingers curl about the axis indicates how spacetime is whirled about due to the spin of the black hole. Using this image, notice that at phases (A) and (C) within the orbit black hole 2 can not impart any effective velocity to black hole 1. However, at phase (B) the dragging of the spacetime caused by black hole 2 will cause black hole 1 to move in the negative z direction, where z is in the direction of the orbital angular momentum (i.e. it will move into the paper in the illustration), and the opposite at phase (D). The same analysis of the effect of the rotation of black hole 1 on black hole 2 shows that with this particular configuration of spins *both* black holes will at each instant have the *same* velocity induced in the direction normal to the orbital plane by the other black hole. In other words, one can think of this as causing the entire orbital plane to oscillate normal to the plane with the orbital frequency, or equivalently, the trajectory of each orbit will be tilted by equal but opposite angles relative to the original orbital plane. This normal-motion by itself does not produce much radiation, however, it does cause the more copious amounts of radiation caused by the circular orbital motion to get blue-shifted in one direction normal to the orbital plane while at the same time being red-shifted in the other. Averaged over one orbit, and ignoring radiation reaction, the net Doppler shift in any one direction is zero. However, as the orbital radius begins to shrink due to radiation reaction, the flux and the magnitude of the Doppler shift increases until about the time of merger. Up until this time the net momentum radiated in the z direction will be a function depending sinusoidally on the phase of the orbit, and slowly increasing in amplitude. Depending on where in the orbit the merger occurs ultimately determines the magnitude and direction of the kick normal to the plane.

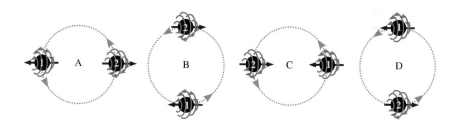

Fig. 5. A depiction of the orbital configuration resulting in the largest kick velocities. The orbital angular momentum points out of the paper in this case, and the spin vectors for each black hole is in the orbital plane within the paper as shown by the solid blue vectors. The grey curved lines illustrate the dragging of spacetime about the black hole caused by its spin. SEE COLOUR PLATES

Of course, the structure of spacetime in the vicinity of two spinning black holes just about to merger will be quite non-trivial, preventing one from unambiguously localizing the positions of the black holes or, when and where the radiation is being produced, so the preceding description of the production of a kick is somewhat cartoonish. However, one can apply it at face value to

come up with an order of magnitude estimate for the kick which is in the
correct ball park, as well as account for the linear dependence of the kick
on the spin a of the black holes and sinusoidal dependence on initial orbital
phase, as follows. When far apart, at any instant in time black hole 1 (2) will
not have any component of orbital angular momentum in the direction of the
spin axis of black 2 (1). Thus, one can approximate the instantaneous velocity
imparted by frame dragging as that which a particle, dropped from rest at
infinity, would have falling toward the black hole. This is a particle on a zero
angular momentum orbit, which will have an instantaneous z velocity of

$$v_z(r,\theta) \approx \frac{2ma\sin(\theta)}{r^2} \qquad (47)$$

where θ is the angle relative to the spin axis of the black hole with spin parameter a and (total) mass m, and r is the distance to the black hole. We
have used the Boyer–Lindquist form of the Kerr metric in the above, and only
kept the term to leading order in r. From this expression one immediately
sees where the linear dependence in a and sinusoidal dependence on the phase
arises. To estimate the maximum possible kick velocity, note that this would
be produced if the Doppler-shifting of the radiation ceases at maximum velocity, i.e. when $\sin(\theta) = 1$. Assume that this occurs when the black holes merge,
and this happens when the two black holes "touch", so when $r \approx 4m$. Thus,

$$v_{zmax} \approx \frac{j}{8}, \qquad (48)$$

where $j \equiv a/m$ is the dimensionless spin parameter. The energy density e of
a gravitational wave is proportional to the square of the wave frequency; thus
the Doppler shifted energy density will be proportional to $(1\pm2v_{zmax})e$, and so
the net momentum density radiated at this moment will be $\delta p = 4v_{zmax}e$. This
accumulates over the last part of an orbit, during which a total of $E = \epsilon M$
($M = 2m$) of energy is emitted in gravitational waves. Therefore the net
momentum radiated in z would be $\delta P = 4v_{zmax}E$, giving an estimate for the
maximum recoil speed $\delta P/M$ as

$$v_{recoil,max} \approx \frac{j\epsilon}{2}. \qquad (49)$$

For a concrete number, take $\epsilon \approx 0.01$ (which is not too unreasonable given that
the net energy emitted in the last orbit/merger/ringdown is around 0.035),
then for an extremal ($j = 1$) black hole this gives $v_{recoil,max} \approx 1500$ km s^{-1}.

Note that a similar line of argument can give an intuitive understanding
of the effective repulsive (attractive) force between binaries with spin axis
aligned (anti-aligned) to the orbital angular momentum, again due to frame
dragging. For empirical formulas giving the net recoil for various spin and
mass configurations see [288, 289].

4.4 Equal Mass, Large Eccentricity, Minimal Spin

An equal mass, zero spin but sizable eccentricity case was in fact the first complete merger event simulated in [97]. Adding eccentricity to the orbit was not intentional, rather this was an artifact of the initial data method, which is as follows. Boosted, highly compact concentrations of scalar field energy are chosen for the initial conditions, which then quickly undergo gravitational collapse and form black holes. Any remnant scalar field energy quickly accretes into the black holes or radiates away from the vicinity of the orbit, leaving behind, for all intents and purposes, a vacuum black hole binary spacetime. For a given initial separation of the scalar field pulses, a single boost parameter k controls the initial data – one scalar field pulse is placed at $(x, y, z) = (d, 0, 0)$ and given a boost $k^i = (0, k, 0)$, while the second is placed at $(-d, 0, 0)$ and given a boost $(0, -k, 0)$. It turns out for sufficiently close separation (as used in the simulations) the resultant black hole binary has non-negligible eccentricity regardless of k. Furthermore, probably due to the scalar field dynamics and accretion, the effective vacuum binary black hole orbit that could be ascribed to the black holes has an apoapsis much further out that the initial scalar field pulses. The consequences are that for the smaller values of k which result in strong interaction of the black holes early on, the black holes have much more kinetic energy than what black holes on a slow, adiabatic inspiral at the same separation would have. This offers an explanation for why the interesting threshold "zoom-whirl" behavior [294, 295] explained in the next paragraph could be observed using this class of initial data, though at the time this was puzzling as it was (incorrectly) assumed the binary was in the adiabatic inspiral regime where any "radiation-reaction" effects would always force the binaries to be closer on average from one orbit to the next.

Imagine what should happen as the boost parameter k is varied. At one extreme, $k = 0$, there will be a head on collision; at the other, $k \approx 1$, the black hole trajectories will be deflected by some amount, though ultimately they will fly apart and separate. At intermediate values of k there should be a significant amount of close-interaction of the black holes, and then they will either merger or separate (to possibly merge at some time in the future). What appears to happen near the threshold value of k between these two distinct end-states is the black holes evolve toward an *unstable near-circular orbit*, remain in that configuration for an amount of time sensitively related to the initial conditions, then either plunge toward coalescence or separate. Specifically, for the *single* class of initial conditions examined in [95, 96], the number of orbits n observed near threshold is found to scale as

$$e^n \propto |k - k^*|^{-\gamma} \qquad (50)$$

with $\gamma \approx 0.34 \pm 0.02$ – see Figs. 6 and 7 that depict this scaling relation for cases that merge, near-threshold orbits, a sample of the gravitational waves and the energy emitted energy as a function of k. Note that due to energy loss via gravitational radiation the threshold cannot be "sharp", i.e. if the time $t_m(k)$

to merger is plotted as a function of k, this will *not* have a discontinuous step at $k = k^*$. There will be a maximum number of orbits N for a given class of initial conditions, and what from a distance might appear like a step function will be resolved into a smooth transition over a region of size $\delta k \approx e^{-N/\gamma}$. Also note that the initial conditions need to be highly fine tuned to obtain even a few whirl-orbits. Thus, when close encounters of near equal mass black holes on hyperbolic or highly eccentric orbits occur in nature (which might occasionally happen in a dense environment such as a globular cluster), it is highly unlikely that it will be a near-immediate-threshold encounter.

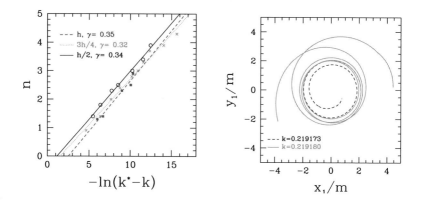

Fig. 6. Left: the number of orbits n versus logarithmic distance of the initial boost parameter k from the immediate merger threshold k^*, for evolutions that did result in a merger. Results from three resolutions are plotted with characteristic mesh spacings h (lowest resolution), $3h/4$ and $h/2$ (highest) to schematically illustrate convergence. For each resolution, a least-squares fit to the data is shown assuming the relation (50). Right: plots of the orbital motion from the two higher resolution simulations ($h/2$) tuned closest to threshold (only the coordinate motion of a single black hole – initially at positive x – is shown for clarity). The dashed curve is the case resulting in a merger, and the curve ends once a common apparent horizon is first detected, while for the solid curve the black holes separate again and here the curve ends when the simulation was stopped.

The behavior just described is very similar to that of equatorial *geodesic* motion on a Kerr background, where geodesics near the threshold of capture approach the unstable circular geodesic orbits of the background spacetime, and exhibit similar scaling behavior of the number of orbits versus distance from threshold as in (50). In that case, the scaling exponent γ is inversely proportional to the Lyapunov (or instability) exponent of the corresponding unstable circular geodesic [296], and in fact numerically has a value quite similar to that in the analogous equal mass scenario; for more information see the discussion in [96]. In contrast to near-equal mass binary black hole encounters in the universe, extreme mass ratio inspirals of a compact object

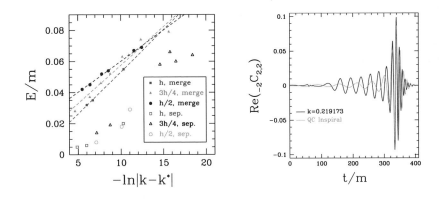

Fig. 7. Left: the total energy radiated in gravitational waves plotted as a function of logarithmic distance from the immediate merger threshold. Data from both super and sub critical cases are shown (and from each of the three characteristic resolutions run), though for clarity only the former have added regression lines. Right: the gravitational waves emitted during a merger event. The real part of the dominant spin weight -2, $\ell = 2$, $m = 2$ spherical harmonic component of Ψ_4 is shown, and for interest the corresponding representation of the signal from the quasi-circular inspiral simulation depicted in Fig. 3 is also shown, time-and-phase-shifted so that the waveforms match at peak amplitude.

into a supermassive black hole are expected to be numerous enough that a significant number of zoom-whirl type orbits will be seen with LISA [51, 52].

4.5 Generic

To date, there have not been any published systematic numerical studies of fully generic initial binary conditions, namely with varying mass ratios, spin magnitudes and orientations, and orbital eccentricity. The reason is simply that this field is still young, though given the rapid rate at which new results have been released over the past couple of years it should not be long before a rather detailed knowledge of a large class of astrophysically relevant merger spacetimes is available.

5 Implications, Prospects and Questions

This chapter concludes with a discussion of some of the implications of current results from the newly uncovered merger phase of the two-body problem and what questions still need to be addressed. As before, the discussion is broken up into the rest-mass dominated regime of relevance to astrophysical black holes, and the kinetic energy dominated regime of the black hole scattering problem.

5.1 Black Holes in our Universe

Even though the merger phase has not yet presented any unexpected or bizarre phenomenology, that there are finally concrete numbers and waveforms associated with an ever growing set of initial conditions allows many consequences of the merger to be seriously explored. The key numbers are the amount of energy and net momentum lost to gravitational waves, and knowledge of the structure of the waves gives data analysts the information to build trust-worthy template banks. Note that the topics discussed in the next several sections hardly exhaust all the possible consequences and applications of black hole mergers, and certainly much of the future work on the "two-body" problem in general relativity will include a thorough examination, numerical and otherwise, of mergers in astrophysical environments.

Consequences of Radiated Energy

An equal mass, non-spinning merger releases close to $\epsilon = 4\%$ of the rest-mass energy of the system into gravitational waves in the last plunge/orbit, merger and ringdown. With spins, depending on the relative alignment, this can increase or decrease by roughly a factor of two. For an unequal mass system with mass ratio $q = M/m$ ($q \geq 1$) the energy will be reduced by slightly less than a factor of q for small q, though approaching a factor of q^2 for large q (see Sect. 4.2). Thus for a "major merger", with q a few or less, a significant amount of the total gravitating energy of the system is effectively lost on a very short time-scale. To get an idea of just how short, define the *light crossing frequency* f_{lc} of a system to be the frequency at which light could cross back and forth between the black holes separated by a distance R, i.e. $f_{lc} = c/2R$. Converting to the units given for the orbital frequency for an equal mass merger in (3), this is

$$ f_{lc} \approx 51\,\text{kHz} \left(\frac{R_s}{R}\right)\left(\frac{M_\odot}{M}\right). \tag{51} $$

Note that this is the *fastest* frequency at which any causal process in the close vicinity of the binary could operate, and by comparison with (3) one can see that near coalescence ($R \to R_s$) the orbital frequency becomes a sizable fraction of this maximum possible frequency. The time-scale over which the final burst of energy is released will therefore be much shorter that almost any other astrophysical process that could be happening close to the binary. A likely non-vacuum environment for a binary is a circumbinary gas disk. Thus, a near-term effect of the passing gravitational waves on the particles in the disk is that essentially instantaneously the central mass they are orbiting will drop by a fraction ϵ [297]; said another way, if they were initially following circular orbits in a potential with mass M, they will suddenly be on eccentric orbits about a potential of mass $M(1-\epsilon)$. Such a rapid perturbation of the disk could set up asymmetric waves and warping in the disk which could conceivably produce a weak but prompt electromagnetic counterpart to the merger event, though no detailed calculations of this have yet been performed. A related phenomena accompanies the secular evolution of the disk as the inspiral and merger occurs [297]. Early on in the inspiral phase the viscous timescale in the disk is shorter than the inspiral rate, allowing the inner edge of the disk with radius roughly twice the binary semi-major axis [298, 299] to "follow" the inspiral. However, eventually the inspiral time becomes much shorter than the viscous time, leading to an essentially non-accreting disk about the final black hole that extends much further than a steady-state accretion disk would (the innermost stable circular orbit). The subsequent inward migration of the disk and turn-on of accretion will produce a strong X-ray afterglow on a timescale of $\approx 7(1 + z)(M/10^6 M_\odot)^{1.32}$ yr (with z the cosmological redshift) that could be seen by future X-ray observatories [297, 300].

Consequences of Radiated Momentum

One of the more significant potential astrophysical consequences of a merger is when asymmetric radiation of linear momentum occurs, resulting in the recoil of the remnant black hole as discussed in Sect. 4. The largest recoil speeds of several thousand kilometers per second for near-equal mass mergers are high enough to be able to eject the remnant from even the most massive galactic halo. If such large kicks are common, it would seem to be in contradiction with the observation of supermassive blackholes in most galaxies, and hierarchical structure formation scenarios [2, 3, 39, 40]. Note that the large recoils require each black hole to be spinning by a fair amount ($a >\approx 0.3$) – X-ray observations of relativistic line broadening in a few AGN suggest spins *are* high, with $a > 0.9$ [304, 305, 306, 307]. As with the effects of energy loss discussed in the previous section, large recoils would also require major mergers, as the kick scales roughly as $1/q^2$ for large mass ratio q [288, 289]. Recent estimates using the effective-one-body model, calibrated with some full numerical results, suggest that for mergers with spin magnitudes of $a = 0.9$, *uniformly* distributed spin configurations and mass ratios $1 \leq q \leq 10$, only about 3% (10%) of mergers result in kicks greater that 1000 km s−1 (500) km s^{-1} [308]. Small kicks could still eject black holes from galaxies in the early universe when halos were much less massive, though the presence of supermassive black holes at the centers of most galaxies might be a more robust consequence of structure formation than naively thought in light of kicks. One study examined the effect of natal kicks in a scenario where supermassive black holes are formed in a primary halo via capture of intermediate mass black holes from surrounding secondary halos, and found that kick velocities imparted to mergers of the intermediate mass black holes had little affect on the growth of the supermassive black hole [301]. Another study following black hole formation through a simplified merger tree model found that even when a high probability of ejection was assigned to merger events, still more than 50% of galaxies today retain their supermassive black holes [309]. An examination of the effect of large recoil velocities on the predicted event rates for LISA to detect supermassive mergers suggested the event rate would drop by *at most* 60% if the seed black holes were light ($\approx 10^2 M_\odot$, from Population III stars) or by at most 15% if the seeds were heavier ($\approx 10^4 M_\odot$, from direct gas collapse of primordial disks)[16] [302].

The study [308] mentioned in the preceding paragraph on the probability of kicks assumed a uniform probability distribution for the spin orientation in the progenitor binary system. However, the distribution of initial spins is likely highly non-uniform – [303] have shown that in gas-rich mergers, torques from the surrounding gas will tend to align the black hole's spin vectors with the net angular momentum vector of the gas, a configuration which results in much more modest recoils of < 200 km s^{-1}. Thus, only a small fraction of supermassive mergers would result in the black hole leaving the host galaxy. For those that do, there could be significant electromagnetic counterparts due to the recoil, as the black hole will drag the inner part (tens of thousands of Schwarzschild radii) of any accretion disk with it [310]. However, given that such recoils would preferentially occur in gas-*poor* environments, accretion-related counterparts might also be correspondingly dim. A search for Doppler shifted emission lines from quasars in the Sloan Digital Sky Survey, which would be a signature of an ejected accretion disk, placed upper limits of incidence of recoiling black holes in quasars at 4% (0.35%) for kicks greater that 500 km s^{-1} (1000) km s^{-1} in the line-of-sight [311]. Note that similar arguments [303] also place doubt on a common explanation that X-shaped jets from radio-loud AGN are the result of spin re-alignment from a recent merger event [312, 313, 314, 315, 316, 317]. Kicks of smaller velocities that temporarily displace the black hole from the galactic center could also have interesting consequences, since this will transfer energy to stars in the nucleus, softening a steep density cusp [318]. Also, modest recoil velocities will have a pronounced effect on the black hole population of globular clusters, effectively depleting the clusters of a large fraction of their black holes and leading to a "rogue" population of wandering black holes in the galactic halo [49, 38, 319].

Implication of Waveform Structure for Detection Efforts

The relative simplicity of the merger waveform, assuming the trend in new results continues and no complicated and lengthy structures in the merger phase occur generically, is on the one hand a boon for gravitational wave astronomy and on the other hand a curse. On the positive side is that the simple transition from inspiral to ringdown should allow the construction of high fidelity hybrid or fully analytic template banks, such as recently presented in [59]. This will ensure, if the circumstances of the sources are consistent with the

---

[16] In both cases configurations were assumed to be most favorable for large kicks, which is probably a significant overestimate.

assumptions of the models, that the waves from the highest possible fraction of events passing earth during the operation of the various instruments will be detected. The downside is that, the less structure and shorter the waves, the more difficult it will be to discriminate between different events, and the less confidence with which one could claim observation of a *particular* event, statistics of source populations, etc. Indeed, in [60, 57] it was shown that high fitting factors can easily be achieved between a numerical source model and some member from a "wrong" template family. Note that this problem is only a significant issue for sources where the part of the waveform dominating the SNR comes from the last few cycles of inspiral, merger and ringdown, i.e. essentially the final burst (for LIGO this will be the merger of tens to hundreds of solar mass black holes, for LISA around 10^7–$10^8 M_\odot$ supermassive black holes). When the inspiral portion of the event is visible to detectors it could be in band for hundreds to thousands of cycles, and in that case even small differences in the phase evolution relative to a given template could drastically affect the SNR. One way to deal with this problem for burst-like sources is to use a small, core template family to search for a given source, then have an expanded control group of template families with systematic deviations from the core family that will be used to place confidence levels and/or error-bars on any conclusions reached with the core family. For example, say one wants to test the hypothesis that all stellar mass mergers occur in environments where the orbits have circularized well before the time of merger. The core template family for this search would thus be a set of zero-eccentricity inspiral events, and the control group would be a set of similar inspirals with eccentricity. The point here is not to go into details of data analysis, rather it is to emphasize how important it will be to investigate non-standard, unexpected or unusual scenarios, not just for the hope of a serendipitous detection of a surprising event, but to strengthen the science that could be done with the usual and more mundane detections. This is perhaps most true for what will be the first triumph of detection of a binary black hole merger – confirmation of the *existence* of black holes. It is easy to lapse into a mental image of a black hole as this dark, concrete object with a surface, rather than *the boundary demarking a region where the geometric structure of spacetime is undergoing gravitational collapse – an intrinsically dynamical regime where space and time itself are funneled to a singular state outside of the grasp of contemporary physical theories.* To claim that such a remarkable scenario exists in the universe demands strong evidence, and part of this evidence would be being able to quantify exactly how distinctive the signatures of mergers of binary black holes within Einstein's theory of gravity are. For example, could compact boson star [320, 321, 322], "gravistar" [323], or other exotic object mergers produce inspiral signals that would be detected using a black hole inspiral template family? How different would a metric theory of gravity have to be to produce observable differences from Einstein's theory (yet be consistent in the weak field)? Could merger events contain the signatures of certain extra-dimension scenarios? The list of such questions is endless, though a reasonable subset will need to be addressed if only to bolster our confidence about what possible future detections could tell us about general relativity and how accurately it describes spacetime.

5.2 The Black Hole Scattering Problem

There is no known natural mechanism in the universe that can accelerate black holes to ultra-relativistic velocities, and hence the black hole scattering problem is largely a thought experiment that can probe a very interesting regime of Einstein's theory. However, over the past few years several ideas have emerged suggesting that an understanding of this problem could have relevance to high-energy particle physics experiments – this will briefly be discussed in the following section. As with rest-mass dominated collisions until recently, full solutions to the metrics describing ultra-relativistic collisions have eluded analytic attempts to obtain them. Most of what is known about this regime comes from studies of the collision of infinite-boosted Schwarzschild black holes, each described by the Aichelberg–Sexl metric [63]. In an impact with zero or small impact parameter, an apparent horizon has been found at the moment of impact [84, 85, 86, 87, 88, 89, 90, 91, 92, 93]; this is not a trivial statement, as the Aichelberg-Sexl metric does not contain an event horizon, and in this extreme case one might imagine that a naked singularity would form, in particular by drawing parallels to collisions of infinite plane gravitational waves [35]. For zero impact parameter, perturbative studies [85, 86, 87, 88] have given a description of the gravitational waves released in the process. Now, as is turning out with merger simulations in the rest-mass dominated regime, it may be that full (presumably numerical) solutions of the scattering problem will only add some details to our understanding of the process, though of course this will not be known until the solutions are discovered. Furthermore, as described in Sect. 2.1, the threshold of immediate merger could have very interesting behavior associated with it, where essentially all of the energy in the spacetime is converted to gravitational waves. Note that in the infinite boost limit the threshold of immediate merger also corresponds to a threshold of black hole *formation*. If, as conjectured by Choptuik [28], the threshold of black hole formation has a universal solution, then at critical impact parameter the structure of spacetime should be described by the Abrahams-Evans axisymmetric critical collapse vacuum solution [324].

The black hole scattering problem will be difficult to simulate numerically. First of all, it is unclear whether generalized harmonic or BSSN evolutions could be used without modification in this regime. Attempted evolution of boosted exact Schwarzschild solutions with Lorentz γ-factors above around 1.7 with the generalized harmonic code used in [228] suggested that at the very least new gauge conditions will be needed for long-term stable evolution. The BSSN code in [256] has been used to evolve boosted Bowen–York black holes with somewhat larger γ factors, though such initial data contains gravitational waves, and apparently a considerable amount with larger boosts [325].

A further issue with evolving highly boosted black holes is the tremendous computational resources that would be required. Consider a single black hole with total energy E, which will be the characteristic length scale of the geometrically non-trivial portion of the spacetime. The black hole would have a rest mass of E/γ, which would be the smallest length scale in the direction transverse to the boost direction. Length contraction in the direction of the boost will compress the horizon by an additional factor of γ in that direction. Furthermore, using a "naive" boost of the Schwarzschild solution, certain components of the metric get scaled by factors of γ^2. Thus, in all, to obtain a numerical solution with a grid-based method will require a mesh spacing a factor of γ^4 smaller than a similar rest-mass dominated problem and obtain similar accuracy. To be well within the kinetic energy dominated regime would require $\gamma \approx 10\text{--}100$, implying around $10^4\text{--}10^8$ times the computational resources. Of course, this is rather simplistic accounting, and certainly with some ingenuity a couple of orders of magnitude could be shaved off the estimated cost.

5.3 High Energy Particle Experiments and Black Hole Collisions

Recently proposed extra-dimension scenarios [326, 327], offer the intriguing possibility that the Planck scale could be within reach of energies attainable by the Large Hadron Collider (LHC) [329, 328, 330, 331, 332, 333]. This implies that the LHC may be able to probe the quantum gravity regime, and that black holes could be produced in substantial quantities by the particle collisions. Similarly, cosmic ray collisions with the earth would produce black holes [334], and this may be detected with current or near-future cosmic ray experiments [335]. In the collision of two particles with super-Planck kinetic energies, gravity dominates the interaction, and thus to a good approximation the collision can be modeled as the ultra-relativistic collision of two black holes [17]. Another intriguing application of ultra-relativistic black hole collisions is in five-dimensional AdS spacetime, and how that might relate to the collision of gold ions at the Relativistic Heavy Ion Collider (RHIC). At RHIC, gold ions are accelerated to Lorentz gamma factors of around 100 before colliding. It is believed that during the collision a quark-gluon plasma (QGP) is formed. The present data supports this idea [336, 337, 338, 339] though there are some puzzles, in particular why the QGP is strongly interacting, behaving almost like an ideal fluid (the energies of RHIC collisions are in the regime where the asymptotic freedom of QCD should be manifest, implying one should have a weakly interacting QGP). One suggested method for deriving properties of the QGP is via the AdS/CFT correspondence of string theory [340, 341, 342]. Specifically, the supposition is that $N = 4$ super-Yang-Mills (SYM) theory at strong coupling, though different in many respects from QCD, can describe

---

[17] Though without a full theory of the physical laws that would operate in this regime, such statements are a bit hand-waving.

some of the features of a "real-world" QGP, and that a practical way of calculating the relevant SYM state is using the AdS/CFT map applied to the corresponding process in 5D AdS spacetime. It has been suggested that the AdS equivalent process is the collision of two black holes [343], and in [344] the quasinormal ringing of a perturbed 5D AdS black hole, which represents the final stage of a black hole collision and may describe aspects of thermalization and collective flow of the QGP, was used to provide in-the-ballpark estimates of the thermalization time and elliptic flow coefficients of an anisotropic heavy ion collision.

Thus there is considerable motivation to study black hole collisions in higher dimensional spacetimes, and in spacetimes with different asymptotic structures in regimes where the asymptotics are expected to affect the physics of the collision (in particular for collisions in AdS the black holes need to be "large" in terms of the length scale imparted by the cosmological constant). Full five (and higher) dimensional numerical simulations of collisions, in particular ultra-relativistic ones, will require computers several generations more powerful than current ones. However, if the analogy between geodesic behavior and the full problem at the threshold of immediate merger described in Sect. 4.4 holds, then for an application to the LHC all that would be needed is a head-on collision simulation, which could be reduced to an 2D simulation[18]. Furthermore, if the threshold geodesic behavior of Myers–Perry solutions [345] are an adequate description of the analogous problem in higher dimensions, it turns out that to a good approximation for dimensions greater than 5 the energy emitted in an ultra-relativistic collision will be given by [346]

$$E(b) = E_0 \left(\Theta(b) - \Theta(b - b^*) \right), \tag{52}$$

where $\Theta(b)$ is the unit step function, b the impact parameter, b^* the critical impact parameter for the geodesic (which is close to the Schwarzschild radius of the equivalent black hole), and E_0 the energy emitted during the head-on collision (estimates of which can be found in [92], for example). In other words, when black holes form, as much energy is lost to gravitational waves as in the head-on collision case, regardless of the impact parameter. This "missing energy", in addition to Hawking radiation emitted when the black holes evaporate, could be used to detect this hypothesized scenario at the LHC.

6 Conclusion

The two-body problem in general relativity is a fascinating, rich problem that is just beginning to be fully revealed by recent breakthroughs in numerical relativity. At the same time, a new generation of gravitational wave detectors

---

[18] Of course, the "catch-22" here is that without doing the full n-dimensional problem it will not be known whether the analogy holds

promise to offer us a view of the universe via the gravitational wave spectrum for the first time. Black hole mergers are a promising source for gravitational waves, and detecting them would provide direct evidence for these remarkable objects, while providing much information about their environments. Suggestions that there might be more than four spacetime dimensions offers the astonishing possibility that black holes could be produced by proton collisions at TeV energies, which will be reached by the Large Hadron Collider, planned to begin operation within a year. Given all this, it is difficult not to be excited about what might be learnt about the universe from the smallest to largest scales during the next decade, and that black hole collisions could have something important to say at both extremes.

Acknowledgments: I would like to thank Alessandra Buonanno, Matthew Choptuik, Gregory Cook, Charles Gammie, David Garfinkle, Steven Gubser, Carsten Gundlach, Luis Lehner, Jeremiah Ostriker, Don Page, David Spergel and Ulrich Sperhake for many stimulating conversations related to some of the discussion presented here.

References

1. R. Narayan, "Black Holes in Astrophysics," New J. Phys. **7**, 199 (2005)
2. M. J. Rees, "Black Hole Models for Active Galactic Nuclei," Ann. Rev. Astron. Astrophys. **22**, 471 (1984).
3. L. Ferrarese and H. Ford, "Supermassive Black Holes in Galactic Nuclei: Past, Present and Future Research," Space Science Reviews, Volume 116, Issue 3-4, pp. 523-624, arXiv:astro-ph/0411247.
4. J. E. McClintock and R. A. Remillard, "Black Hole Binaries," Compact Stellar X-ray Sources," eds. W.H.G. Lewin and M. van der Klis, Cambridge University Press, arXiv:astro-ph/0306213.
5. R. Schodel *et al.*, "A Star in a 15.2 year orbit around the supermassive black hole at the center of the Milky Way," Nature **419**, 694 (2002).
6. A. M. Ghez, S. Salim, S. D. Hornstein, A. Tanner, M. Morris, E. E. Becklin and G. Duchene, "Stellar Orbits Around the Galactic Center Black Hole," Astrophys. J. **620**, 744 (2005)
7. K. Gebhardt *et al.*, "A Relationship Between Nuclear Black Hole Mass and Galaxy Velocity Dispersion," Astrophys. J. **539**, L13 (2000)
8. E. E. Flanagan and S. A. Hughes, "The basics of gravitational wave theory," New J. Phys. **7**, 204 (2005)
9. J. Weber, "Evidence for discovery of gravitational radiation," Phys. Rev. Lett. **22**, 1320 (1969).
10. K. S. Thorne, "Gravitational Wave Research: Current Status And Future Prospects," Rev. Mod. Phys. **52**, 285 (1980).
11. R. A. Hulse and J. H. Taylor, "Discovery of a pulsar in a binary system," Astrophys. J. **195**, L51 (1975).
12. A. Abramovici et al., " LIGO: The Laser interferometer gravitational wave observatory", Science **256**, 325 (1992)

13. B. Caron et al., "The Virgo Interferometer", Class. Quant. Grav. **14**, 1461 (1997)

14. H. Lück et al., "The Geo-600 Project", Class. Quant. Grav. **14**, 1471 (1997)

15. M. Ando et al., "Stable Operation of a 300-m Laser Interferometer with Sufficient Sensitivity to Detect Gravitational Wave Events Within our Galaxy", Phys. Rev. Lett. **86**, 3950 (2001)

16. P. Astone *et al.*, "First Cooling Below 0.1-K Of The New Gravitational Wave Antenna 'Nautilus' Of The Rome Group," Europhys. Lett. **16**, 231 (1991).

17. P. Astone *et al.*, "Long term operation of the Rome 'Explorer' cryogenic gravitational wave detector," Phys. Rev. D **47**, 362 (1993).

18. M. Cerdonio *et al.*, Class. Quant. Grav. **14**, 1491 (1997).

19. E. Mauceli *et al.*, "The ALLEGRO gravitational wave detector: Data acquisition and analysis," Phys. Rev. D **54**, 1264 (1996)

20. D. G. Blair, E. N. Ivanov, M. E. Tobar, P. J. Turner, F. van Kann and I. S. Heng, "High sensitivity gravitational wave antenna with parametric transducer readout," Phys. Rev. Lett. **74**, 1908 (1995).

21. K. Danzmann, "LISA - an ESA cornerstone mission for a gravitational wave observatory," Class. Quant. Grav. **14**, 1399 (1997).

22. M. Maggiore, "Gravitational wave experiments and early universe cosmology," Phys. Rept. **331**, 283 (2000)

23. C. Cutler and K. S. Thorne, "An overview of gravitational-wave sources," arXiv:gr-qc/0204090.

24. S. W. Hawking and R. Penrose, "The Singularities of gravitational collapse and cosmology," Proc. Roy. Soc. Lond. A **314**, 529 (1970).

25. S.W. Hawking and G.F.R. Ellis, G. F. R., *The Large Scale Structure of Space-Time*, Cambridge University Press (1973).

26. B. K. Berger, "Numerical Approaches To Space-Time Singularities," Living Rev. Rel. **1**, 7 (1998).

27. E. Poisson and W. Israel, "Internal structure of black holes," Phys. Rev. D **41**, 1796 (1990).

28. M. W. Choptuik, "Universality And Scaling In Gravitational Collapse Of A Massless Scalar Field," Phys. Rev. Lett. **70**, 9 (1993).

29. C. Gundlach, "Critical phenomena in gravitational collapse - Living Reviews," Living Rev. Rel. **2**, 4 (1999)

30. W. Israel, "Event Horizons In Static Vacuum Space-Times," Phys. Rev. **164**, 1776 (1967).

31. B. Carter, "Axisymmetric Black Hole Has Only Two Degrees of Freedom," Phys. Rev. Lett. **26**, 331 (1971).

32. B. Carter, "Has the black hole equilibrium problem been solved?," arXiv:gr-qc/9712038.

33. A. Ashtekar and B. Krishnan, "Isolated and dynamical horizons and their applications," Living Rev. Rel. **7**, 10 (2004) [arXiv:gr-qc/0407042].

34. R. Arnowitt, S. Deser and C.W. Misner, in *Gravitation: An Introduction to Current Research*, ed. L. Witten, New York, Wiley (1962)

35. K. A. Khan and R. Penrose, "Scattering of two impulsive gravitational plane waves," Nature **229**, 185 (1971).

36. K. Belczynski, V. Kalogera, F. A. Rasio, R. E. Taam and T. Bulik, "On the rarity of double black hole binaries: Consequences for gravitational-wave detection," Astrophys.J. **662**, 504 (2007)

37. S. F. Portegies Zwart and S. McMillan, "Black hole mergers in the universe," Astrophys. J.**528** L17 (2000)

38. R. M. O'Leary, F. A. Rasio, J. M. Fregeau, N. Ivanova and R. O'Shaughnessy, "Binary Mergers and Growth of Black Holes in Dense Star Clusters," Astrophys. J. **637**, 937 (2006)

39. S. D. M. White and M. J. Rees, "Core condensation in heavy halos: A Two stage theory for galaxy formation and clusters," Mon. Not. Roy. Astron. Soc. **183**, 341 (1978).

40. V. Springel *et al.*, "Simulating the joint evolution of quasars, galaxies and their large-scale distribution," Nature **435**, 629 (2005)

41. D. Merritt and M. Milosavljevic, "Massive Black Hole Binary Evolution" Living Rev. Relativity **8**, 8 (2005)

42. M. Colpi, S. Callegari, M. Dotti, S. Kazantzidis and L. Mayer, "On the inspiral of Massive Black Holes in gas-rich galaxy mergers," 2007 STScI Spring Symposium: Black Holes, eds. M. Livio & A. M. Koekemoer, Cambridge University Press arXiv:0706.1851 [astro-ph].

43. P. C. Peters and J. Mathews, "Gravitational radiation from point masses in a Keplerian orbit," Phys. Rev. **131**, 435 (1963).

44. P. C. Peters, "Gravitational Radiation and the Motion of Two Point Masses", Phys. Rev. **136**, B1224 (1964)

45. L. Wen, "On the Eccentricity Distribution of Coalescing Black Hole Binaries Driven by the Kozai Mechanism in Globular Clusters," Astrophys. J. **598**, 419 (2003) [arXiv:astro-ph/0211492].

46. S. F. Portegies Zwart and S. McMillan, "Simulating young star clusters with primordial binaries," arXiv:astro-ph/0411188.

47. K. Gultekin, M. Coleman Miller and D. P. Hamilton, "Three-body dynamics with gravitational wave emission," Astrophys. J. **640**, 156 (2006)

48. I. Mandel, D. A. Brown, J. R. Gair and M. C. Miller, "Rates and Characteristics of Intermediate-Mass-Ratio Inspirals Detectable by Advanced LIGO," Astrophys. J. **681**, 1431 (2008)

49. M. C. Miller and D. P. Hamilton, "Four-body effects in globular cluster black hole coalescence," arXiv:astro-ph/0202298.

50. S. J. Aarseth, "Post-Newtonian N-body simulations," Mon. Not. Roy. Astron. Soc. **378**, 285 (2007) [arXiv:astro-ph/0701612].

51. J. R. Gair, L. Barack, T. Creighton, C. Cutler, S. L. Larson, E. S. Phinney and M. Vallisneri, "Event rate estimates for LISA extreme mass ratio capture sources," Class. Quant. Grav. **21**, S1595 (2004)

52. P. Amaro-Seoane, J. R. Gair, M. Freitag, M. Coleman Miller, I. Mandel, C. J. Cutler and S. Babak, "Astrophysics, detection and science applications of intermediate- and extreme mass-ratio inspirals," arXiv:astro-ph/0703495.

53. L. Blanchet, "Gravitational radiation from post-Newtonian sources and inspiralling compact binaries," Living Rev. Rel. **5**, 3 (2002)

54. A. Buonanno and T. Damour, "Effective one-body approach to general relativistic two-body dynamics," Phys. Rev. D **59**, 084006 (1999)

55. A. Buonanno, G. B. Cook and F. Pretorius, "Inspiral, merger and ring-down of equal-mass black-hole binaries," arXiv:gr-qc/0610122.

56. J. G. Baker, J. R. van Meter, S. T. McWilliams, J. Centrella and B. J. Kelly, "Consistency of post-Newtonian waveforms with numerical relativity," arXiv:gr-qc/0612024.

57. Y. Pan *et al.*, "A data-analysis driven comparison of analytic and numerical coalescing binary waveforms: Nonspinning case," arXiv:0704.1964 [gr-qc].

58. M. Hannam, S. Husa, U. Sperhake, B. Brugmann and J. A. Gonzalez, "Where post-Newtonian and numerical-relativity waveforms meet," arXiv:0706.1305 [gr-qc].

59. A. Buonanno, Y. Pan, J. G. Baker, J. Centrella, B. J. Kelly, S. T. McWilliams and J. R. van Meter, "Toward faithful templates for non-spinning binary black holes using the effective-one-body approach," arXiv:0706.3732 [gr-qc].

60. T. Baumgarte, P. Brady, J. D. E. Creighton, L. Lehner, F. Pretorius and R. De-Voe, "Learning about compact binary merger: The interplay between numerical relativity and gravitational-wave astronomy," arXiv:gr-qc/0612100.

61. A. Buonanno and T. Damour, "Transition from inspiral to plunge in binary black hole coalescences," Phys. Rev. D **62**, 064015 (2000) [arXiv:gr-qc/0001013].

62. T. Damour and A. Nagar, "Faithful Effective-One-Body waveforms of small-mass-ratio coalescing black-hole binaries," arXiv:0705.2519 [gr-qc].

63. P. C. Aichelburg and R. U. Sexl, "On the Gravitational field of a massless particle," Gen. Rel. Grav. **2**, 303 (1971).

64. S. A. Teukolsky, "Perturbations of a rotating black hole. 1. Fundamental equations for gravitational electromagnetic, and neutrino field perturbations," Astrophys. J. **185**, 635 (1973).

65. R. Ruffini and M. Sasaki, "On A Semirelativistic Treatment Of The Gravitational Radiation From A Mass Thrusted Into A Black Hole", Prog.Theor.Phys. **66**, 1627 (1981)

66. T. C. Quinn and R. M. 93, "An axiomatic approach to electromagnetic and gravitational radiation reaction of particles in curved spacetime," Phys. Rev. D **56**, 3381 (1997)

67. Y. Mino, M. Sasaki and T. Tanaka, "Gravitational radiation reaction to a particle motion," Phys. Rev. D **55**, 3457 (1997)

68. S. A. Hughes, "Computing radiation from Kerr black holes: Generalization of the Sasaki-Nakamura equation," Phys. Rev. D **62**, 044029 (2000) [Erratum-ibid. D **67**, 089902 (2003)]

69. K. Glampedakis, S. A. Hughes and D. Kennefick, "Approximating the inspiral of test bodies into Kerr black holes," Phys. Rev. D **66**, 064005 (2002) [arXiv:gr-qc/0205033].

70. K. Glampedakis and S. Babak, "Mapping spacetimes with LISA: Inspiral of a test-body in a 'quasi-Kerr' field," Class. Quant. Grav. **23**, 4167 (2006)

71. J.R. Gair , D.J. Kennefick and S.L. Larson, "Semi-relativistic approximation to gravitational radiation from encounters with black holes", Phys.Rev. **D72**, 084009 (2005), *Erratum-ibid.* **D74**, 109901 (2006)

72. F. D. Ryan, "Gravitational waves from the inspiral of a compact object into a massive, axisymmetric body with arbitrary multipole moments," Phys. Rev. D **52**, 5707 (1995).

73. S. W. Hawking, "Gravitational radiation from colliding black holes," Phys. Rev. Lett. **26**, 1344 (1971).

74. C.V. Vishveshwara, Nature **227**, 936 (1970).

75. M. Davis, R. Ruffini, W. H. Press and R. H. Price, "Gravitational radiation from a particle falling radially into a schwarzschild black hole," Phys. Rev. Lett. **27**, 1466 (1971).

76. W. Press, Astrophys J. Letters **170**, L105 (1971).
77. S. Chandrasekhar and S. Detweiler, Proc. R. Soc. Lond. A **344**, (1975) 441.
78. E. W. Leaver, Phys. Rev. D **34**, 384 (1986).
79. K. D. Kokkotas and B. G. Schmidt, "Quasi-normal modes of stars and black holes," Living Rev. Rel. **2**, 2 (1999)
80. F. Echeverria, Phys. Rev. D **40**, 3194 (1997).
81. E. Berti, V. Cardoso and C. M. Will, "On gravitational-wave spectroscopy of massive black holes with the space interferometer LISA," Phys. Rev. D **73**, 064030 (2006)
82. R. H. Price and J. Pullin, Phys. Rev. Lett. **72**, 3297 (1994) [arXiv:gr-qc/9402039].
83. R. H. Price, "Nonspherical perturbations of relativistic gravitational collapse. 1. Scalar and gravitational perturbations," Phys. Rev. D **5**, 2419 (1972).
84. R.Penrose, *unpublished* (1974)
85. P. D. D'Eath, "High Speed Black Hole Encounters And Gravitational Radiation," Phys. Rev. D **18**, 990 (1978).
86. P. D. D'Eath and P. N. Payne, "Gravitational Radiation In High Speed Black Hole Collisions. 1. Perturbation Treatment Of The Axisymmetric Speed Of Light Collision," Phys. Rev. D **46**, 658 (1992).
87. P. D. D'Eath and P. N. Payne, "Gravitational Radiation In High Speed Black Hole Collisions. 2. Reduction To Two Independent Variables And Calculation Of The Second Order News Function," Phys. Rev. D **46**, 675 (1992).
88. P. D. D'Eath and P. N. Payne, "Gravitational radiation in high speed black hole collisions. 3. Results and conclusions," Phys. Rev. D **46**, 694 (1992).
89. D. M. Eardley and S. B. Giddings, "Classical black hole production in high-energy collisions," Phys. Rev. D **66**, 044011 (2002) [arXiv:gr-qc/0201034].
90. H. Yoshino and Y. Nambu, "Black hole formation in the grazing collision of high-energy particles", Phys.Rev. **D67**, 024009 (2003)
91. E. Berti, M. Cavaglia and L. Gualtieri, " Gravitational energy loss in high-energy particle collisions: Ultrarelativistic plunge into a multidimensional black hole". Phys. Rev. **D69**, 124011 (2004)
92. H. Yoshino and S. Rychkov, "Improved analysis of black hole formation in high-energy particle collisions", Phys.Rev. **D71**, 104028 (2005)
93. V. Cardoso, E. Berti and M. Cavaglia, "What we (don't) know about black hole formation in high-energy collisions", Class.Quant.Grav. **22**, L61-R84 (2005)
94. M.W. Choptuik, *private communication*
95. F. Pretorius, " Simulation of binary black hole spacetimes with a harmonic evolution scheme", Class. Quant. Grav. **23**, S529 (2006)
96. F. Pretorius and D. Khurana, "Black hole mergers and unstable circular orbits," Class. Quant. Grav. **24**, S83 (2007)
97. F. Pretorius, "Evolution of Binary Black Hole Spacetimes", Phys. Rev. Lett. **95**, 121101 (2005)
98. M. Campanelli, C.O. Lousto, P. Marronetti and Y. Zlochower, "Accurate Evolutions of Orbiting Black-Hole Binaries Without Excision", Phys. Rev. Lett. **96**, 111101 (2006)
99. J. G. Baker, J. Centrella, D. Choi, M. Koppitz and J. van Meter, "Gravitational Wave Extraction from an Inspiraling Configuration of Merging Black Holes", Phys. Rev. Lett. **96**, 111102 (2006)
100. O. A. Reula, "Hyperbolic Methods For Einstein's Equations," Living Rev. Rel. **1**, 3 (1998).

101. L. Lehner, "Numerical relativity: A review," Class. Quant. Grav. **18**, R25 (2001)
102. C. Bona, T. Ledvinka, C. Palenzuela and M. Zacek, "General-covariant evolution formalism for Numerical Relativity," Phys. Rev. D **67**, 104005 (2003) [arXiv:gr-qc/0302083].
103. C. Bona, T. Ledvinka, C. Palenzuela and M. Zacek, "General-covariant constraint-free evolution system for Numerical Relativity," arXiv:gr-qc/0209082.
104. C. Gundlach, J. M. Martin-Garcia, G. Calabrese and I. Hinder, "Constraint damping in the Z4 formulation and harmonic gauge," Class. Quant. Grav. **22**, 3767 (2005).
105. S.G.Hahn and R.W. Lindquist, Ann. Phys. **29**, 304 (1964).
106. C.W. Misner, "Wormhole Initial Conditions", Phys. Rev. **118**, 1110 (1960) Ann. Phys. **29**, 304 (1964)
107. L. Smarr, "The Structure of General Relativity with a Numerical Illustration: The Collision of Two Black Holes", *Univ. of Texas at Austin Ph.D. Thesis* (1975)
108. K.R. Eppley, *Princeton Ph.D. Thesis* (1977)
109. L. Smarr, in *Sources of Gravitational Radiation*, ed. L. Smarr, Seattle, Cambridge University Press (1979)
110. P. Anninos, D. Hobill, E. Seidel, L. Smarr and W. Suen, "Collision of Two Black Holes", Phys. Rev. Lett. **71**, 2851 (1993)
111. P. Anninos, K. Camarda, J. Masso, E. Seidel, W. M. Suen, M. Tobias and J. Towns, "3-D numerical relativity at NCSA," arXiv:gr-qc/9412059.
112. P. Anninos, K. Camarda, J. Masso, E. Seidel, W. M. Suen and J. Towns, "Three-dimensional numerical relativity: The Evolution of black holes," Phys. Rev. D **52**, 2059 (1995)
113. J.W. York, Jr., in *Sources of Gravitational Radiation*, ed. L. Smarr, Seattle, Cambridge University Press (1979)
114. J. Winicour, "Characteristic Evolution and Matching," arXiv:gr-qc/0102085.
115. R. Gomez *et al.*, "Stable characteristic evolution of generic 3-dimensional single-black-hole spacetimes," Phys. Rev. Lett. **80**, 3915 (1998)
116. N. T. Bishop, R. Gomez, L. Lehner and J. Winicour, "Cauchy-characteristic extraction in numerical relativity," arXiv:gr-qc/9705033.
117. R. Gomez, R. L. Marsa and J. Winicour, "Black hole excision with matching," Phys. Rev. D **56**, 6310 (1997) [arXiv:gr-qc/9708002].
118. W. Barreto, A. Da Silva, R. Gomez, L. Lehner, L. Rosales and J. Winicour, "The 3-dimensional Einstein-Klein-Gordon system in characteristic numerical relativity," Phys. Rev. D **71**, 064028 (2005)
119. T. Piran, "Numerical Codes for Cylindrical Relativistic Systems", J. Comp. Phys. **35**, 254 (1980)
120. B. Gustafsson, H. Kreiss and J. Oliger, *Time Dependent Problems and Difference Methods*, New York, John-Wiley & Sons, Inc. 1995
121. C. Bona and J. Masso, "Hyperbolic evolution system for numerical relativity", Phys. Rev. Lett. **68**, 1097 (1992)
122. S. Frittelli and O. A. Reula, Commun. Math. Phys. **166** 221 (1994).
123. C. Bona, J. Masso, E. Seidel and J. Stela, "A New formalism for numerical relativity," Phys. Rev. Lett. **75**, 600 (1995)

124. A. Abrahams, A. Anderson, Y. Choquet-Bruhat and J. W. . York, "Einstein And Yang-Mills Theories In Hyperbolic Form Without Gauge Fixing," Phys. Rev. Lett. **75**, 3377 (1995)

125. H. Friedrich, "Hyperbolic Reductions For Einstein's Equations," Class. Quant. Grav. **13**, 1451 (1996).

126. S. Frittelli and O. A. Reula, "First-order symmetric-hyperbolic Einstein equations with arbitrary fixed gauge," Phys. Rev. Lett. **76**, 4667 (1996)

127. A. Abrahams, A. Anderson, Y. Choquet-Bruhat and J. W. . York, "Geometrical hyperbolic systems for general relativity and gauge theories," Class. Quant. Grav. **14**, A9 (1997)

128. Y. Choquet-Bruhat and J. W. . York, "Well Posed Reduced Systems for the Einstein Equations," arXiv:gr-qc/9606001.

129. C. Bona, J. Masso, E. Seidel and J. Stela, "First order hyperbolic formalism for numerical relativity," Phys. Rev. D **56**, 3405 (1997)

130. M. S. Iriondo, E. O. Leguizamon and O. A. Reula, "Einstein's equations in Ashtekar's variables constitute a symmetric hyperbolic system," Phys. Rev. Lett. **79**, 4732 (1997)

131. G. Yoneda and H. a. Shinkai, "Symmetric hyperbolic system in the Ashtekar formulation," Phys. Rev. Lett. **82**, 263 (1999)

132. A. Anderson and J. W. . York, "Fixing Einstein's equations," Phys. Rev. Lett. **82**, 4384 (1999)

133. G. Yoneda and H. a. Shinkai, "'Constructing hyperbolic systems in the Ashtekar formulation of general relativity," Int. J. Mod. Phys. D **9**, 13 (2000)

134. A. Arbona, C. Bona, J. Masso and J. Stela, "Robust evolution system for Numerical Relativity," Phys. Rev. D **60**, 104014 (1999)

135. M. Alcubierre, B. Brugmann, M. A. Miller and W. M. Suen, "A conformal hyperbolic formulation of the Einstein equations," Phys. Rev. D **60**, 064017 (1999)

136. S. Frittelli and O. A. Reula, "Well-posed forms of the 3+1 conformally-decomposed Einstein equations," J. Math. Phys. **40**, 5143 (1999)

137. L. E. Kidder, M. A. Scheel and S. A. Teukolsky, "Extending the lifetime of 3D black hole computations with a new hyperbolic system of evolution equations," Phys. Rev. D **64**, 064017 (2001)

138. H. a. Shinkai and G. Yoneda, "Adjusted ADM systems and their expected stability properties," Class. Quant. Grav. **19**, 1027 (2002)

139. Y. Choquet-Bruhat, "Non strict and strict hyperbolic systems for the Einstein equations," arXiv:gr-qc/0111017.

140. H. Friedrich, "Conformal Einstein evolution," Lect. Notes Phys. **604**, 1 (2002)

141. K. Alvi, "First-order symmetrizable hyperbolic formulations of Einstein's equations including lapse and shift as dynamical fields," Class. Quant. Grav. **19**, 5153 (2002)

142. C. Bona, L. T. and C. Palenzuela, "A 3+1 covariant suite of numerical relativity evolution systems," Phys. Rev. D **66**, 084013 (2002)

143. L. T. Buchman and J. M. Bardeen, "A hyperbolic tetrad formulation of the Einstein equations for numerical relativity," Phys. Rev. D **67**, 084017 (2003) [Erratum-ibid. D **72**, 049903 (2005)]

144. G. Nagy, O. E. Ortiz and O. A. Reula, "Strongly hyperbolic second order Einstein's evolution equations," Phys. Rev. D **70**, 044012 (2004)

145. O. A. Reula, "Strongly hyperbolic systems in General Relativity," arXiv:gr-qc/0403007.

146. G. Y. Chee and Y. Guo, "Symmetric hyperbolic system in self-dual teleparallel gravity," Phys. Rev. D **70**, 044009 (2004).
147. V. Paschalidis, A. Khokhlov and I. D. Novikov, "Well-posed constrained evolution of 3+1 formulations of General Relativity," Phys. Rev. D **75**, 024026 (2007)
148. V. Paschalidis, "Mixed Hyperbolic - Second-Order Parabolic Formulations of General Relativity," rXiv:0704.2861 [gr-qc].
149. B. Brugmann, "Adaptive mesh and geodesically sliced Schwarzschild spacetime in 3+1 dimensions," Phys. Rev. D **54**, 7361 (1996)
150. G. B. Cook *et al.* [Binary Black Hole Challenge Alliance Collaboration], "Boosted three-dimensional black-hole evolutions with singularity excision," Phys. Rev. Lett. **80**, 2512 (1998)
151. M. A. Scheel, T. W. Baumgarte, G. B. Cook, S. L. Shapiro and S. A. Teukolsky, "Numerical evolution of black holes with a hyperbolic formulation of general relativity," Phys. Rev. D **56**, 6320 (1997)
152. C. Bona, J. Masso, E. Seidel and P. Walker, "Three dimensional numerical relativity with a hyperbolic formulation," arXiv:gr-qc/9804052.
153. M. A. Scheel, T. W. Baumgarte, G. B. Cook, S. L. Shapiro and S. A. Teukolsky, "Treating instabilities in a hyperbolic formulation of Einstein's equations," Phys. Rev. D **58**, 044020 (1998)
154. M. Alcubierre, G. Allen, B. Brugmann, E. Seidel and W. M. Suen, "Towards an understanding of the stability properties of the 3+1 evolution equations in general relativity," Phys. Rev. D **62**, 124011 (2000)
155. B. Brugmann, "Numerical relativity in 3+1 dimensions," Annalen Phys. **9**, 227 (2000)
156. B. Brugmann, "Binary Black Hole Mergers in 3d Numerical Relativity," Int. J. Mod. Phys. D **8**, 85 (1999)
157. K. Camarda and E. Seidel, "Numerical evolution of dynamic 3D black holes: Extracting waves," Phys. Rev. D **57**, 3204 (1998)
158. G. Allen, K. Camarda and E. Seidel, "Black hole spectroscopy: Determining waveforms from 3D excited black holes," arXiv:gr-qc/9806036.
159. K. Camarda and E. Seidel, "Three-dimensional simulations of distorted black holes. I. Comparison with axisymmetric results," Phys. Rev. D **59**, 064019 (1999)
160. M. Alcubierre *et al.*, "Towards a stable numerical evolution of strongly gravitating systems in general relativity: The conformal treatments," Phys. Rev. D **62**, 044034 (2000)
161. M. Alcubierre, W. Benger, B. Brugmann, G. Lanfermann, L. Nerger, E. Seidel and R. Takahashi, "The 3D grazing collision of two black holes," Phys. Rev. Lett. **87**, 271103 (2001)
162. H. J. Yo, T. W. Baumgarte and S. L. Shapiro, "Improved numerical stability of stationary black hole evolution calculations," Phys. Rev. D **66**, 084026 (2002)
163. B. Kelly, P. Laguna, K. Lockitch, J. Pullin, E. Schnetter, D. Shoemaker and M. Tiglio, "A cure for unstable numerical evolutions of single black holes: adjusting the standard ADM equations," Phys. Rev. D **64**, 084013 (2001)
164. M. A. Scheel, L. E. Kidder, L. Lindblom, H. P. Pfeiffer and S. A. Teukolsky, "Toward stable 3D numerical evolutions of black-hole spacetimes," Phys. Rev. D **66**, 124005 (2002) [arXiv:gr-qc/0209115].

165. M. Tiglio, L. Lehner and D. Neilsen, "3D simulations of Einstein's equations: symmetric hyperbolicity, live gauges and dynamic control of the constraints," Phys. Rev. D **70**, 104018 (2004)

166. U. Sperhake, K. L. Smith, B. Kelly, P. Laguna and D. Shoemaker, "Impact of densitized lapse slicings on evolutions of a wobbling black hole," Phys. Rev. D **69**, 024012 (2004)

167. D. Shoemaker, K. Smith, U. Sperhake, P. Laguna, E. Schnetter and D. Fiske, "Moving black holes via singularity excision," Class. Quant. Grav. **20**, 3729 (2003)

168. M. Anderson and R. A. Matzner, "Extended Lifetime in Computational Evolution of Isolated Black Holes," Found. Phys. **35**, 1477 (2005)

169. M. Alcubierre *et al.*, "Dynamical evolution of quasi-circular binary black hole data," Phys. Rev. D **72**, 044004 (2005)

170. L. T. Buchman and J. M. Bardeen, "Schwarzschild Tests of the WEBB Tetrad Formulation for Numerical Relativity," Phys. Rev. D **72**, 124014 (2005)

171. P. Diener *et al.*, "Accurate evolution of orbiting binary black holes," Phys. Rev. Lett. **96**, 121101 (2006)

172. T. W. Baumgarte and S. L. Shapiro, "Numerical relativity and compact binaries," Phys. Rept. **376**, 41 (2003)

173. B. Bruegmann, W. Tichy and N. Jansen "Numerical simulation of orbiting black holes", Phys.Rev.Lett. **92** 211101, (2004)

174. Y. Choquet-Bruhat and T. Ruggeri, "Hyperbolicity Of The 3 + 1 System Of Einstein Equations," Commun. Math. Phys. **89**, 269 (1983).

175. O. Brodbeck, S. Frittelli, P. Hubner and O. A. Reula, "Einstein's equations with asymptotically stable constraint propagation," J. Math. Phys. **40**, 909 (1999)

176. O. Sarbach and M. Tiglio, "Exploiting gauge and constraint freedom in hyperbolic formulations of Einstein's equations," Phys. Rev. D **66**, 064023 (2002)

177. T. W. Baumgarte and S.L. Shapiro, "Numerical integration of Einstein's field equations", Phys. Rev. **D59**, 024007 (1999)

178. A. P. Gentle, N. D. George, A. Kheyfets and W. A. Miller, "The constraints as evolution equations for numerical relativity," Class. Quant. Grav. **21**, 83 (2004)

179. S. Bonazzola, E. Gourgoulhon, P. Grandclement and J. Novak, "A constrained scheme for Einstein equations based on Dirac gauge and spherical coordinates", Phys.Rev. **D70**, 104007 (2004)

180. L. Andersson and V. Moncrief, "Elliptic-hyperbolic systems and the Einstein equations," Annales Henri Poincare **4**, 1 (2003)

181. M. Holst, L. Lindblom, R. Owen, H. P. Pfeiffer, M. A. Scheel and L. E. Kidder, "Optimal Constraint Projection for Hyperbolic Evolution Systems," Phys. Rev. D **70**, 084017 (2004) [arXiv:gr-qc/0407011].

182. C. R. Evans and J. F. Hawley, "Simulation of magnetohydrodynamic flows: A Constrained transport method," Astrophys. J. **332**, 659 (1988).

183. D. L. Meier, "Constrained Transport Algorithms for Numerical Relativity. I. Development of a Finite Difference Scheme," Astrophys. J. **595**, 980 (2003)

184. P.J.E. Peebels, *Principles of Physical Cosmology*, Princeton University Press (1993).

185. G. Calabrese, L. Lehner and M. Tiglio, "Constraint-preserving boundary conditions in numerical relativity," Phys. Rev. D **65**, 104031 (2002)

186. B. Szilagyi, B. G. Schmidt and J. Winicour, "Boundary conditions in linearized harmonic gravity," Phys. Rev. D **65**, 064015 (2002)

187. G. Calabrese, J. Pullin, O. Sarbach, M. Tiglio and O. Reula, "Well posed constraint-preserving boundary conditions for the linearized Einstein equations," Commun. Math. Phys. **240**, 377 (2003)

188. S. Frittelli and R. Gomez, "Boundary conditions for hyperbolic formulations of the Einstein equations," Class. Quant. Grav. **20**, 2379 (2003)

189. S. Frittelli and R. Gomez, "Einstein boundary conditions for the 3+1 Einstein equations," Phys. Rev. D **68**, 044014 (2003)

190. C. Gundlach and J. M. Martin-Garcia, "Symmetric hyperbolicity and consistent boundary conditions for second-order Einstein equations," Phys. Rev. D **70**, 044032 (2004)

191. C. Bona, T. Ledvinka, C. Palenzuela-Luque and M. Zacek, "Constraint-preserving boundary conditions in the Z4 Numerical Relativity formalism," Class. Quant. Grav. **22**, 2615 (2005)

192. O. Sarbach and M. Tiglio, "Boundary conditions for Einstein's field equations: Analytical and numerical analysis," J. Hyperbol. Diff. Equat. **2**, 839 (2005)

193. L. E. Kidder, L. Lindblom, M. A. Scheel, L. T. Buchman and H. P. Pfeiffer, "Boundary conditions for the Einstein evolution system," Phys. Rev. D **71**, 064020 (2005)

194. D. N. Arnold and N. Tarfulea, "Boundary conditions for the Einstein-Christoffel formulation of Einstein's equations," arXiv:gr-qc/0611010.

195. M. C. Babiuc, H. O. Kreiss and J. Winicour, "Constraint-preserving Sommerfeld conditions for the harmonic Einstein equations," arXiv:gr-qc/0612051.

196. L. T. Buchman and O. C. A. Sarbach, "Towards absorbing outer boundaries in General Relativity," Class. Quant. Grav. **23**, 6709 (2006) [arXiv:gr-qc/0608051].

197. O. Rinne, "Stable radiation-controlling boundary conditions for the generalized harmonic Einstein equations," Class. Quant. Grav. **23**, 6275 (2006) [arXiv:gr-qc/0606053].

198. H. O. Kreiss, O. Reula, O. Sarbach and J. Winicour, "The Einstein equations, boundaries and integration by parts," arXiv:0707.4188 [gr-qc].

199. L. T. Buchman and O. C. A. Sarbach, "Improved outer boundary conditions for Einstein's field equations," Class. Quant. Grav. **24**, S307 (2007) [arXiv:gr-qc/0703129].

200. M. C. Babiuc, H. O. Kreiss and J. Winicour, "Constraint-preserving Sommerfeld conditions for the harmonic Einstein equations," Phys. Rev. D **75**, 044002 (2007).

201. O. Rinne, L. Lindblom and M. A. Scheel, "Testing outer boundary treatments for the Einstein equations," arXiv:0704.0782 [gr-qc].

202. H. Kreiss and J. Oliger, "Methods for the Approximate Solution of Time Dependent Problems", *Global Atmospheric Research Programme, Publications Series No. 10.* (1973)

203. C. W. Misner, "Over the Rainbow: Numerical Relativity beyond Scri+," arXiv:gr-qc/0512167.

204. J. R. van Meter, D. R. Fiske and C. W. Misner, "Excising das All: Evolving Maxwell waves beyond Scri," Phys. Rev. D **74**, 064003 (2006)

205. C. W. Misner, "Hyperboloidal slices and artificial cosmology for numerical relativity," arXiv:gr-qc/0409073.

206. G. Calabrese, C. Gundlach and D. Hilditch, "Asymptotically null slices in numerical relativity: Mathematical analysis and spherical wave equation tests," Class. Quant. Grav. **23**, 4829 (2006)
207. R.M. Wald, *General Relativity*, The University of Chicago Press (1984).
208. S. Brandt and B. Brugmann, "Black hole punctures as initial data for general relativity," Phys. Rev. Lett. **78**, 3606 (1997)
209. M. Hannam, S. Husa, D. Pollney, B. Brugmann and N. O'Murchadha, "Geometry and Regularity of Moving Punctures," arXiv:gr-qc/0606099.
210. M. Hannam, S. Husa, B. Brugmann, J. A. Gonzalez, U. Sperhake and N. O. Murchadha, "Where do moving punctures go?," arXiv:gr-qc/0612097.
211. J. Thornburg, P. Diener, D. Pollney, L. Rezzolla, E. Schnetter, E. Seidel and R. Takahashi, "Are moving punctures equivalent to moving black holes?," arXiv:gr-qc/0701038.
212. T. W. Baumgarte and S. G. Naculich, "Analytical Representation of a Black Hole Puncture Solution," Phys. Rev. D **75**, 067502 (2007)
213. J. D. Brown, "Puncture Evolution of Schwarzschild Black Holes," arXiv:0705.1359 [gr-qc].
214. G. B. Cook, "Initial Data for Numerical Relativity," Living Rev. Rel. **3**, 5 (2000)
215. H. P. Pfeiffer, "The initial value problem in numerical relativity," arXiv:gr-qc/0412002.
216. E. Gourgoulhon, "Construction of initial data for 3+1 numerical relativity," arXiv:0704.0149 [gr-qc].
217. S. Nissanke, "Post-Newtonian freely specifiable initial data for binary black holes in numerical relativity," Phys. Rev. D **73**, 124002 (2006) [arXiv:gr-qc/0509128].
218. N. Yunes, W. Tichy, B. J. Owen and B. Bruegmann, "Binary black hole initial data from matched asymptotic expansions," Phys. Rev. D **74**, 104011 (2006)
219. N. Yunes and W. Tichy, "Improved initial data for black hole binaries by asymptotic matching of post-Newtonian and perturbed black hole solutions," Phys. Rev. D **74**, 064013 (2006) [arXiv:gr-qc/0601046].
220. B. J. Kelly, W. Tichy, M. Campanelli and B. F. Whiting, "Black hole puncture initial data with realistic gravitational wave content," arXiv:0704.0628 [gr-qc].
221. T. DeDonder, *La Gravifique Einsteinienne*, Gunthier-Villars, Paris (1921); T. DeDonder, *The Mathematical Theory of Relativity*. Massachusetts Institute of Technology, Cambridge, MA (1927).
222. V. Fock, *The Theory of Space Time and Gravitation*, Pergamon Press, New York (1959)
223. Y. Four'es-Bruhat, "Théorème d'éxistence pour certains systèmes d'équations aux dérivées partielles non linéaires", *Acta Math.*, **88** 141 (1952); Y. Bruhat, "The Cauchy problem" in *Gravitation: An Introduction to Current Research*, ed. L. Witten, New York, Wiley (1962)
224. A. E. Fischer and J. E. Marsden, "The Einstein evolution equations as a first-order quasi-linear symmetric hyperbolic system," Commun. Math. Phys., **28** 1 (1972)
225. J. Renn and T. Sauer, "Heuristics and mathematical representation in Einstein's search for a gravitational field equation", in *History of General Relativity, volume 7 of Einstein Studies*, eds. H. Goenner, J. Renn, and J. Ritter, Birkhäuser, Cambridge, MA (1998).

226. D. Garfinkle, "Harmonic coordinate method for simulating generic singularities", Phys.Rev. **D65**, 044029 (2002)

227. B. Szilagyi and J. Winicour, "Well-Posed Initial-Boundary Evolution in General Relativity", Phys.Rev. **D68**, 041501 (2003)

228. F. Pretorius, "Numerical Relativity Using a Generalized Harmonic Decomposition", Class. Quant. Grav. **22** 425, (2005)

229. L. Lindblom, M. A. Scheel, L. E. Kidder, R. Owen and O. Rinne, "A New Generalized Harmonic Evolution System", Class.Quant.Grav. **23**, S447 (2006)

230. M. C. Babiuc, B. Szilagyi and J. Winicour, "Harmonic Initial-Boundary Evolution in General Relativity," Phys.Rev. **D73**, 064017 (2006)

231. B. Szilagyi, D. Pollney, L. Rezzolla, J. Thornburg and J. Winicour "An explicit harmonic code for black-hole evolution using excision", gr-qc/0612150

232. C. Bona and J. Masso, "Harmonic Synchronizations of Space-time," Phys. Rev. D **38**, 2419 (1988).

233. C. Bona, J. Masso and J. Stela, "Numerical black holes: A Moving grid approach," Phys. Rev. D **51**, 1639 (1995)

234. M. Shibata and T. Nakamura, "Evolution of three-dimensional gravitational waves: Harmonic slicing case", Phys. Rev. **D52**, 5428 (1995).

235. M. Alcubierre, "The appearance of coordinate shocks in hyperbolic formalisms of General Relativity," Phys. Rev. D **55**, 5981 (1997)

236. M. Alcubierre and J. Masso, "Pathologies of hyperbolic gauges in general relativity and other field theories," Phys. Rev. D **57**, 4511 (1998)

237. S. D. Hern, "Coordinate Singularities in Harmonically-sliced Cosmologies," Phys. Rev. D **62**, 044003 (2000) [arXiv:gr-qc/0001070].

238. M. Alcubierre, "Are gauge shocks really shocks?," Class. Quant. Grav. **22**, 4071 (2005)

239. D. Garfinkle, C. Gundlach and D. Hilditch, "Comments on Bona-Masso type slicing conditions in long-term black hole evolutions," arXiv:0707.0726 [gr-qc].

240. H. Friedrich, "On the Hyperbolicity of Einstein's and Other Gauge Field Equations", Commun. Math. Phys **100**, 525 (1985)

241. R. Owen, "Constraint Damping in First-Order Evolution Systems for Numerical Relativity," arXiv:gr-qc/0703145.

242. B. Szilagyi, D. Pollney, L. Rezzolla, J. Thornburg and J. Winicour, "An explicit harmonic code for black-hole evolution using excision," Class. Quant. Grav. **24**, S275 (2007) [arXiv:gr-qc/0612150].

243. L.Lindblom, "Generalized Harmonic Gauge Drivers" as presented at the *18th International Conference on General Relativity and Gravitation*

244. T. Nakamura, K. Oohara and Y. Kojima, Prog. Theor. Phys. Suppl. **90**, 1 (1987)

245. A. Lichnerowicz, J. Math. Pure Appl. **23**, 37 (1944), J. W. York, Phys. Rev. Lett. **26**, 1656 (1971).

246. M. Alcubierre and B. Brugmann, "Simple excision of a black hole in 3+1 numerical relativity," Phys. Rev. D **63**, 104006 (2001)

247. P. Laguna and D. Shoemaker, "Numerical stability of a new conformal-traceless 3+1 formulation of the Einstein equation," Class. Quant. Grav. **19**, 3679 (2002)

248. J. G. Baker, J. Centrella, D. I. Choi, M. Koppitz and J. van Meter, "Binary black hole merger dynamics and waveforms," Phys. Rev. D **73**, 104002 (2006)

249. A. P. Gentle, "The BSSN formulation is a partially constrained evolution system," arXiv:0707.0339 [gr-qc].

250. O. Sarbach, G. Calabrese, J. Pullin and M. Tiglio, "Hyperbolicity of the BSSN system of Einstein evolution equations," Phys. Rev. D **66**, 064002 (2002)

251. H. Beyer and O. Sarbach, "On the well posedness of the Baumgarte-Shapiro-Shibata-Nakamura formulation of Einstein's field equations," Phys. Rev. D **70**, 104004 (2004)

252. C. Gundlach and J. M. Martin-Garcia, "Hyperbolicity of second-order in space systems of evolution equations," Class. Quant. Grav. **23**, S387 (2006)

253. C. Gundlach and J. M. Martin-Garcia, "Well-posedness of formulations of the Einstein equations with dynamical lapse and shift conditions," Phys. Rev. D **74**, 024016 (2006)

254. M. Alcubierre, B. Brugmann, P. Diener, M. Koppitz, D. Pollney, E. Seidel and R. Takahashi, "Gauge conditions for long-term numerical black hole evolutions without excision," Phys. Rev. D **67**, 084023 (2003)

255. F. Herrmann, D. Shoemaker and P. Laguna, "Unequal-Mass Binary Black Hole Inspirals," arXiv:gr-qc/0601026.

256. U. Sperhake, "Binary black-hole evolutions of excision and puncture data," arXiv:gr-qc/0606079.

257. L. E. Kidder, M. A. Scheel, S. A. Teukolsky, E. D. Carlson and G. B. Cook, "Black hole evolution by spectral methods," Phys. Rev. D **62**, 084032 (2000)

258. M. Boyle, L. Lindblom, H. Pfeiffer, M. Scheel and L. E. Kidder, "Testing the Accuracy and Stability of Spectral Methods in Numerical Relativity," Phys. Rev. D **75**, 024006 (2007)

259. P. Grandclement and J. Novak, "Spectral Methods for Numerical Relativity," arXiv:0706.2286 [gr-qc].

260. Y. Zlochower, J. G. Baker, M. Campanelli and C. O. Lousto, "Accurate black hole evolutions by fourth-order numerical relativity," Phys. Rev. D **72**, 024021 (2005)

261. http://www.cactuscode.org

262. E. Schnetter, S. H. Hawley and I. Hawke, "Evolutions in 3D numerical relativity using fixed mesh refinement," Class. Quant. Grav. **21**, 1465 (2004)

263. http://www.physics.drexel.edu/olson/paramesh-doc/Users_manual/amr.html

264. PAMR (Parallel Adaptive Mesh Refinement) and AMRD (Adaptive Mesh Refinement Driver) libraries (http://laplace.physics.ubc.ca/Group/Software.html)

265. D. Neilsen, E. W. Hirschmann, M. Anderson and S. L. Liebling, "Adaptive mesh refinement and relativistic MHD," arXiv:gr-qc/0702035.

266. P. Marronetti, W. Tichy, B. Bruegmann, J. Gonzalez, M. Hannam, S. Husa and U. Sperhake, "Binary black holes on a budget: Simulations using workstations," Class. Quant. Grav. **24**, S43 (2007)

267. D. Brown, O. Sarbach, E. Schnetter, M. Tiglio, P. Diener, I. Hawke and D. Pollney, "Excision without excision: the relativistic turducken," arXiv:0707.3101 [gr-qc].

268. B. Imbiriba et al., "Evolving a puncture black hole with fixed mesh refinement," Phys. Rev. D **70**, 124025 (2004) [arXiv:gr-qc/0403048].

269. S. Husa, J. A. Gonzalez, M. Hannam, B. Bruegmann and U. Sperhake, "Reducing phase error in long numerical binary black hole evolutions with sixth order finite differencing," arXiv:0706.0740 [gr-qc].

270. M. Campanelli, C. O. Lousto and Y. Zlochower, "The last orbit of binary black holes," Phys. Rev. D **73**, 061501 (2006) [arXiv:gr-qc/0601091].

271. J. G. Baker, M. Campanelli, F. Pretorius and Y. Zlochower, "Comparisons of binary black hole merger waveforms," Class. Quant. Grav. **24**, S25 (2007) [arXiv:gr-qc/0701016].

272. H. P. Pfeiffer, D. A. Brown, L. E. Kidder, L. Lindblom, G. Lovelace and M. A. Scheel, "Reducing orbital eccentricity in binary black hole simulations," arXiv:gr-qc/0702106.

273. S. Husa, M. Hannam, J. A. Gonzalez, U. Sperhake and B. Brugmann, "Reducing eccentricity in black-hole binary evolutions with initial parameters from post-Newtonian inspiral," arXiv:0706.0904 [gr-qc].

274. G. B. Cook and H. P. Pfeiffer, "Excision boundary conditions for black hole initial data," Phys. Rev. D **70**, 104016 (2004) [arXiv:gr-qc/0407078].

275. J. M. Bowen and J. W. . York, "Time asymmetric initial data for black holes and black hole collisions," Phys. Rev. D **21**, 2047 (1980).

276. E. Berti, V. Cardoso, J. A. Gonzalez, U. Sperhake, M. Hannam, S. Husa and B. Bruegmann, "Inspiral, merger and ringdown of unequal mass black hole binaries: A multipolar analysis," arXiv:gr-qc/0703053.

277. J. G. Baker, J. Centrella, D. I. Choi, M. Koppitz, J. R. van Meter and M. C. Miller, "Getting a kick out of numerical relativity," Astrophys. J. **653**, L93 (2006) [arXiv:astro-ph/0603204].

278. J. A. Gonzalez, U. Sperhake, B. Bruegmann, M. Hannam and S. Husa, "Total recoil: the maximum kick from nonspinning black-hole binary inspiral," Phys. Rev. Lett. **98**, 091101 (2007) [arXiv:gr-qc/0610154].

279. M. Fitchett, *MNRAS* **203**, 1049 (1983).

280. M. Campanelli, C. O. Lousto and Y. Zlochower, "Gravitational radiation from spinning-black-hole binaries: The orbital hang up," Phys. Rev. D **74**, 041501 (2006) [arXiv:gr-qc/0604012].

281. M. Campanelli, C. O. Lousto and Y. Zlochower, "Spin–orbit interactions in black-hole binaries," Phys. Rev. D **74**, 084023 (2006) [arXiv:astro-ph/0608275].

282. M. Campanelli, C. O. Lousto, Y. Zlochower, B. Krishnan and D. Merritt, "Spin Flips and Precession in Black-Hole-Binary Mergers," Phys. Rev. D **75**, 064030 (2007) [arXiv:gr-qc/0612076].

283. F. Herrmann, I. Hinder, D. Shoemaker, P. Laguna and R. A. Matzner, "Gravitational recoil from spinning binary black hole mergers," arXiv:gr-qc/0701143.

284. M. Koppitz, D. Pollney, C. Reisswig, L. Rezzolla, J. Thornburg, P. Diener and E. Schnetter, "Getting a kick from equal-mass binary black hole mergers," arXiv:gr-qc/0701163.

285. D. Pollney *et al.*, "Recoil velocities from equal-mass binary black-hole mergers: a systematic investigation of spin–orbit aligned configurations," arXiv:0707.2559 [gr-qc].

286. M. Campanelli, C. O. Lousto, Y. Zlochower and D. Merritt, "Large Merger Recoils and Spin Flips From Generic Black-Hole Binaries," Astrophys. J. **659**, L5 (2007) [arXiv:gr-qc/0701164].

287. D. I. Choi, B. J. Kelly, W. D. Boggs, J. G. Baker, J. Centrella and J. van Meter, "Recoiling from a kick in the head-on case," arXiv:gr-qc/0702016.

288. J. G. Baker, W. D. Boggs, J. Centrella, B. J. Kelly, S. T. McWilliams, M. C. Miller and J. R. van Meter, "Modeling kicks from the merger of non-precessing black-hole binaries," arXiv:astro-ph/0702390.

289. M. Campanelli, C. O. Lousto, Y. Zlochower and D. Merritt, "Maximum gravitational recoil," arXiv:gr-qc/0702133.

290. B. Vaishnav, I. Hinder, F. Herrmann and D. Shoemaker, "Matched Filtering of Numerical Relativity Templates of Spinning Binary Black Holes," arXiv:0705.3829 [gr-qc].

291. J. D. Schnittman et al., "Anatomy of the binary black hole recoil: A multipolar analysis," arXiv:0707.0301 [gr-qc].

292. B. Brugmann, J. A. Gonzalez, M. Hannam, S. Husa and U. Sperhake, "Exploring black hole superkicks," arXiv:0707.0135 [gr-qc].

293. L. E. Kidder, "Coalescing binary systems of compact objects to postNewtonian 5/2 order. 5. Spin effects," Phys. Rev. D **52**, 821 (1995)

294. C. Cutler, D. Kennefick and E. Poisson, "Gravitational radiation reaction for bound motion around a Schwarzschild black hole," Phys. Rev. **D50**, 3816 (1994).

295. K. Glampedakis and D. Kennefick, "Zoom and whirl: Eccentric equatorial orbits around spinning black holes and their evolution under gravitational radiation reaction," Phys. Rev. **D66**, 044002 (2002)

296. N. Cornish and J. Levin, "Lyapunov timescales and black hole binaries", Class.Quant.Grav. **20**, 1649 (2003)

297. M. Milosavljevic and E. S. Phinney, "The Afterglow of Massive Black Hole Coalescence," Astrophys. J. **622**, L93 (2005)

298. A. I. Macfadyen and M. Milosavljevic, "An Eccentric Circumbinary Accretion Disk and the Detection of Binary Massive Black Holes," arXiv:astro-ph/0607467.

299. K. Hayasaki, S. Mineshige and H. Sudou, "Binary Black Hole Accretion Flows in Merged Galactic Nuclei," arXiv:astro-ph/0609144.

300. M. Dotti, R. Salvaterra, A. Sesana, M. Colpi and F. Haardt, "On the search of electromagnetic cosmological counterparts to coalescences of massive black hole binaries," Mon. Not. Roy. Astron. Soc. **372**, 869 (2006)

301. M. Micic, T. Abel and S. Sigurdsson, "The Role of Primordial Kicks on Black Hole Merger Rates," arXiv:astro-ph/0512123.

302. A. Sesana, "Extreme recoils: impact on the detection of gravitational waves from massive black hole binaries," arXiv:0707.4677 [astro-ph].

303. T. Bogdanovic, C. S. Reynolds and M. C. Miller, "Alignment of the spins of supermassive black holes prior to coalescence," arXiv:astro-ph/0703054.

304. K. Iwasawa et al., "The variable iron K emission line in MCG-6-30-15," Mon. Not. Roy. Astron. Soc. **282**, 1038 (1996)

305. A. C. Fabian et al., "A long hard look at MCG-6-30-15 with XMM-Newton," Mon. Not. Roy. Astron. Soc. **335**, L1 (2002)

306. C. S. Reynolds and M. A. Nowak, "Fluorescent iron lines as a probe of astrophysical black hole systems," Phys. Rept. **377**, 389 (2003)

307. L. W. Brenneman and C. S. Reynolds, "Constraining Black Hole Spin Via X-ray Spectroscopy," Astrophys. J. **652**, 1028 (2006)

308. J. D. Schnittman and A. Buonanno, "The Distribution of Recoil Velocities from Merging Black Holes," arXiv:astro-ph/0702641.

309. J. D. Schnittman, "Retaining Black Holes with Very Large Recoil Velocities," arXiv:0706.1548 [astro-ph].

310. A. Loeb, "Observable Signatures of a Black Hole Ejected by Gravitational Radiation Recoil in a Galaxy Merger," arXiv:astro-ph/0703722.

311. E. W. Bonning, G. A. Shields and S. Salviander, "Recoiling Black Holes in Quasars," arXiv:0705.4263 [astro-ph].

312. R.D. Ekers, R. Fanti, C. Lari and P. Parma, Nature **276**, 588 (1978)

313. J. Dennett-Thorpe, P. A. G. Scheuer, R. A. Laing, A. H. Bridle, G. G. Pooley and W. Reich, "Jet reorientation in AGN: two winged radio galaxies," Mon. Not. Roy. Astron. Soc. **330**, 609 (2002)

314. D. Merritt and R. D. Ekers, "Tracing black hole mergers through radio lobe morphology," Science **297**, 1310 (2002) [arXiv:astro-ph/0208001].

315. S. Komossa, "Observational evidence for supermassive black hole binaries," AIP Conf. Proc. **686**, 161 (2003)

316. D. Vir Lal and A. Pramesh Rao, "GMRT observations of X-shaped radio sources," Mon. Not. Roy. Astron. Soc. **374**, 1085 (2007)

317. C. C. Cheung, "FIRST 'Winged' and 'X'-shaped Radio Source Candidates," arXiv:astro-ph/0701278.

318. D. Merritt, M. Milosavljevic, M. Favata, S. A. Hughes and D. E. Holz, "Consequences of gravitational radiation recoil," Astrophys. J. **607**, L9 (2004)

319. K. Holley-Bockelmann, K. Gultekin, D. Shoemaker and N. Yunes, "Gravitational Wave Recoil and the Retention of Intermediate Mass Black Holes," arXiv:0707.1334 [astro-ph].

320. R. Ruffini and S. Bonazzola, "Systems of selfgravitating particles in general relativity and the concept of an equation of state," Phys. Rev. **187**, 1767 (1969).

321. M. Colpi, S. L. Shapiro and I. Wasserman, "Boson Stars: Gravitational Equilibria of Selfinteracting Scalar Fields," Phys. Rev. Lett. **57**, 2485 (1986).

322. C. Palenzuela, L. Lehner and S. L. Liebling, "Orbital Dynamics of Binary Boson Star Systems," arXiv:0706.2435 [gr-qc].

323. C. B. M. Chirenti and L. Rezzolla, "How to tell a gravastar from a black hole," arXiv:0706.1513 [gr-qc].

324. A. M. Abrahams and C. R. Evans, "Critical behavior and scaling in vacuum axisymmetric gravitational collapse," Phys. Rev. Lett. **70**, 2980 (1993).

325. U.Sperhake, *private communication*

326. N. Arkani-Hamed, S. Dimopoulos and G. Dvali, "The hierarchy problem and new dimensions at a millimeter", Phys. Lett. **B 429**, 263 (1998)

327. L. Randall and R. Sundrum, "Large Mass Hierarchy from a Small Extra Dimension", Phys. Rev. Lett. **83**, 3370 (1999)

328. S.B. Giddings and S. Thomas, "High energy colliders as black hole factories: The end of short distance physics", Phys. Rev. **D 65**, 056010 (2002)

329. T. Banks and W. Fischler, "A Model for High Energy Scattering in Quantum Gravity", hep-th/9906038

330. S. Dimopoulos and G. Landsberg, "Black Holes at the Large Hadron Collider", Phys. Rev. Lett. **87** 161602, (2001)

331. P. Kanti, "Black Holes in Theories with Large Extra Dimensions: a Review", Int.J.Mod.Phys. **A19**, 4899 (2004)

332. D. M. Gingrich, "Black hole cross-section at the large hadron collider", Int.J.Mod.Phys. **A21**, 6653 (2006)

333. D. M. Gingrich, " Effect of charged partons on black hole production at the large hadron collider", hep-ph/0612105

334. J.L. Feng and A. D. Shapere, "Black hole production by cosmic rays", Phys. Rev. Lett. **88** 021303, (2002)

335. G. Landsberg, "Black Holes at Future Colliders and Beyond: a Topical Review", J.Phys. **G32**, R337 (2006)

336. I. Arsene *et al.* [BRAHMS Collaboration], "Quark gluon plasma and color glass condensate at RHIC? The perspective from the BRAHMS experiment," Nucl. Phys. A **757**, 1 (2005)

337. K. Adcox *et al.* [PHENIX Collaboration], "Formation of dense partonic matter in relativistic nucleus nucleus collisions at RHIC: Experimental evaluation by the PHENIX collaboration," Nucl. Phys. A **757**, 184 (2005)

338. B. B. Back *et al.*, "The PHOBOS perspective on discoveries at RHIC," Nucl. Phys. A **757**, 28 (2005)

339. J. Adams *et al.* [STAR Collaboration], "Experimental and theoretical challenges in the search for the quark gluon plasma: The STAR collaboration's critical assessment of the evidence from RHIC collisions," Nucl. Phys. A **757**, 102 (2005) [arXiv:nucl-ex/0501009].

340. J. M. Maldacena, "The large N limit of superconformal field theories and supergravity," Adv. Theor. Math. Phys. **2**, 231 (1998) [Int. J. Theor. Phys. **38**, 1113 (1999)]

341. E. Witten, "Anti-de Sitter space and holography," Adv. Theor. Math. Phys. **2**, 253 (1998)

342. S. S. Gubser, I. R. Klebanov and A. M. Polyakov, "Gauge theory correlators from non-critical string theory," Phys. Lett. B **428**, 105 (1998)

343. H. Nastase, "The RHIC fireball as a dual black hole," arXiv:hep-th/0501068.

344. J. J. Friess, S. S. Gubser, G. Michalogiorgakis and S. S. Pufu, "Expanding plasmas and quasinormal modes of anti-de Sitter black holes," JHEP **0704**, 080 (2007)

345. R. C. Myers and M. J. Perry, "Black Holes In Higher Dimensional Space-Times," Annals Phys. **172**, 304 (1986).

346. C.Merrick, "Black Holes in Large Extra Dimensions at the LHC through Simulation", Princeton University Junior-thesis Paper (2007).

Index

aberration effect, 6
acceleration searches, 55
accretion
  and emission lines, 271
  and X-ray afterglow, 347
  conditions for, 141
  in mass transfer, 161
  relation to equilibrium spin period,
    143
  super-Eddington-limited, 168
accretion disks, 137, 143, 265
  chemical composition, 283
  hyper-accreting, 254, 259
  innermost regions, 273
  viscous timescale, 347
accretion-induced collapse, 178
accretion-induced magnetic field decay,
  173
accretor phase, 87
active galactic nuclei, 275, 305
adaptive mesh refinement, 332
Aichelberg-Sexl metric, 350
Alfven radius, 87, 139, 142
alternative theories of gravity, 44, 349
  future limits, 69
Amati correlation, 253
angular momentum transfer, 85
Arnowitt-Deser-Misner formulation,
  318, 326
arrival time, 57, 100
  barycentric, 58
  topocentric, 58
assymetric radiation

of linear momentum, 347
asynchronicity
  and unipolar inductor model, 295
Australian Telescope Compact Array,
  112

baryonic mass
  difference from gravitational, 158
BATSE experiment, 245
Baumgarte-Shapiro-Shibata-Nakamura
  formalism, 315, 329
BeppoSAX satellite, 245
binary hardening, 213
binary mergers
  as gamma-ray burst source, 248
  of compact objects, 250
binary pulsars
  classification, 136, 137
binary systems
  and supra-thermal star production,
    230
  classification, 127, 129
  conservative case in evolution, *see*
    conservative case
  double line', 13
  encounter with neutron stars, 220
  equipotential surfaces, 153
  interactions with single stars, 207,
    208
  merger lifetime, 255
  merger rate, 191
  neutron-star-white-dwarf, 12, 15, 50,
    170, 175
  of two black holes, 225